AF471346

Lecture Notes in Earth Sciences

Vol. 1: Sedimentary and Evolutionary Cycles. Edited by U. Bayer and A. Seilacher. VI, 465 pages. 1985.

Vol. 2: U. Bayer, Pattern Recognition Problems in Geology and Paleontology. VII, 229 pages. 1985.

Vol. 3: Th. Aigner, Storm Depositional Systems. VIII, 174 pages. 1985.

Vol. 4: Aspects of Fluvial Sedimentation in the Lower Triassic Buntsandstein of Europe. Edited by D. Mader. VIII, 626 pages. 1985.

Vol. 5: Paleogeothermics. Edited by G. Buntebarth and L. Stegena. II, 234 pages. 1986.

Vol. 6: W. Ricken, Diagenetic Bedding. X, 210 pages. 1986.

Vol. 7: Mathematical and Numerical Techniques in Physical Geodesy, Edited by H. Sünkel. IX, 548 pages. 1986.

Vol. 8: Global Bio-Events. Edited by O.H. Walliser. IX, 442 pages. 1986.

Vol. 9: G. Gerdes, W.E. Krumbein, Biolaminated Deposits. IX, 183 pages. 1987.

Vol. 10: T.M. Peryt (Ed.), The Zechstein Facies in Europe. V, 272 pages. 1987.

Vol. 11: L. Landner (Ed.), Contamination of the Environment. Proceedings, 1986. VII, 190 pages. 1987.

Vol. 12: S. Turner (Ed.), Applied Geodesy. VIII, 393 pages. 1987.

Vol. 13: T.M. Peryt (Ed.), Evaporite Basins. V, 188 pages. 1987.

Vol. 14: N. Cristescu, H.I. Ene (Eds.), Rock and Soil Rheology. VIII, 289 pages. 1988.

Vol. 15: V.H. Jacobshagen (Ed.), The Atlas System of Morocco. VI, 499 pages. 1988.

Vol. 16: H. Wanner, U. Siegenthaler (Eds.), Long and Short Term Variability of Climate. VII, 175 pages. 1988.

Vol. 17: H. Bahlburg, Ch. Breitkreuz, P. Giese (Eds.), The Southern Central Andes. VIII, 261 pages. 1988.

Vol. 18: N.M.S. Rock, Numerical Geology. XI, 427 pages. 1988.

Vol. 19: E. Groten, R. Strauß (Eds.), GPS-Techniques Applied to Geodesy and Surveying. XVII, 532 pages. 1988.

Vol. 20: P. Baccini (Ed.), The Landfill. IX, 439 pages. 1989.

Vol. 21: U. Förstner, Contaminated Sediments. V, 157 pages. 1989.

Vol. 22: I.I. Mueller, S. Zerbini (Eds.), The Interdisciplinary Role of Space Geodesy. XV, 300 pages. 1989.

Vol. 23: K.B. Föllmi, Evolution of the Mid-Cretaceous Triad. VII, 153 pages. 1989.

Vol. 24: B. Knipping, Basalt Intrusions in Evaporites. VI, 132 pages. 1989.

Vol. 25: F. Sansò, R. Rummel (Eds.), Theory of Satellite Geodesy and Gravity Field Theory. XII, 491 pages. 1989.

Vol. 26: R.D. Stoll, Sediment Acoustics. V, 155 pages. 1989.

Vol. 27: G.-P. Merkler, H. Militzer, H. Hötzl, H. Armbruster, J. Brauns (Eds.), Detection of Subsurface Flow Phenomena. IX, 514 pages. 1989.

Lecture Notes in Earth Sciences

Edited by Somdev Bhattacharji, Gerald M. Friedman, Horst J. Neugebauer and Adolf Seilacher

27

G.-P. Merkler H. Militzer H. Hötzl H. Armbruster J. Brauns (Eds.)

Detection of Subsurface Flow Phenomena

Springer-Verlag
Berlin Heidelberg GmbH

Editors

Dr. Georg-Paul Merkler
Institute of Applied Geology, University of Karlsruhe
Kaiserstr. 12, D-7500 Karlsruhe, FRG

Prof. Dr. Heinz Militzer
Bergakademie Freiberg
DDR-9200 Freiberg, GDR

Prof. Dr. Heinz Hötzl
Institute of Applied Geology, University of Karlsruhe
Kaiserstr. 12, D-7500 Karlsruhe, FRG

Dipl.-Ing. Heinrich Armbruster
Head of the Section Groundwater, Problems with Structures
Bundesanstalt für Wasserbau (BAW)
Kußmaulstr. 17, D-7500 Karlsruhe 21, FRG

Prof. Dr.-Ing. Josef Brauns
Institute of Soil and Rock Mechanics, University of Karlsruhe
Kaiserstr. 12, D-7500 Karlsruhe, FRG

DOI 10.1007/978-3-540-46834-9

Originally published by Springer-Verlag Berlin Heidelberg New York in 1989
MyCopy version of the original edition 1989
2132/3140-543210 – Printed on acid-free paper
www.springer.com/mycopy

PREFACE

It is increasingly necessary to develop industrial and hydraulic engineering constructions under unfavourable geological or geotechnical conditions. Furthermore, it becomes more and more important to build effectively and economically and to find optimal solutions for a long-term steady function of the constructions. This emphatically demands exhaustive information on the structural situations and engineering parameters of local site assessments by areal investigations of the sites and the petrophysical parameters in situ. This requires, however, the use of geophysical techniques. During the last two or three decades international applied geophysics has systematically developed new possibilities for site investigations for the determination of petrophysical parameters in situ as well as for observation of the system building and site. As in "New techniques in engineering", geophysical methods make it possible to develop areal models of subsurface conditions of building sites, to quantify relevant engineering parameters in situ, as well as to analyze the long-term behaviour of the buildings, which are influenced by internal or external factors. With regard to the broad spectrum of applied geophysics, there are few methods, that especially favour application in engineering and groundwater studies. These methods are distinguished by a relatively simple measuring technique and good measuring progress, e.g. the geoelectrical self-potential method, the geoelectrical resistivity method as well as a newly developed devices for geothermic measurements. There exist numerous publications, broadly scattered in the technical literature, concerning the theoretical bases and applications of these methods, but until now, there have been only a few meetings to exchange experience and results on an international level. This was the aim of the symposium "Detection of Subsurface Flow Phenomena by Self-Potential/Geoelectrical and Thermometric Methods", held in Karlsruhe from 14-18 March 1988. An outstanding part of the symposioum was represented by the results of a research project, coordinated by the University of Karlsruhe (Department of Geology and Institute of Soil and Rock Mechanics) and the Federal Waterway Engineering and Research Institute (BAW), Karlsruhe. Regarding the subject "Experiments to ascertain the relations between hydraulic potentials in the underground and the geoelectrical and thermic potentials set off by these", the research work took four years.

The project was sponsered by the Volkswagen Foundation/Hannover. The goal was to develop and test objective techniques for detecting leakages in dams, locating, demarcating and designating quantitatively inhomogeneous spheres in dams with the aim of detecting damage and subsurface flow phenomena as soon as possible. The symposium consisted of a three-day lecture meeting with about 40 papers and a summarizing respectively closing roundtable discussion, a visit to the laboratories and to the in situ constructions within the area of BAW developed in the frame of the research project. This included a technical excursion to the Rhine-Staustufe Iffezheim with its very impressive waterway constructions and an excursion to the Geophysical Observatory near Schiltach (Black Forest). The Observatory belongs to the Universities of Karlsruhe and Stuttgart.

Approximately 80 scientists from 15 countries participated the symposium. They were welcomed by the Rector of the University, Professor Dr. A. Kunle and the representative of the Federal Ministry of Traffic, Dr. G. Schröder. Professor Dr. H. Hötzl elucidated the scientific problems and the economical importance of the project as a speaker of the research group.

The following papers dealt with the fundamental aspects of geoelectrical and thermometric measurements, with the theory of these methods, the state and developing tendencies concerning devices, data acquisition, processing and interpretation as well as noise effects. It became clear that the solution of the complex scientific-technical problems of waterway constructions and environmental protection requires broad, interdisciplinary cooperation and international collaboration. Thus it would be possible to minimize the personnel, temporal and economic efforts.

The intended cooperation of geoscientists, engineering geologists, building engineers and representatives of other disciplines make it possible, not only to exchange experiences and results relating to international problems unsolved until now, but also to determine new guidelines with regard to the scientific organization of further investigations.

Thus in order to inform all interested parties of the main topics of the symposium and to advance international cooperation in the future, the present review includes a part of the papers and reports of the excursions recommended by the participants of the meeting, which have been divided into the following topics:

- Introduction to engineering-geophysical problems and attempts at their solution;

- Geoelectrical self-potential measurements;

- Geoelectrical resistivity measurements;

- Geothermic measurements;

- Case histories;

- Some topics of the roundtable discussion;

- Reports concerning the excursions.

The editors wish to thank very much all those, who contributed to the success of the symposium and to the publication of the present report. Finally they venture the note, that the authors theirselves are responsible for the content of their papers.

H. Hötzl; G.-P. Merkler; H. Militzer

Karlsruhe, Dec. 1988.

CONTENTS

INTRODUCTION

MEASUREMENTS OF SELF-POTENTIAL

SPECIAL APPLICATIONS OF GEOELECTRICAL MEASUREMENTS

GEOTHERMICAL MEASUREMENTS

APPLICATION OF GEOELECTRICAL MEASUREMENTS IN ENGINEERING AND ROUND TABLE DISCUSSION

EXCURSION

Introduction

EFFECT OF LEAKS IN DAMS AND TRIALS TO DETECT LEAKAGES BY GEOPHYSIKAL MEANS

by H. Armbruster[1], J. Brauns[2], W. Mazur[2] and G.P. Merkler[3]

Abstract

A research project was carried out over a period of 3.5 years, dealing with the hydraulics of leaks in dam seals and with geophysical methods to detect and to localize these leaks. The main object of the investigations was a large model dam (H = 3.5 m, L = 20 m, V = 600 m^3) in a huge open pit which was sealed with HDPE plates to form a watertight basin and an "impervious base" of the dam. The dam body was made of sand and had an upstream sealing face which was constructed with a number of artifical leaks to be operated under full reservoir conditions. The instrumentation of the dam body consisted of a number of piezometer and temperature gauges for observing the seepage processes and the related temperature changes. A net of self-potential electrodes was installed in the downstream face of the dam which were observed along with the numerous tests performed under varying conditions. A thermo-camera was also installed and was used to observe the infrared thermo-reflection of the downstream face of the dam during the leakage tests.

The tests showed that even small leaks in sealing faces can lead to considerable water losses and extensive percolations of embankment dams. Local infiltrations through leaks in an upstream sealing spread out widely into the dam body and do not result in local leakages in the downstream faces of homogeneous embankment dams. Additional safety elements like toe drains are urgently to be recommended.

The geophysical measurements show that the self-potentials react immediately to changes in the seepage conditions, at least in certain regions; but a generalization of these results is not yet possible. Also the temperature measurements show significant reactions, which are substantially influenced by seasonal and day/night fluctuations.

The investigations show that the geophysical methods applied can be useful for the detection of seepages through dams, but only under certain favourable conditions. More research work and systematic investigations seem to be required.

---

[1]Bundesanstalt für Wasserbau
[2]Institut für Bodenmechanik und Felsmechanik der Universität
[3]Lehrstuhl für Angewandte Geologie der Universität
all in Karlsruhe (Federal Republic of Germany)

Lecture Notes in Earth Sciences, Vol. 27
G.-P. Merkler et al. (Eds.)
Detection of Subsurface Flow Phenomena

1. Introduction

The safety of dams is (after DIN 19700) determined essentially by:

- the stability in the cases of static, hydraulic and dynamic loads
- the efficiency of seals and its joints to the subsoil and to adjacent buildings
- the foundation of the dam
- the harmless drainage of seepage water
- the correct design and dimension of buildings joining to the dam.

In the case of inefficiency of a seal - for example by leaks - an uncontrolled and dangerous seepage can result. Sometimes such seepages cause a failure of the dam and there are enough examples of damages in Germany in recent times (Katzwang, Kirchheim, Bostalsee an others).

The reasons of such failures are various. In one case the thin dam sealing (asphaltic concrete lining) was sensitive to local strains and developed cracks or holes. In another case some mishap occured in connection with a plastic sheet used as seal. In a third case the junction between a sealing layer and a solid structure penetrating the dam was the neuralgic point (joint between an asphaltic facing and an intake tower failed). Another possible reason are imperfections in dam seals caused by air or water bubbles in asphaltic facings if such bubbles burst. Cracks may also develop in asphaltic cores of water retaining dams, when very fast cooling causes too much shrinkage or contraction of the material. In another case nearly 200 joints of a 2 km long diaphragm wall opened due to temperature changes of the water which was stored in a reservoir created by a 2 km long surrounding dam.

Experiences of this sort cause our interest in the effect of leaks in thin sealing elements of dams and led the German Ministry of Transport to look for new methods for the early recognition of dam leakages. For this purpose the Ministry supports research projects of others by providing materials, instruments and staff. One of such common research projects was carried out by the University of Karlsruhe and the Federal Waterways Engineering and Research Institute in the period from 04/01/1984 to 10/31/1987 to investigate the leakage problem and to develop and test methods for detecting percolated zones in dams. The project included hydraulic and geophysical measurements (thermic and geoelectrical) on models in the laboratory and in particular on a large scale model dam.

The authors would like to take this opportunity to thank the Minister of Transport for his financial support and his agreement to this publication, which presents part of the results of the research work.

2. Dam Model

In 1984 a large scale dam model (height: 3.6m; length: top 22.5m, toe 13.4m; width: top 1.1m, toe 17.2m; volume: $600m^3$) was built on the premises of the Federal Waterways Engineering and Research Institute in Karlsruhe. It is placed in a pit with a HDPE plastics sheet (thickness 2.5mm) forming the reservoir and the impervious foundation. The material of the dam is uniform sand (0.2 to 2 mm, $k_f=2\cdot10^{-4}$m/s). A short toe drain made of gravel (2 to 16 mm) has been provided in order to assure stability even when full percolation occurs. On the downstream slope a grass cover was provided in order to simulate natural conditions and to protect the slope against erosion.

The dam is instrumented with piezometer tubes and temperature cells in various profiles and elevations. Along the downstream surface, electrodes for self-potential measurements are installed. The temperatures of the dam surface are observed by means of an infrared camera. The water seeping through the dam is collected in a ditch along the downstream dam toe and measured by means of a calibrated weir. The acquisition of all data is controlled automatically by a computer which also serves for the evaluation of the data. The conditions of an open air model require the measurements of precipitation, air and water temperatures and the use of a filter system to keep the water clean.

After a first test without a dam sealing, in order to obtain reference values for comparing the later data from tests with a leaky dam seal, the upstream sealing membrane was installed and sealed against the HDPE sheet along the line of contact. 35 prefabricated concrete elements with defined leak openings (3 cm by 80 cm each) were incorporated in the upstream seal in three rows at different elevations (level 1 to 3). The leak openings were covered with movable plates so that each leak could be opened or closed when the reservoir was full as required in the testing programme.

Figure 1 shows a view of the model dam and the water reservoir, Figure 2 gives a cross-section.

3. Hydraulic Test Results

Unsealed Dam

A first test was run in September 1984 without a dam sealing. For steady conditions of impoundments at 1/3, 2/3 and 3/3 of reservoir height, the seepage quantities corrected to a reference water temperature of 20°C were 0.14, 0.61 and 1.55 l/s, respectively. The phreatic lines corresponded very well with those predicted according to [1] or [2].

Figure 1

View of the model dam and reservoir

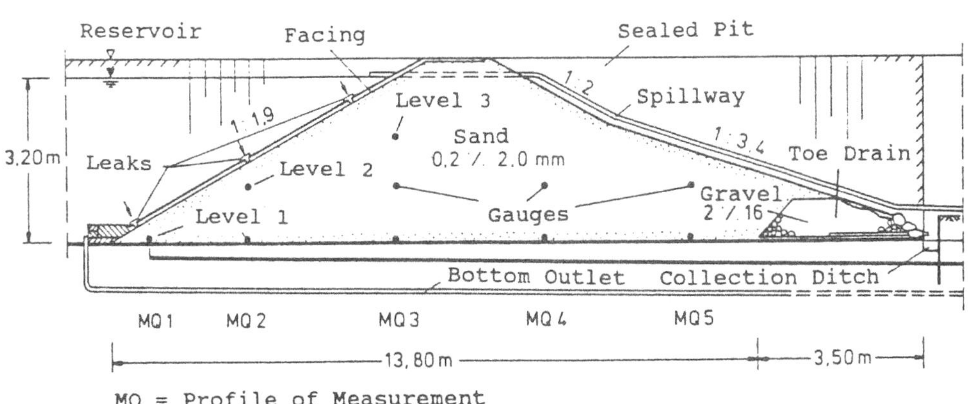

Figure 2

Cross section of the model dam

Sealed Dam with Leaks (Leakage Tests)

From 1985 to 1987 more than 30 tests have been run with varying conditions of leak positions and leak sizes. In these tests development in time and space and the dimension of seepages (volume and geometrical form of the saturated portion of the dam, position and size of the seepage line, seepage quantity) have been regarded.

Figure 3 shows for example a section of the model dam within development of percolation after opening two leaks in the centre portion of the dam at low level. Figure 4 shows the corresponding development in time of the quantity of water emerging from the dam and development in time of the saturated sectional area. In this example with a degree of damage of about 2‰ for steady conditions the seepage quantity is about 30% of that one of the unsealed dam.

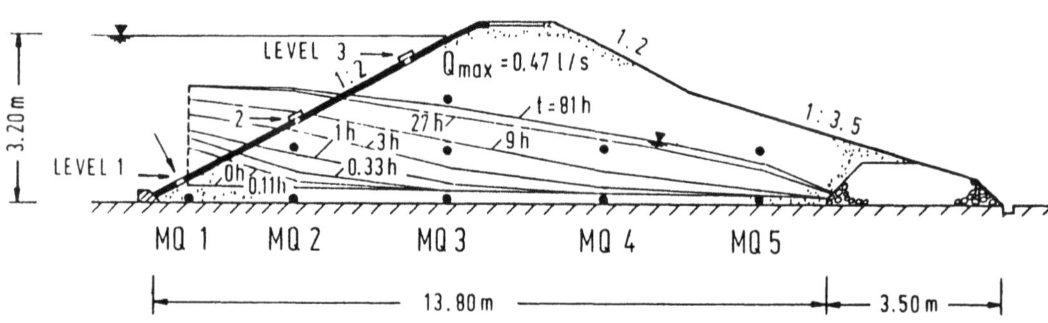

Figure 3

Section of the model dam - development of percolation (saturated zone) after opening two leaks in the centre portion of the dam at low level (level 1)

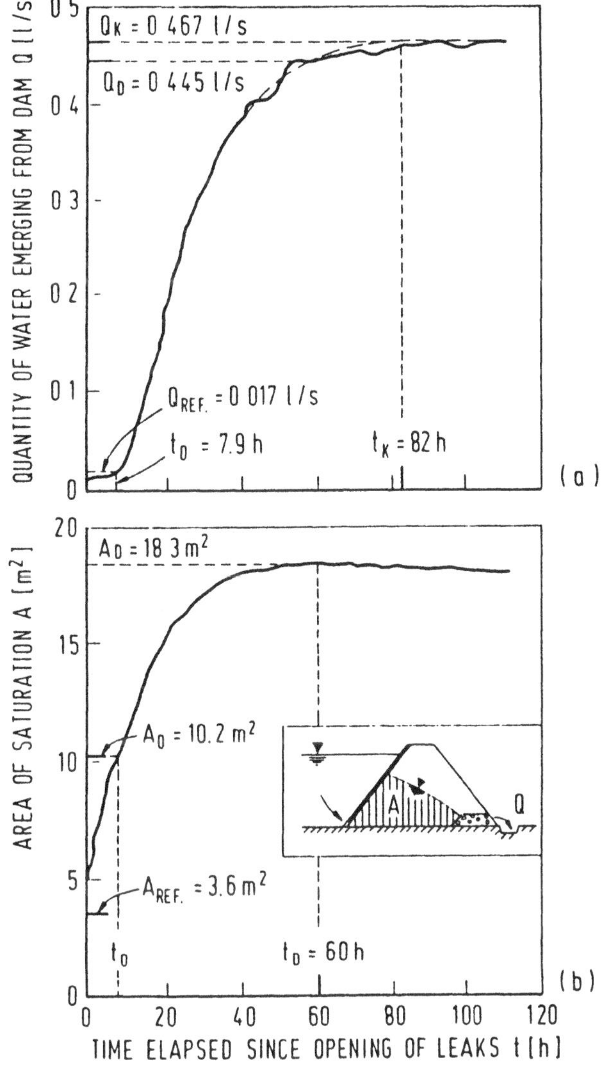

Figure 4

Development of (a) the quantity of water emerging from the dam and of (b) the saturated sectional area for two open leaks in centre portion of dam at low level

Q_{REF}, A_{REF}	Values of Q and A before opening of leaks (due to imperfections in joint between dam facing and lining of pit with HDPE sheet)
t_0	First increase of Q after opening of leaks
t_D	Piezometer readings become constant
t_A	Capillary effects have developed

The results of all leakage tests are shown in figure 5 (seepage quantity Q and the ratio Q/Q_0 versus degree of damage κ for stady state). Within a range of κ-values covered in the tests (0.1°/∘∘ to 10°/∘∘), the relative seepage quantity varies between 10 and 80%. That means that even small leaks in sealing faces can lead to considerable water losses and extensive percolations of embankment dams.

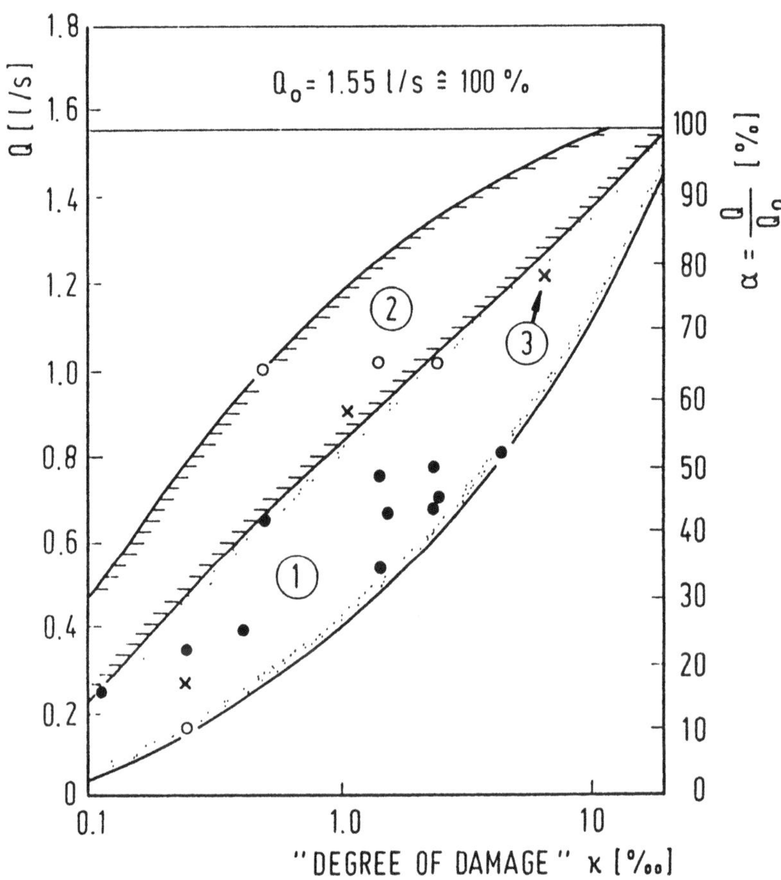

Figure 5

Seepage quantity (steady conditions) as a function of "degree of damage"

(1) Leaks at a low level (2) Leaks at a high level
(3) Open leaks at both low and high levels
Q_0 Reference seepage quantity for dam without sealing

Another important result of the tests is that local infiltrations through leaks in an upstream sealing spread out widely into the dam body and do not result in local leakages in the downstream faces of homogeneous embankment dams. The maximum of inclination of the seepage line observed in direction parallel to the dam axis (with a local leak in centre of the sealing face) was 6° (steady state).

A comparison of field testings and laboratory testings showed that the head loss in the leak itself (thickness of 10 cm) is nominal only for leaks less than 0.5 to 1 mm in width.

4. Results of Thermal Measurements

The investigation using thermometric methods were performed in order to find out how the changes in the hydraulic conditions and the temperature fields relate to one another (thermal behaviour of a dam in connection with loading with water and in the course of water infiltration through a leak). The final aim of these efforts is to use thermometric methods as a mean for leakage detection.

So during the whole period of tests (from 1984 to 1987) thermal measurements of soil, reservoir water, air and dam surface temperatures were made by different kinds of observation systems:

- temperature cells firmly installed with a cable connection
- temperature cells inside the reservoir
- quicksilver thermometers
- thermovision system (infrared camera measuring the radiation intensity in the infrared range without contact).

The distribution of temperatures in dams depend on different influences so that we can distinguish the following cases:

- dry dams without any water loading
- dry dams with a water loading but without any water percolation

(dams with undamaged upstream sealings, sealings without any leak)

- dams with water percolation as a result of storage and damaged sealings (sealing with leaks).

The distribution of temperatures in a dam body without any water loading is similar to that in the earth body below the ground surface, dependent on the surrounding climate (solar radiation, wind, precipitation etc.) with its daily and seasonal changes of the air temperature. The temperature curves (variation with time) are somewhat sinus-shaped whereas the amplitudes are maximal near the ground surface; besides damping a retardation is to be observed with depth. As regards the temperature field in the section of the dam, a typical situation is shown in figure 6 (a) (symmetrical temperature distribution increasing with depth in a steady temperature condition).

As soon as the reservoir is filled with water, but the upstream sealing is impervious, conductive temperature effects are superimposed, depending on the temperature of the water. The influence of the aerial zone (downstream slope face) corresponds to the climate. The influence of the water zone (upstream slope face) corresponds to the water temperature. In the example shown here, the temperature conditions changed drastically: the reservoir was filled with water with a temperature of 19°C while air temperature was 6°C (figure 6 (b): distribution of temperature 16 days after filling the reservoir, conductive heat transport).

In the case of a water percolated dam, there is - in addition to conductive heat transport - also convection and (to a neglectable extent) dispersion. While the water seeps through a dam body as a result of a leak in the upstream facing of the dam, the temperatures of the water and the soil of the dam converge. The main part of the convection will take place in the percolated, saturated zone and a small portion in the capillary zone. The change in the temperature of a dam with seepage depends to a

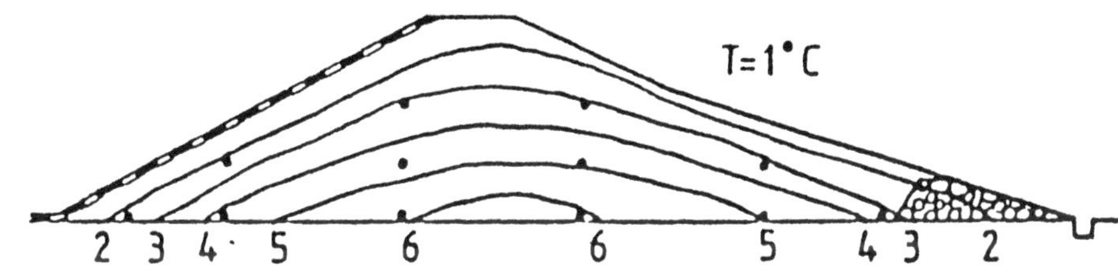

a) without water (12.3.87)

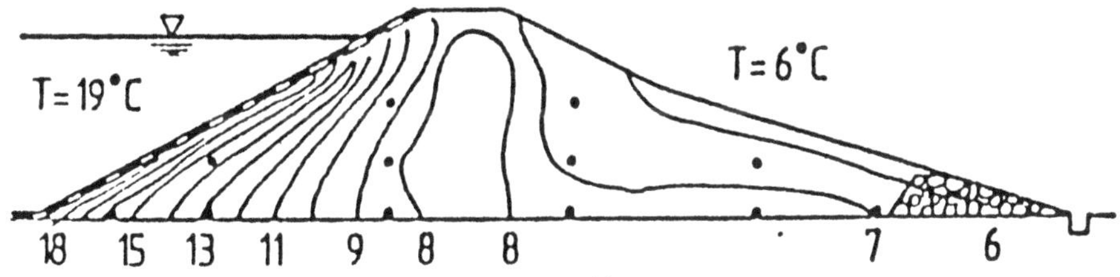

b) with warm water without leakage (30.3.87)

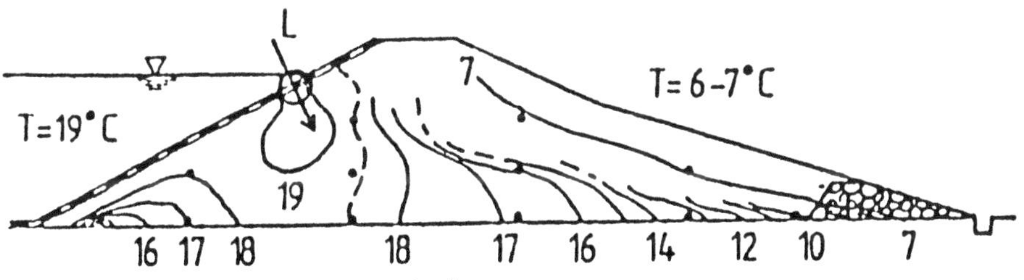

c) 48 hours after opening the leakage

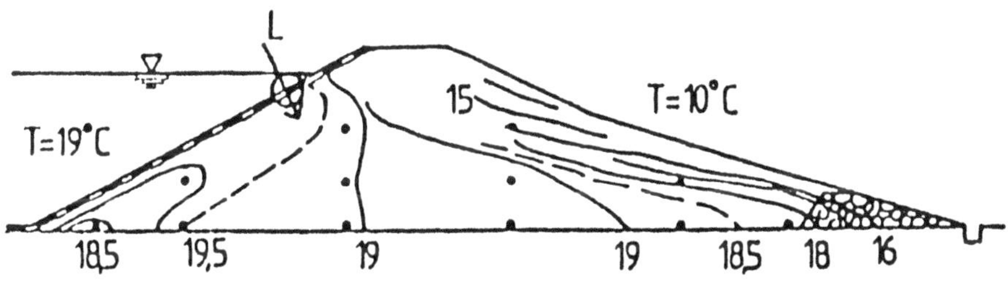

d) 14 days after opening the leakage

Figure 6

Section of the model dam - distribution of temperatures

large extent on the velocity with which the water passes the dam. This on the other hand depends on the leak (height, size), on geohydraulic properties (dam material, sealings, subsoil) and on geometric parameters. The temperature variations in a percolated dam are rather complex, if all variations in the boundary conditions are taken into account. Even under the simplifying conditions of the open air model tests, one has to study the situation for a while until all effects are understood.

Figures 6 (c) and (d) show the distribution of temperature in the dam as it results from the convective heat transport two days after opening two leaks centered at a high level and two weeks after opening the leaks.

As can be seen from figure 6, surface measurements of temperature can hardly indicate even a local leak in such type of dam which has an upstream facing and a homogeneous body, since the infiltration of water spreads out widely and does not result in a local temperature anomaly in the downstream surface of the dam. On the other hand, dams showing veins of percolation due to inhomogenities in their inner structure may show such temperature anomalies and, thus, detection by surface temperature measurements may be possible.

5. Results of Self-Potential Measurements

The measurements of self-potential were made over a long period of the research project to find out the relations between and influences of the hydraulic and geoelectrical fields and to have some possibility for detecting seepage as a result of leaks in an early state of water infiltration.

At the downstream facing of the dam there have been installed up to 40 unpolarizable electrodes (Cu - $CuSO_4$) to measure the self-potentials.

The main results of all the self-potential measurements are:

- self-potential field on the surface of a dam reacts to seepage processes inside the dam in a measurable range
- changes in self-potential fields can restrict only to a part of the downstream facing of a dam
- influences of changings in temperature and other effects (precipitation, evapotranspiration etc.) superimpose to influences of seeping water, so that, in the case of dominance of such influences, extensive data analysis is necessary, for example smoothing and filtering of the data with numerical algorithms
- to explain the fundamental phenomena and to solve the problems in measuring, fundamental and systematic laboratory investigations are needed.

One direct result of the self-potential measurements is that there was a strong temperature effect leading to pronounced day/night-ondulations of the self-potential values under normal weather conditions as they prevail over the major portion of the year in our region of the world.

Last we will give an example of a self-potential reading made during a leakage test. Figure 7 shows the correlation between hydraulic and geoelectrical measurements for a test in November 1985 with a vertical leak in the sealing face of the dam and with more or less constant temperature conditions over the entire duration of the test. Besides the diagrams of hydraulic heads and water level in the reservoir measured by pressure gauges and in addition to the diagram of quantity Q of water emerging from the dam, the curves of selfpotential of some electrodes are plotted in the figure.

The electrical potentials in the upper zone of the downstream face of the dam remarkably decrease as the water infiltrates into the dam body. The same electrodes react again as soon as the infiltration rate increases. The electrical potentials in the toe region of the dam - on the other hand - were observed as not re-

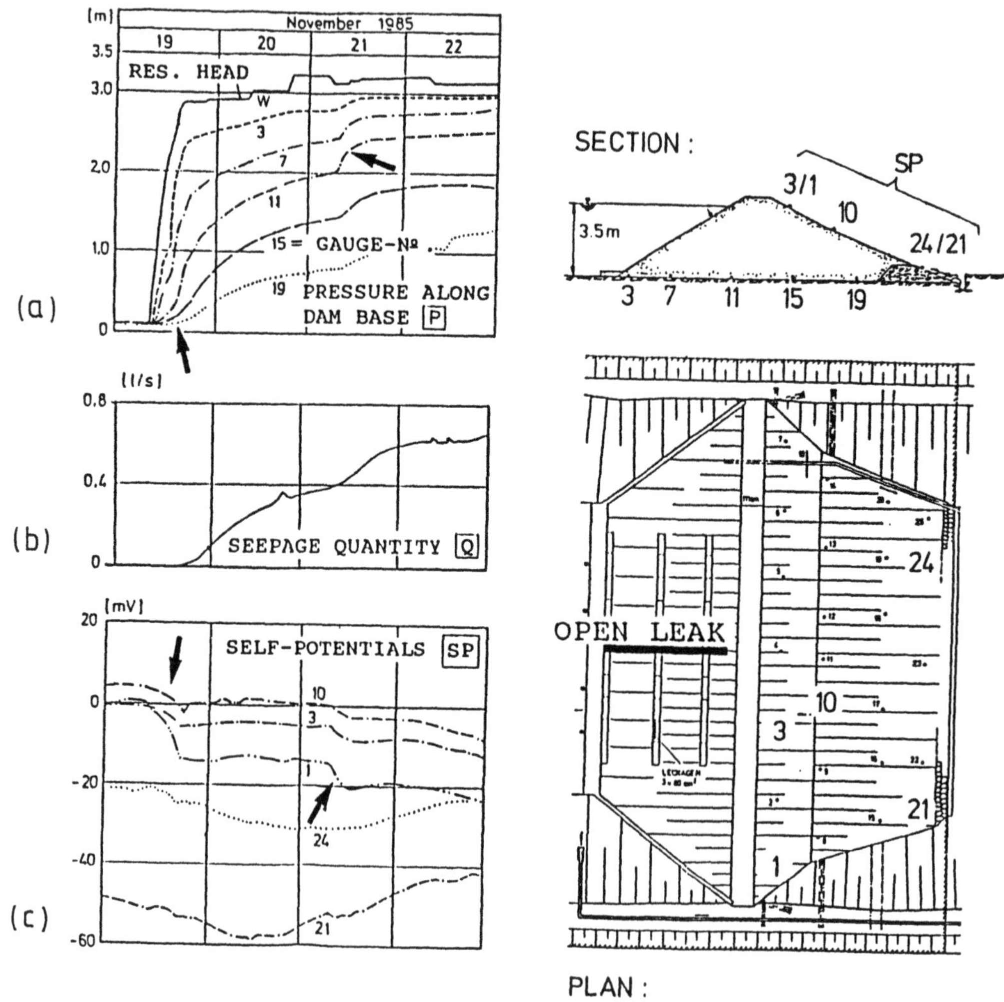

Figure 7

Correlation between hydraulic and self-potential measurements
(a) Hydraulic heads and water level W measured by pressure gauges
(b) Development of seepage quantity Q emerging from the dam
(c) Self-potential values of the electrodes 1, 3, 10, 21 and 24

acting on the changes in the hydraulic conditions. As regards the pattern of the distribution of potentials along the surface of observations, we have not found a key to transduce the electrical field into the pattern of flow.

6. Conclusion

The results of the investigations briefly described in this contribution give strong support to the opinion that embankment dams must be designed so that seepages, which may occur when thin sealing elements malfunction, can safely be drained from the dam body. The need for a "second line of defence" (for example a toe drain) is, thus, emphasized. Even small leaks in sealing faces can lead to considerable water losses and extensive percolations of embankment dams.

The geophysical measurements show that the self potentials react immediately to changes in the seepage conditions, at least in certain regions; but a generalisation of these results is not yet possible. Also the temperature measurements show significant reactions, which are substantially influenced by seasonal and day/night fluctuations.

The investigations show that temperature and - to some degree - self-potential measurements can be useful for the detection of seepages through dams, but only under certain favourable conditions and when applied in certain indicative points. More research work and systematic investigations seem to be required to elaborate the conditions under which such measurement techniques can be applied with success.

Bibliography

[1] KOZENY, J. - Hydraulik. Springer Verlag, Wien, 1963.

[2] POLUBARINOVA-KOCHINA, P.Ya. - Theory of Ground Water Movement. Princeton University Press, Princeton, 1962.

THE PROJECT OF THE VOLKSWAGEN FOUNDATION "GEOELECTRICS/THERMOMETRY"

H. Armbruster[1], A. Blinde[2] and H. Hötzl[3]

1 Aim of the Project

If there is a difference in the hydraulic potential between two points in the subsoil, water flows through the subsoil, which is assumed to be permeable. If this difference in the potential can be maintained, after a certain time a constant potential field exists, which has a measurable value at every point of the aquifer. This hydraulic potential field depends not only on hydraulic, geometric and geohydraulic parameters but also on boundary conditions. Due to the transport of water, which is always related to the transport of heat and physical chemical substances, the existing chemical, physical, thermal and geoelectrical potential fields of the aquifer are changed, which otherwise depend on other parameters and boundary conditions than the hydraulic field. These potential fields of the subsoil are, on the other hand, related to the potential of the surface, so that a hydraulic change in the subsoil also produces a change in the potential fields of the surface. The aim of this project was to relate two of the potential fields of the surface, namely geoelectric field B (Fig. 1) and thermal field C to the causal hydraulic field A in the subsoil and this should be able to be recordered analytically.

1) Bundesanstalt für Wasserbau, Karlsruhe, FRG
2) Institut für Boden- und Felsmechanik, Universität Karlsruhe, FRG
3) Lehrstuhl für Angewandte Geologie, Universität Karlsruhe, FRG

Lecture Notes in Earth Sciences, Vol. 27
G.-P. Merkler et al. (Eds.)
Detection of Subsurface Flow Phenomena

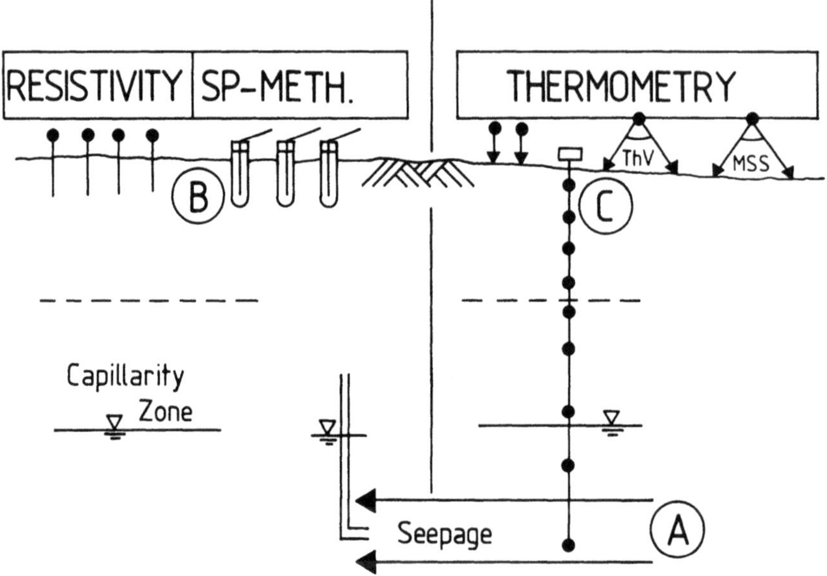

Fig. 1: Seepage (hydraulic pot.field) causes two pot.fields on the surface (Thv = Thermovision camera, MSS = Multispectral Scanner)

At the same time it should be clarified as to which relationships exist between fields B and C caused by hydraulic field A and which, however, cannot possibly be directly formulated (Fig. 2).

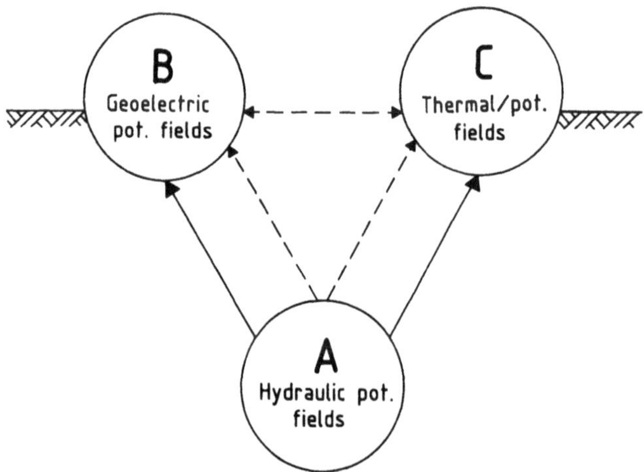

Fig. 2: The relations between the 3 involved pot. fields

The very ambitious aim could only be tackled in stages and embedded in a number of projects from various institutes, each with their own main emphasis.

2 Contributing Institutes

2.1 Federal Waterways Engineering and Research Institute (BAW)

The BAW is the central research institute of the Federal German Ministry of Transport (BMV) in the field of waterways (canals, rivers). It thus also deals with problems regarding the groundwater flow in and around constructions (e.g. dams, locks, weirs, etc.), which are influenced by sealing elements and any possible defects of these. The Ministry is responsible for dam stretches extending approximately 700 km , the stability of which is influenced by the groundwater flow. The BAW has thus been working for years in the field of groundwater hydraulics and leakage research, above all using thermal methods.

2.2 Institute for Applied Geology (AGK) at the University of Karlsruhe

This institute, being part of the Institute of Geology, has its main emphasis in the field of groundwater hydraulics, where above all for

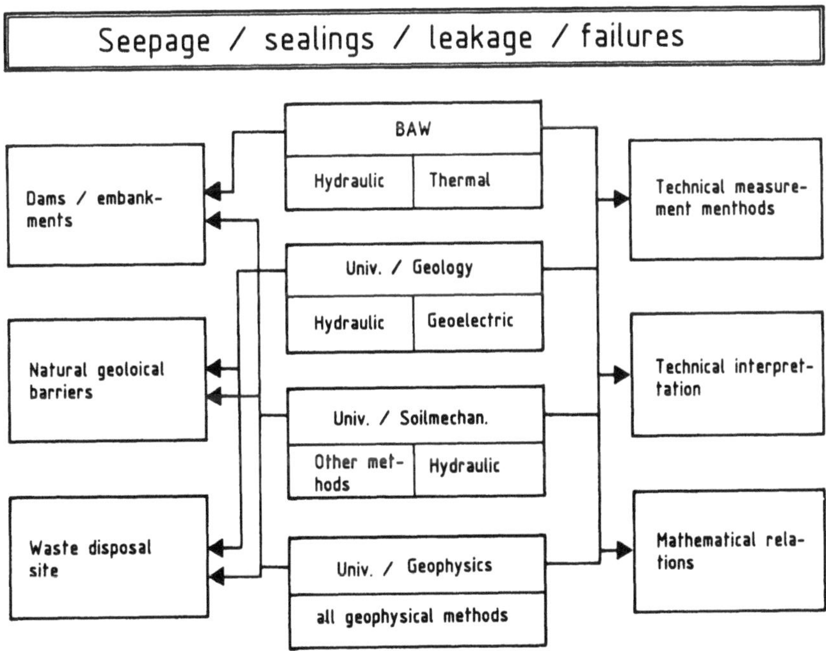

Fig. 3. The scientific institutes and their common problems

years work has been carried out on the research of water conductivity of fissured rock and granular soil with tracers and geoelectric methods.

2.3 Institute of Soil and Rock Mechanics (IBF), Department for Embankment Dam and Foundation Engineering, University of Karlsruhe

This department of the IBF has its emphasis in the field of groundwater hydraulics, where for years theoretical and practical work has been carried out, above all on the hydraulic effects of leakages in dam sealings. This department, which has in the meantime been renamed, still works today in the field of deposits.

2.4 Other Institutes

Within the framework of the co-operation on a scientific (and, in individual cases, also a commercial) basis, the know-how of other institutes could also partly be used. These include the Geophysical Institute of the University of Karlsruhe, the Soil Mechanics Institute of the College of Groningen (Netherlands), Geothermik Consult in Hannover and Spacetec in Freiburg, both FRG.

2.5 Co-operation

The co-operation of the three institutes mainly involved can be seen in Fig. 3. The title is intended to emphasize once more the multiplicity of the problem. Each institute contributed its own specific experience, which had partly been gained in other research projects.

The most important of these were a project of the IBF on leakage research in dams, commissioned by the BMV (sect. 2.1) in co-operation with the BAW, and a project of the AGK together with the IBF to investigate flow processes, commissioned by the German Research Community (DFG).

3 Measuring Technics

3.1 Hydraulics

The main problem lies in determing the geoelectric and thermal fields. The initial sizes are the hydraulic potentials (field A) of which the absolute sizes and changes, however, must be known.

The measurement of the hydraulic potentials was carried out on models by means of measuring the stand pipe reflecting level in tubes, the ends of which were equipped with a filter stone. To automate this process by means of a two-way cock, pressure pickups were attached, of which the electric signal could be called up on a computer scanner. The measurements followed after the tubes had been deaerated and the pressure pickups had been calibrated both manually (controls) and automatically.

To measure the natural hydraulic fields, observation pipes with a defined filter section were used, of which the water levels were measured with light perpendiculars (manually or self-recording) with a pressure transducer (automatic call-up).

3.2 Geoelectrics

The measurement of resistances in the soil was carried out with metal probes, which were inserted up to 15 cm into the ground and in general were wired according to the WENNER layout. The measurements were not automated.

The measurements of the self-potentials were carried out with various probe systems, which were tested at the same time. Various unpolarizable probes were used, in the open air mostly on a $CuSO_4$-Cu basis. All measurements were initially carried out manually and later on the models by computer via a scanner.

3.3 Geothermics

For the measurements in the soil (in the laboratory and in the open air) various measuring systems were tested, which work partly on a semiconductor basis and partly on a resistance basis. The measurements were carried out manually at first and later by computer via a scanner.

For measuring the surface temperatures in the laboratoy thermovision cameras were used and for the measurements in the open air a multispectral scanner, which recorded the thermal spectrum (medium-infrared) as well as visible light in two wavelength ranges.

Additional measurements, such as air and water temperatures, rainfall, wind speed, chemical parameters, moisture, etc., required additional, usually manual measuring systems.

4 Physical Models

4.1 Models for Preliminary Studies

Only part of the research was carried out on these models, which were generally built for other purposes, i.e. not all three fields were systematically measured at the same time. All the models consisted of sand (partly with supporting bodies of gravel) and water flowed through them.

1. On the half-truncated cone (Fig. 4a) only the spacial spreading of water from a circular leak was measured.

2. On the ring dam, hydraulic fields and temperature fields on the surface were examined (Fig. 4b).

3. In the small plexiglass channel (Fig. 4c) a dam was built and water flowed through it, whereby the hydraulic and self-potential parameters were examined (Armbruster and Merkler 1983).

4. In a circular plexiglass tube (Fig. 4d) placed horizontally and in which water flowed through, the hydraulic and self-potential field at various hydraulic gradients could be measured.

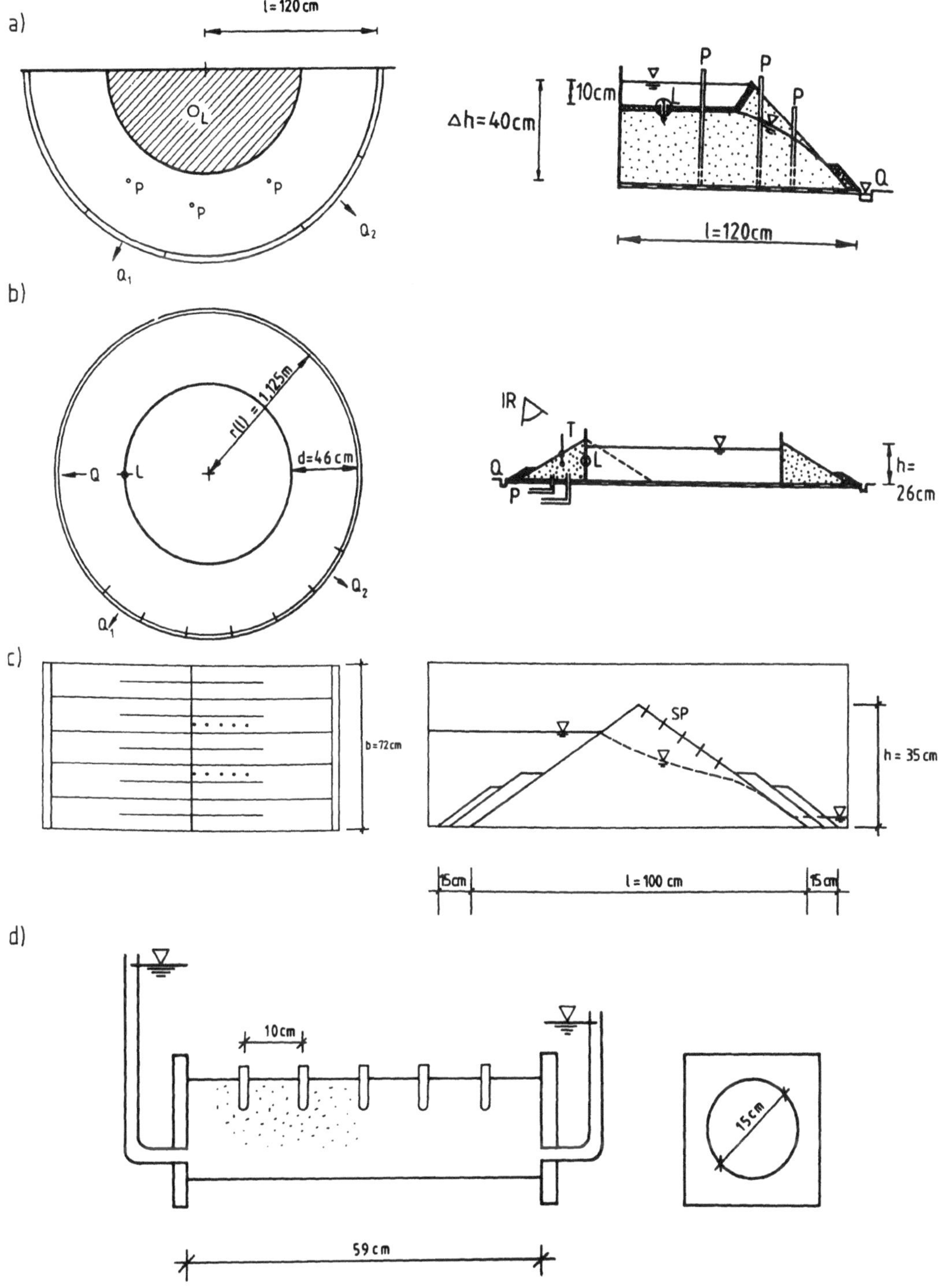

Fig. 4. The Models for preliminary studies (a ÷ d) see text
P = Piezometer, T = Temperature, Q = discharge, L = leakage,
SP = self-potentials, iR = infrared

4.2 Model of a Large Channel

The large channel in the laboratories of the BAW has plexiglass walls to avoid natural electric voltages from corrosion. It is 6 m long, 2 m wide and 1.5 m high, so that three-dimensional flow processes can develop (Fig. 5).

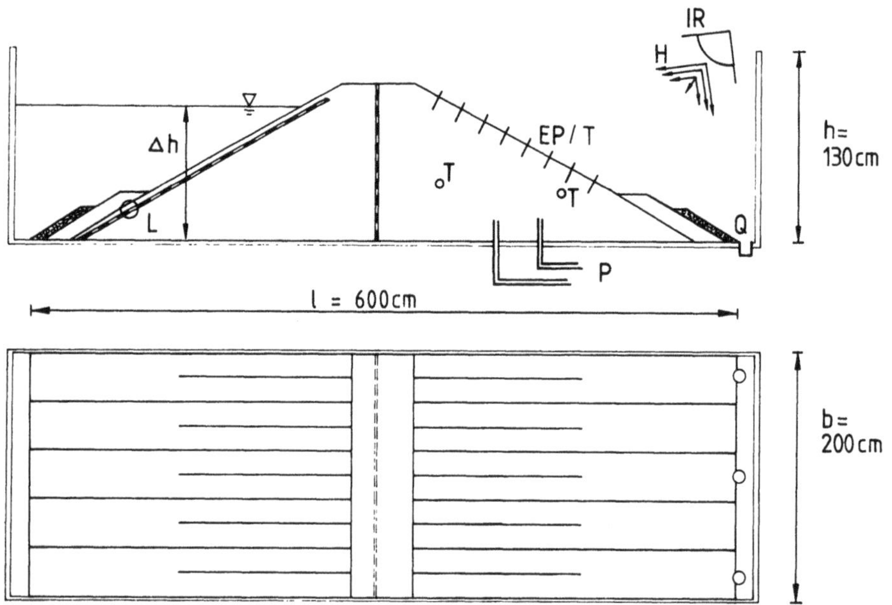

Fig. 5. The large plexiglass model
SP = Self Potentials, T = Temp., IR = infrared, H = Heating

The model was specially built for the research commission mentioned and thus allows the measurements of all three field sizes by means of measuring cells inside the model (hydraulics, thermics) and on the surface (resistivities, self-potentials, non-destructive thermics). Through a relatively simply modification, various hydraulic basic situations such as:

1. Seepage of dams;

2. Flow to a drain (at right angles/longitudinally);

3. Leakage flows with seals destroyed in some areas or with intact, but slightly permeable sealing elements;

4. Underseepage of constructions, etc.

can be produced. Due to the possibility of heating the flowing medium in addition, temperature effects can be increased. The measurements of all sizes can be carried out manually, but presently, however, they are for the most part automated (call-up, storage, processing).

4.3 Model of a Dam in the Open

This dam is outside in the grounds of the BAW, almost 4 m high, with a basic width of approx. 20 m, but only about 20 m long (Fig. 6). It was built for the research of the hydraulic effect of leakages, commissioned by the BMV (Brauns 1986; Armbruster et al. 1987).

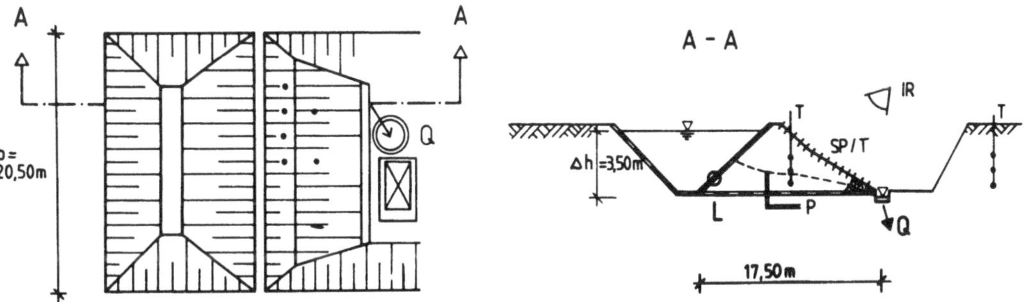

Fig. 6: The model dam in the open

It was available for the research project, i.e. the hydraulic, geo-electric and temperature measurements were generally carried out at the same time. As with the large channel there were measuring cells inside (hydraulics, thermics) and on the surface (self-potentials, resistances, non-destructive thermics), however, in a coarser grid than in the channel.

Since the modification of the dam was not easy to accomplish, the change in the hydraulic field as laid down in the project was achieved by using reclosable leakage system of the surface sealing (30 openings, different arrangement and sizes). The measurements were taken under actual natural conditions (sun, rain, snow, day/night). The measurements were mostly automated (call-up, storage, processing). Thermal measurements were also taken, even when there was no seepage.

5 Investigations in Natural Fields

5.1 Dams in Canals and River Barrages

Commissioned by the Ministry of Transport, the BAW examined, in part together with the AGK, a large number of dam points in German canals (lateral dykes) and dammed rivers (lateral dykes and dams across valleys)(Armbruster and Merkler 1982). These dams have, in addition to varying hydraulic and geometric boundary conditions, also varying sealing and drain elements, so that the measuring results can be systemized. The dam points examined showed that the seepage was in part not as planned, so that some had to be repaired (measurements before repairs/after repairs). The measurements included the hydraulic field at points (observation tubes), geoelectrics (resistances with varying distances between the electrodes, self-potentials) over the whole area and thermics, partly longitudinal verticals (in observation tubes), partly over the whole area with the infrared measuring device.

5.2 Sealings

The effect of sealings was examined either indirectly in repaired areas of the dams as in section 5.1 or directly in a project, which had been sponsored beforehand by the DFG (Blinde et al. 1983). Here, a diaphragm around a planned deposit was initially incompletely closed and thus simulated a leak flow.

5.3 Subsidences

Measurements near wells (radial flow) and drains (stable flow) are planned, but have not as yet been started systematically.

6 Progress of the Project/Theoretical Work

The project has the aforementioned large complex of physical models, on which measurements are carried out systematically, in order to obtain the qualitative (Phase I) and quantitative (Phase II) correlations of the required fields. The correlations found will be checked with the analytical and numerical models (Phase III) being worked at the same time, whereby Phases II and III will correct each other. The finally determined correlations will be checked (Phase IV) under real conditions (dams, sealings, subsidence, etc.), which are already partially available. Thus, nearly all phases overlap, which naturally greatly impedes the systematic processing of the data measured.

References

Armbruster H., Merkler G. P. (1982
Möglichkeiten der Leckstellenortung an Dämmen, Geotechn. H. 1

Armbruster H., Merkler G. P. (1983)
Measurement of subsoil flow phenomena by thermic and geoelectric methods. Bull. Int. Assoc. Eng. Geol., No. 26 - 27, Paris

Armbruster H., Degen F.-P., Faber S., Mazur W., Merkler G. P. (1987)
Durchsickerung von Dämmen und Deichen bei Dichtungsleckagen und Methoden zur Erkennung von Sickervorgängen. Bericht des IBF an das BMV (unpublished)

Blinde A., Hötzl H., Merkler G. P. (1983)
Determination of the disaggregation anisotropy of rock masses by means of geophysical investigation methods. Bull. Int. Assoc. Eng. Geol., Nr. 26-27, Paris

THE APPLICATION OF GEOELECTRICAL AND THERMAL MEASUREMENTS TO LOCATE DAM LEAKAGES

H. Armbruster[1], A. Blinde[2], J. Brauns[2], H.D. Döscher[1], H. Hötzl[3], G.P. Merkler[3]

1 Introduction

The dam system of the constantly banked-up waterways in FRG (700 km in length) vary with regard to their construction (geometry, soil), to the sealing systems (vertical, horizontal seals), to the draining (lateral ditches, drains, groundwater) and above all to their age, which automatically reduces the knowledge of the aforementioned points. Thus, methods are being sought, which give non-destructive results over the whole area required, providing information on

- the general stability of the dam;
- the stability of individual sections of a dam, which for certain reasons appear to be endangered (prognoses);
- the stability of repaired sections of a dam (controls).

For relatively short dams, which are usually dams across the valleys, measuring systems can already be installed during the construction phase, which also control the completed dam. However, in very much longer lateral dykes of canals and banked-up rivers monitoring is difficult, even if they are newly constructed. However, with canal dams up to a 100-years-old, new methods have to be found.

1) Bundesanstalt für Wasserbau, Karlsruhe, FRG
2) Institut für Boden- und Felsmechanik, Universität Karlsruhe, FRG
3) Lehrstuhl für Angewandte Geologie, Universität Karlsruhe, FRG

Lecture Notes in Earth Sciences, Vol. 27
G.-P. Merkler et al. (Eds.)
Detection of Subsurface Flow Phenomena

2 Measuring Procedures to Locate Leakages

There are basically five groups of dam monitoring systems which are suitable for locating leakages (Table 1).

Group	System	Classification
I	Observations, with simple auxiliary if necessary	Non-destructive methods
II	The system is part of the dam construction	
III	Detection by assistance of scheduled systems, installed during work	
IV	Detection with systems working on the surface of the dam	
V	Detection with systems fitted supplementarily in the dam	Destructive methods

Tabl. 1: Detection methods for recognizing dam failures

Within these five groups a number of individual procedures can be described (Armbruster and Merkler 1982), which with varying efforts or expense provide informations about the current load of the dam and enable a safe calculation of the stability to be made. These calculations require the following parameters:

- external geometry: obtained by geodetic measurements;

- structure of the subsoil and the dam; obtained by direct methods (boreholes, excavating, bore probing, slot probing) and by indirect methods (driving rods, pressure probing); supported in addition by surface measurements (radar, resistance, self-potential, seismics);

- Soil mechanical parameters: obtained directly from soil probes and tests in the borehole from probing (driving, pressure, wing, isotope probes) and indirectly from surface measurements;

- geohydraulic parameters or loads: obtained directly from tests in the borehole (isotopes, Lefranc tests, diagnosis effect tests) and indirectly from measurements of the water level or pressure, water quantities, temperatures or from tracer tests;

- external loads: only necessitate measurements in special cases.

Most of the above measuring procedures are routine, which however often only enable information to be obtained about certain points. In the following, two of these methods are described in more detail.

3 Surface Measurements

The procedures for surface measurements are used under two aspects:

1. A danger point is suspected in the dam. The measuring procedure should invalidate or confirm the suspicion and in the latter case give the extent of the danger point and enable further specific measurements to be made.

2. The procedures are used as a precautionary measure for dam safety. If there are any indications of danger points, these indications must be followed up.

The measuring procedures required must consequently be as quick as possible and cover a complete area, with regard to both the execution and the evaluation.

3.1 Geoelectrics

3.1.1 Resistivity Method

The <u>measurement of the electrical resistance</u> of the soil against an artifically produced potential is an old tried and trusted method. The electrodes are inserted 10 - 15 cm in the dam in a grid formation, whereby the distance between the electrodes, which relates the information to depth, is variable. The apparent specific resistance measured over a larger area, which depends on the soil and its water content, including the chemical composition of the water, is entered in layouts as isolines and provides inter alia the following information:

- The soil/water conditions are stable over larger areas at a defined depth (or vary in some areas);

- The resistance measured corresponds to the existing soil (or does not);

- Horizontal seals exist (or are missing);

- The water levels (where the soil is known) are higher/lower than expected);

- The height and depth of the preferred directions of water in the aquifer could be observed (or not).

Resistance measurements are simple to carry out and to interpret. The choice of suitable distances between the electrodes (e.g. 2.5 m; 5 m; 7.5 m; 10 m) makes it possible to use the same electrodes several times using the WENNER layout. With appropriate preparation a dam area of 1 - 5 hectometers of a dam 10 m in height can be measured per day with one measuring group.

3.1.2 Self-Potential Method

The measurement of the naturally existing electrical potential of individual points against a reference electrode (self-potentials) has become possible with a high precision by the use of modern high-resistance voltmeters. For this purpose non-polarizable electrodes of varying types (usually ceramic plugs with a $CuSO_4$ solution around a Cu rod are used, which must have good contact with the soil. Here, the potential measured is that which occurs during water transport in capillary tubes (HELMHOLTZ). The polarity sign of this potential and its amount (generally $<$ 100 mV) give indications as to the height and direction of the hydraulic gradients in the subsoil. The choice of distance between the electrodes is only important with regard to the quality of the information about a whole area.

The potentials measured over a larger area, entered on the layout as isolines, or the difference of the potentials of neighbouring points (gradient method), also plotted on the layout, provide inter alia the following information:

- the flow conditions in the subsoil are stable over a larger area (or not);

- the flow has an ascending or descending character;

- the strongest flow is either deep and strong or higher than in the neighbouring area;

- the direction of the flow (water conductivity) could be observed;

- the toe of the dam is under stronger flow pressure (or not);

- the water flows stably to a draining ditch (or not);

- the waterflow to or through seals is stable (or not).

Self-potential measurements are simple to carry out (there are problems in gravelly dams, when placing the electrodes); when interpreting the information specific knowledge of certain aspects is necessary, if it is not possible to perform measurements with all the electrodes simultaneously. The self-potentials depend on the temperature; disturbances from other fields (e.g. electric cables in the ground, which are not earthed) are frequent. With a sufficient quantity of electrodes a measuring team can achieve a similar daily workload as when using the resistivity method.

3.2 Thermography

Under thermography the non-destructive imaging recording of the surface temperature of the dam and of the bordering areas is to be understood. This takes place by:

- measuring the temperature radiation at particular points with the help of a pyrometer (case A);

- measuring the temperature radiation over a whole area with the help of thermovision equipment on the ground or in the air, which scans the area line by line and forms a picture. The picture, which can be seen on the monitor and be stored as individual pictures or be recorded on tape, therefore consists of a number of picture elements (pixel). The information available is thus a localized pixel with a defined temperature radiation (case B);

- measuring the temperature radiation over a whole area and other radiations of the electromagnetic spectrum with the help of a multispectral scanner on the ground or in the air, whereby the recording technique corresponds to case B. Here, several pieces of information on the reflection radiation per pixel are simply stored, which are then available for an evaluation (case C).

In all three cases, if absolute temperature values are required, the termal reflection radiation has to be converted with the help of known emission coefficients and/or reflection emitters.

The principle of the evaluation of these recordings, which with the help of electronic processing of the data are available as pictures or plans, during indentification touches on thermal anomalies, the causes of which are to be found in the subsoil or below the water surface (e.g. in the lateral ditch or in the tail water of dams across valleys).

It is clear that moisture of the surface due to a leakage or even water flowing out will show itself clearly in a thermal measurement. This applies to a lesser extent to a deep-lying flow at a point which brings moisture to the surface in cohesive soil in a capillary manner.

The heat is also conducted to the surface from leakage flow areas with a different temperature, which is generally visible with a thermal solution of = 0.2 K (the geometric solution depends on the distance).

In case C, in addition to the temperature, the radiation of other wavelength can be used. Reference is made here to the literature (Hager et al. 1985; Armbruster and Merkler 1983; Armbruster et al. 1985). Due to the recording of the area at different times, case A is not suitable as a method to be applied during the day in the open air. After the anomalies have been identified in a difficult evaluation process, these points of anomaly are examined further using other methods, i.e. the extent of specifically recorded leakage areas will be ascertained.

When using an aeroplane, this method is complex. However, over 100 km can be flown per day. Evaluation methods are still being developed.

4 Examples of Multimeasurements in Dams

4.1 Dam Across the Valley of a River Barrage

The dam across the valley of a river barrage was built in a former river bed, without diverting the river. For this purpose, a supportive construction consisting of coarse material was piled up perpendicular to the river (Fig. 1).

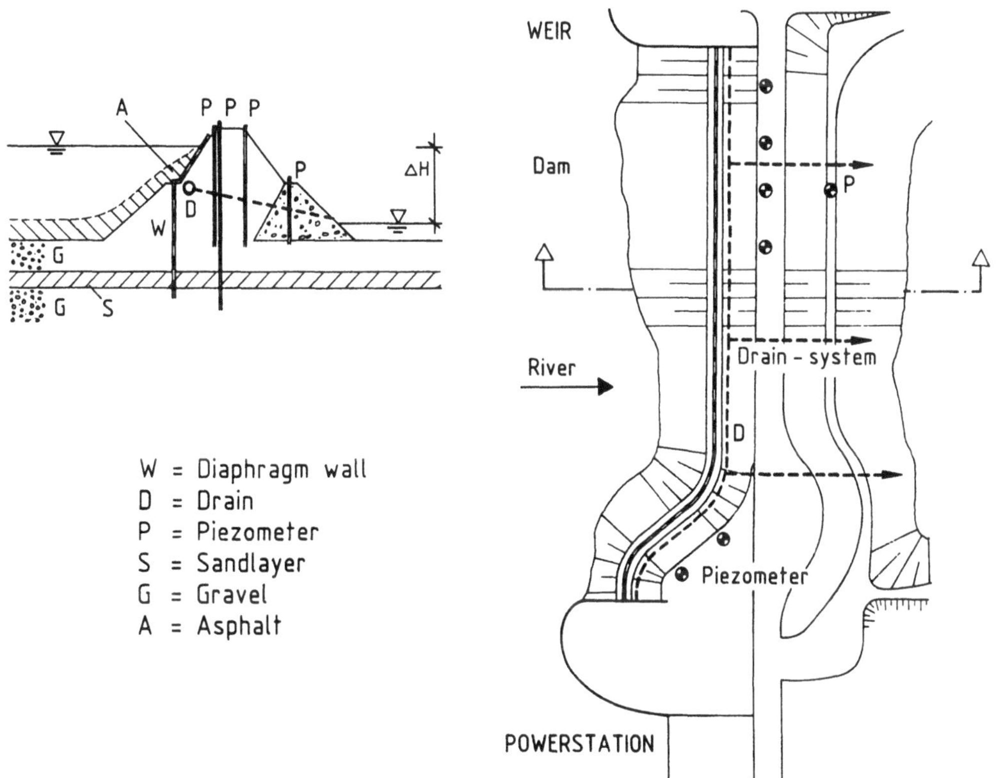

Fig. 1: dam across the valley (part of a barrage system)

More sandy gravel was poured into the head water of the supportive construction, which at the beginning filled up the spaces in the supporting construction, after which it was possible to build up the dam layer for layer under dry conditions, since the damming did not take place until after completion. The lower meters of the dam thus consist of sandy gravel with a lesser proportion of sand. The sealing consists of a single phase diaphragm wall, which integrates in a lower lying fine sand layer, which connects to the asphalt concrete of the surface sealing. The control system of the sealings (drain with control outlets into the tail water) originally planned here was deleted due to a construction error, so that a complete control system was installed afterwards:

- Control of the hydraulic potential field through about 40 observation tubes, which are arranged in five sections perpendicular to the dam axle and along the sealing wall (Fig. 1). The filter area = measuring section of some of these tubes is below the fine sand layer, so that the potential distribution in the dam can be measured. The measurement enabled the observation of the pressures of individual points, the pressure distribution in the solid body of the dam, the changes in pressure in the course of time and the pressure reduction through the seals to be observed (Armbruster 1987).

- Control of the thermal potential field in the dam by measuring the water temperatures in the observation tubes. These measurements are only possible in the saturated area of the dam, depending on the level of the tail water. The measurements enabled the observation of the changes in temperature in vertical sections, on horizontal surfaces of different heights and in the course of time to be carried out. Thus, they supply a much more differentiated picture of the seepage flow processes than the hydraulic measurements, as it is possible to measure an infinite number of temperatures over the depth (in general, measurements are taken 1 m apart). In addition, temperature cells were installed in the non-saturated part of the dam (Armbruster 1983, 1987).

- Control of the deformations on the surface through measuring pins (settlements) inside the dam through a number of settlements gauges at varying depths (withdrawal of material).

- Control of the dam construction by taking a sample when observation tubes are drilled.

- Control of permeability through tracer tests. Points of insertion are either the head water or tubes nearby, withdrawal points for water samples are the tailwater tubes. The naturally available chemicals in the river water were used as tracers. The testing took place weekly.

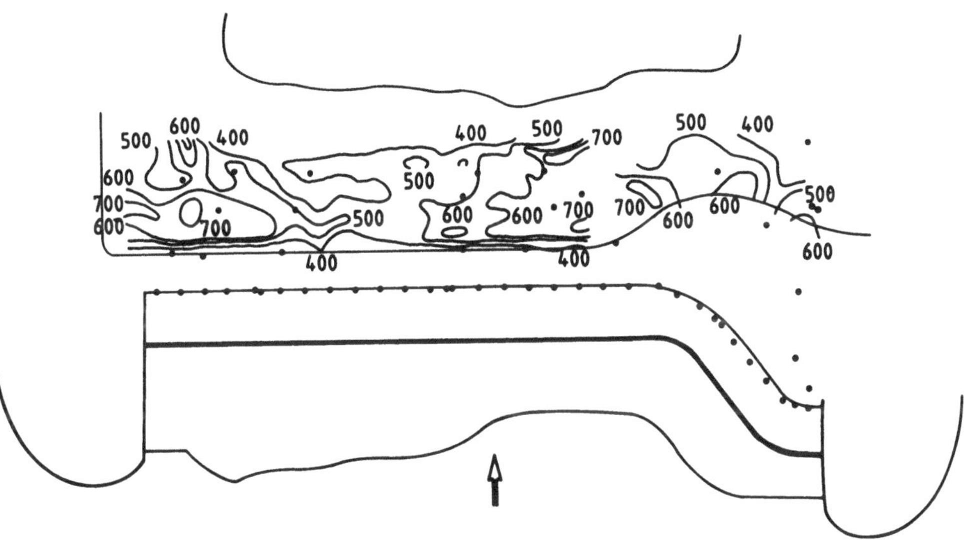

Fig. 2: Resistivity map (Ωm), e = 10 m (WENNER)

- Control of the <u>homogeneity of the pile-up of the dam</u> by resistance measurements (Fig. 2). The isolines of the resistances show a relatively even pile-up of the dam. Two different distances between electrodes were used for measuring.

- Control of the <u>stability of the seepage flow</u> through self-potential measurements (Fig. 3).

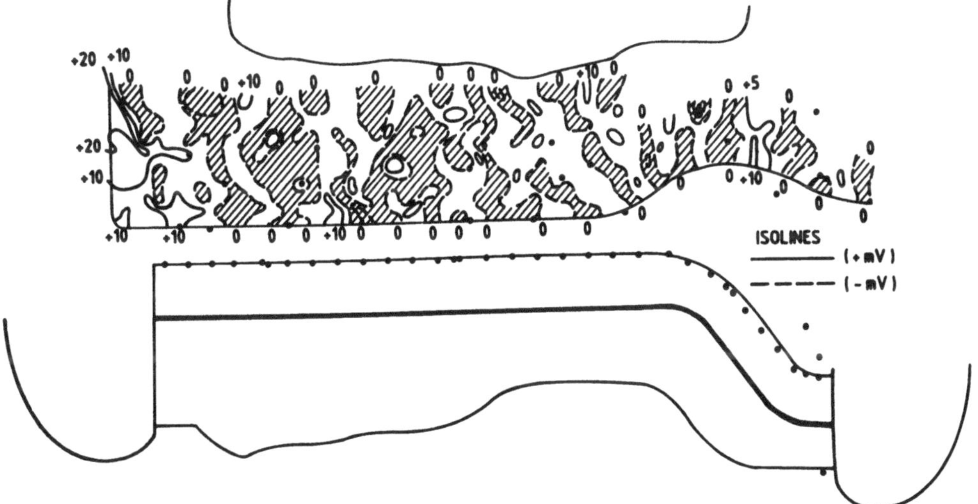

Fig. 3: Self-potentials map (mV), shaded areas are negative

The self-potentials do not show a stable seepage flow, but there is a general tendency all over the measured dam area. The greater differences near the sealings (evident through the temperature measurements) cannot be measured, since they are under water. The piling-up of the very wide dam with permeable material prevents a concentrated flow, even if there were a larger leak in the seal in the water.

- Control of the underseepage and through-seepage by infrared pictures of the dam and tail water. Two pictures were taken with the multispectral scanner 6 years apart and these show that there are no through-seepage anomalies (temperature pictures of the embankment surface), but that the underseepage is not stable (Fig. 4).

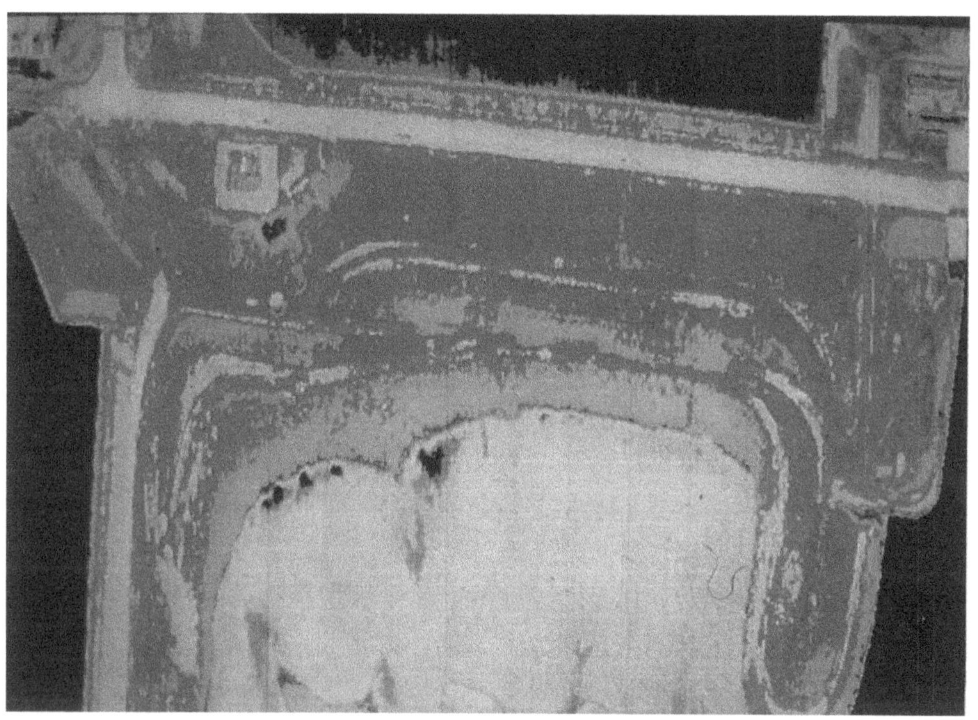

Fig. 4: Surface temperatures of the dam, imaged by aeroplane (infrared scanning)

The varying emergence of seepage water produces different temperatures, which are clearly visible and enable conclusions to be drawn regarding dam areas with stronger underseepage (Armbruster et al. 1985).

4.2 Lateral Dyke of a River Barrage

It was possible to erect the lateral dyke of the river under dry conditions on the old river bank. For this purpose, after the contact surface was cleared, the vertical sealing was built as a diaphragm wall, which cuts the uppermost meters (7 - 12 m) of the coarse, permeable gravel, without reaching the slightly permeable layer. The core seal (Fig. 5) subsequently drawn-up with the dam connects to the vertical wall. The solid body of the dam consists of sandy gravel.

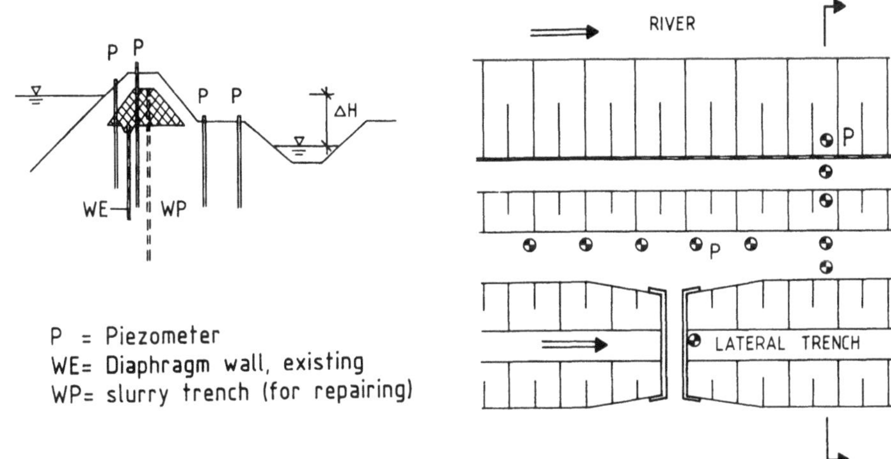

Fig. 5: Lateral dam along the river

The planned underseepage is collected in a lateral ditch, which also takes smaller rerouted streams and leads into the tail water of the river barrage. A special control system (in addition to water level and outflow measurements in the lateral ditch) was not planned.

The observations of the dam by walking along it once a week showed relatively soon after the damming that there were greater seepages at several points into the lateral ditch. As a result, at these points a traverse line of observation tubes was inserted in each of the drill holes and in addition over a longer stretch about fifteen 5-m deep observation tubes were rammed in along the toe of the dam and further measurements were taken. The following control facilities were thus provided:

- Control of the hydraulic potentials in a cross-section of the dam along the toe of the dam. Due to the uniformly short tubes, practically only the phreatic line could be ascertained and any alteration in it in the course of time (no potential field). It fluctuates with the water level of the lateral ditch and the water level of the river, which, depending on the distance to the dam construction, can increase at hightide of the river. The highest water levels occured in the area of a leakage in the sealing system (Fig.6).

- Control of the thermal potentials at the toe of the dam at a depth of about 2 - 5 m (longitudinal section possible) and in a cross-section (potential field measurable below the phreatic line). The temperatures change with the water temperature of the river and of the groundwater. In the leakage area the river has the greater influence (Armbruster 1983). The extent should be ascertained through the measurements in the pipes along the toe (Fig. 6).

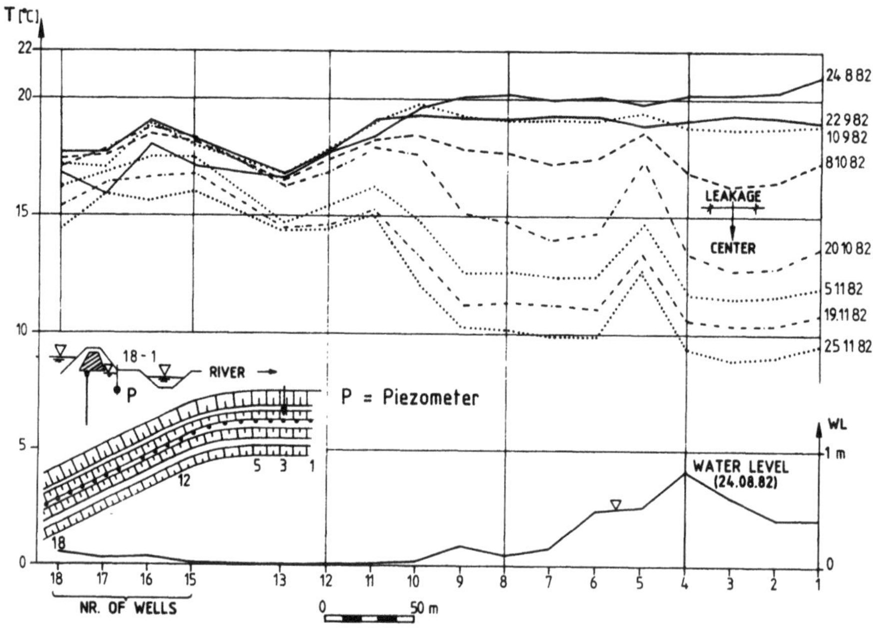

Fig. 6: Temperatures along the dam toe in the depth of 2 m (a) water level in the piezometer along the dam (b)

- Control of the dam construction by taking samples when drilling the five pipes of the cross-section. The driven pipes at the toe of the dam provide few clues.

- Control of the permeability as in section 4.1.

- Control of the piling-up of the dam and the core seal by taking resistance measurements with small (e = 2.5 m) and medium-sized distances between the electrodes (e = 5 - 10 m). The resistivities (Fig. 7) clearly show areas with cohesive cores and less cohesive cores. This was confirmed by slot probes.

- Control of the seepage flow through self-potentials. The self-potentials (Fig. 8) clearly show areas with ascending water, which pene-

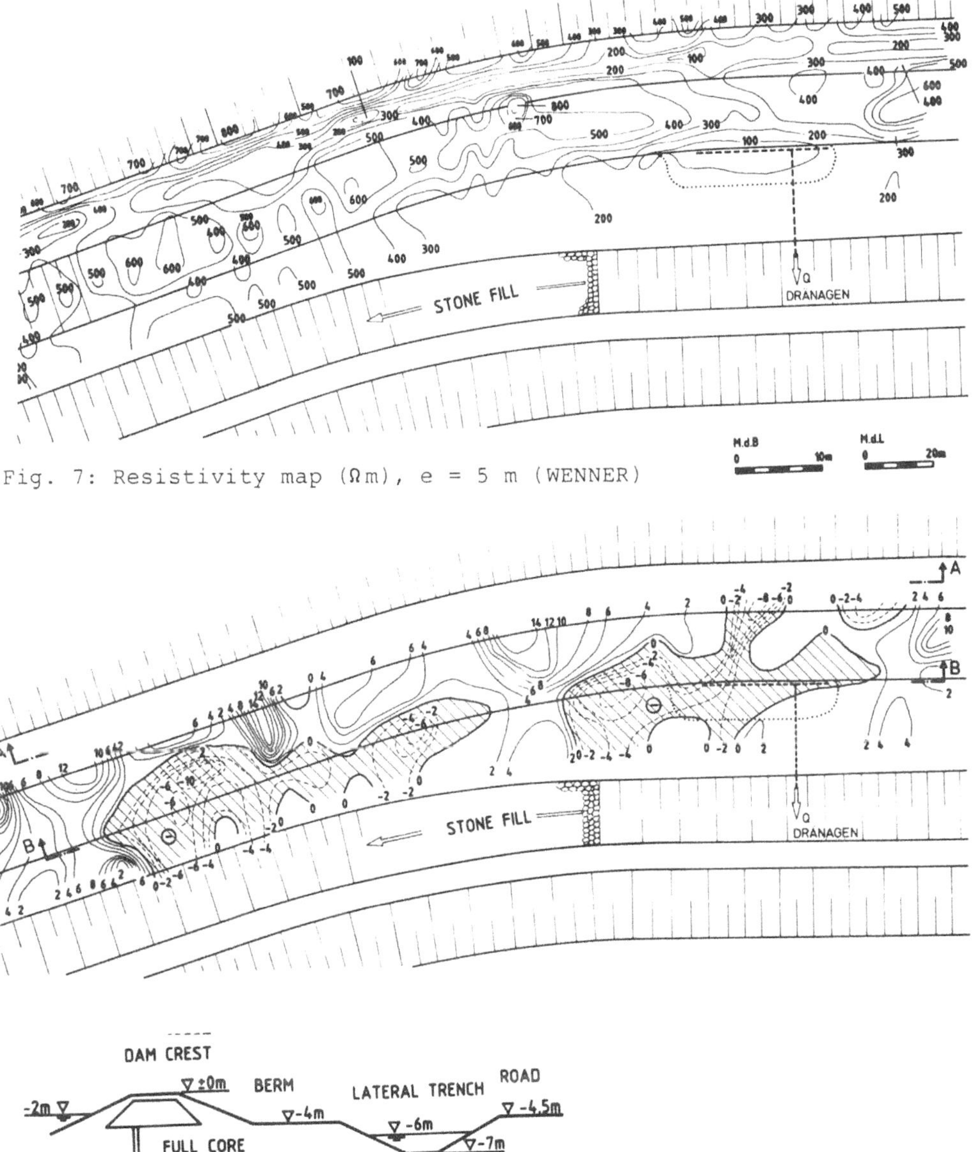

Fig. 7: Resistivity map (Ωm), e = 5 m (WENNER)

Fig. 8. Self-potentials map (mV), e = 5 m, shifted regions are negative

trates thorugh the leakage of the sealing and that the water tension is released in the gravel body. In measurements after later repairs through a diaphragm of the crest of the dam, these areas disappear.

- Control of the underseepage and through-seepage by measuring the water levels in the lateral ditch, by measuring the outflow in defined cross-sections and by the ovservation of gravitational water outflows in the embankment of the lateral ditch. In addition, the temperatures of the dam surface (through-seepage), the lateral ditch embankment (water outflow) and the water in the lateral ditch (underseepage) were measured by means of infrared pictures twice in 6 years. The pictures show greater temperature anomalies in the leakage area (Fig. 9) at the toe of the dam and in the lateral ditch (embankment and water), which later disappeared after repairs.

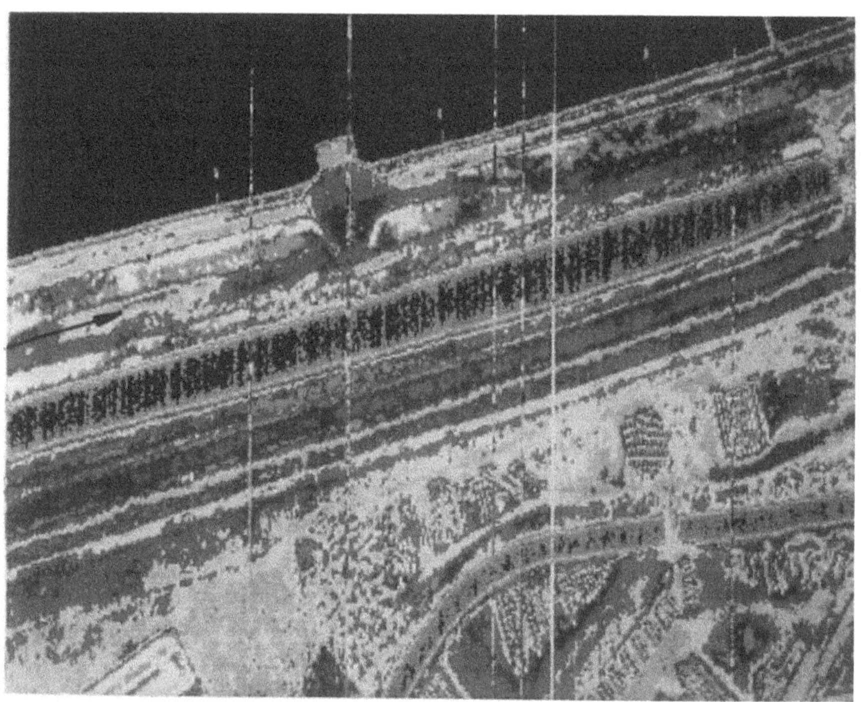

Fig. 9: Surface temperatures, imaged by aeroplane (infrared scanning)

4.3 Lateral Dyke of Channel

Channels are generally equipped with continuous sealings, as they are narrow, and these include the canal bottom and both lateral dykes and generally cut off any connection to the groundwater.

In the area of South German canal, of which the dams consisted of silty sands with a continuous surface sealing (soft seal), wet patches have existed at the toe of the dam since the canal was constructed, the cause of which was to be investigated by taking measurements. The few and very short observation tubes (Fig. 10) did not give a clear picture of the leakage flow.

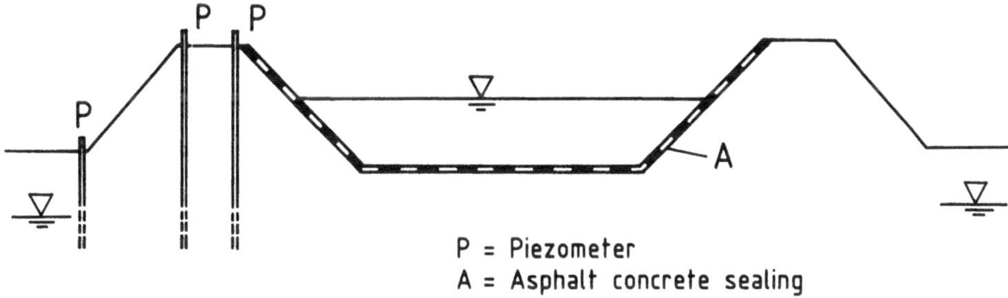

Fig. 10: Lateral dams of a channel

The measurements were concentrated on surface measurements:

- A control of the hydraulic potentials was only possible at a few points. A lowering of the banked-up water level of the canal led to a drying of the pipes, which led to the conclusion that there was a leakage in the water changing area.

- A control of the thermal potential fields over a length of 600 m inside the dam by measurements with non-fixed soil thermometers led to the discovery of leakage areas (Armbruster and Merkler 1982), including the level of the leak.

- The control of the pile-up of the dam by resistance measurements provided no acceptable results, as the dam pile-up was even and it was not impaired by the leakage of the surface seal.

- The control of the seepage flow through self-potentials indicated that there were areas with stronger seepage (at the leakages). Due to the very low flow speeds in the cohesive soil these anomalies are not very distinctive.

- The control of the through-seepage of the dam (since there is no lateral ditch, the underseepage cannot be measured) with the help of infrared pictures naturally showed the thermal anomalies in the wet area and indicated other areas, which were followed up with the other measurements (above all thermally in the soil).

5 Evaluation of the Results

The above examples show that measurements must be adapted according to the position of the sealing systems, the draining, the construction material of the dam and the subsoil. Each type of measurement has specific disadvantages or advantages.

The geoelectric measurements, which are relatively simple to perform, require sound information on the dam construction for their interpretation. They give good indications of the anomalies, which must then be investigated further.

The complex infrared measurements are well-suited for dams with open lateral ditches and dams with tail water and, in addition, for dams with cohesive soil with little plant growth. It is even more so the case here that the temperature anomalies only indicate suspicious points, which must then be investigated by looking directly at the spot and using simple aids and then, if suspicions continue, with further measurements.

References

Armbruster H. (1983)
Messung von thermischen Feldern zur Überwachung von Dämmen. Symp. Meßtechnik der DGEG, Munich, FRG

Armbruster H. (1987)
The distribution of temperature in water percolated earth dams of a river barrage. 9th Eur. Conf. Soil mechanics and foundation engineering, Dublin

Armbruster H., Merkler G. P. (1982)
Möglichkeiten der Leckstellenortung an Erddämmen. Geotechnik 1/82

Armbruster H., Merkler G. P. (1983)
Measurement of subsoil flow phenomena by thermic and geoelectric methods. Bull. Int. Assoc. Eng. Geol. (Paris) 26 - 27

Armbruster H., Döscher H. D., Sartori M. (1985)
The infrared thermography, a control system for the efficiency of sealing elements and drains. ICOLD, Lausanne

Merkler G. P., Blinde A., Armbruster H., Döscher H. D. (1985)
Field investigations for the assessment of permeability and identification of leakages in dams and dam foundations. 15th Congr. Large Dams, Lausanne, Q 58, R 7

Measurements of Self-Potential

Data Quality For Engineering Self-Potential Surveys

Robert F. Corwin
Consulting Geophysicist
El Cerrito, California, USA

Abstract

Self-potential data quality depends on survey configuration and procedures, equipment selection and maintenance, recognition of sources of error and noise, and use of appropriate data reduction techniques. Survey procedures that minimize movement of the base reference electrode and account for electrode drift and polarization help to reduce the effects of cumulative error. Input impedance of the measuring instrument must be considerably greater than electrode contact resistance to avoid loading of the measurement circuit, and cable insulation integrity for prevention of ground loops is critical.

Noise sources may be natural or artificial, and may be constant or vary with time. Recognition of noise potentials is important to avoid their being misinterpreted as anomalies generated by sources of interest. The use of a telluric current monitor to record time-varying potentials in the survey area, inspection of site plans, and careful field observation assist in noise source recognition. Magnetic or electromagnetic survey data also are helpful for detection of metallic noise sources.

1 Introduction

Self-potential (SP) data obtained in support of engineering investigations often are more severely affected by error and noise than are SP data for mineral or geothermal exploration. The severity of these effects is due to the relatively low SP anomaly (signal) levels generated by most sources of engineering interest and to the prevalence of artificial noise sources in the developed areas in which most engineering geophysical investigations are conducted.

Standard geophysical reference texts such as Heiland (1940), Parasnis (1966), Semenov (1974), and Telford et al. (1976) give brief descriptions of SP field procedures. However, unlike most other geophysical methods, widely accepted data quality control standards and procedures have not been established for the SP method. Therefore, this chapter discusses in some detail the effects on SP data quality of survey procedures, equipment, data reduction techniques, and natural and artificial noise sources.

Lecture Notes in Earth Sciences, Vol. 27
G.-P. Merkler et al. (Eds.)
Detection of Subsurface Flow Phenomena

Because SP is a passive, potential field technique (like magnetics or gravity), there is no way of changing source parameters to vary depth of investigation or to help differentiate signal from noise. As for other potential field methods, smoothing and filtering techniques can be applied to SP data. However, it is preferable to recognize and remove error and noise to as great an extent as possible before using such techniques.

In the following discussions, "error" is defined as the irreproducible component of a given SP reading associated with the data acquisition process, while "noise" in general is a potential generated by a source that is not of interest for the purposes of the investigation. In some cases (for example, time-varying noise generated by natural or artificial sources) these definitions will unavoidably overlap. Potentials considered as noise for engineering SP surveys (e.g., those generated by corrosion or telluric currents) may be the desired signal for other applications.

2 Survey Configurations

A variety of survey configurations are used to obtain SP field data. While all of the configurations described below have been used successfully, their susceptibility to random and systematic error differs. Descriptions of the most commonly used survey configurations, along with discussions of their advantages and disadvantages, are given below. For all survey configurations, SP sign convention requires the negative terminal of the measuring voltmeter to be connected to the electrode at the survey base station and the positive terminal to the electrode at the measurement station.

The gradient configuration (also called the dipole, leapfrog, or fixed-electrode configuration; Telford et al. 1976) utilizes two electrodes and a connecting wire of fixed length equal to the measurement station separation. The voltmeter is connected between one of the electrodes and one end of the connecting wire. Measurements are made by moving this dipole along the survey line, with the trailing electrode occupying the station of the previous leading electrode. Reversal of electrode positions between stations (the so-called leapfrog technique) helps to reduce cumulative error caused by electrode polarization. The SP value at a given station is obtained by successive addition of individual dipole readings. Sometimes the measured dipole readings, rather than their sum, may be plotted. If the dipole length is small relative to the anomaly wavelength, such plots represent essentially the gradient or derivative of the total SP field.

An advantage of the gradient configuration is the relatively short connecting wire that minimizes exposure to damage. Also, because no wire retrieval is involved, it is not necessary to expend time re-traversing the survey line. However, this configuration is extremely sensitive to spurious "anomalies" generated by cumulative error. Components of reading error, discussed in more detail later, include soil contact effects, electrode polarization, and time-varying potentials. The effects of electrode polarization can be reduced by use of the "leapfrog" procedure mentioned above, but the other two components can be difficult to quantify or correct. These components add a random error to each measured value, and because these values are added together in the data reduction process the errors can accumulate to significant levels, often several tens of mV or more.

Because of the small electrode separation, the observed magnitude of time-varying potentials for the gradient configuration generally is small. For example, a telluric current variation (discussed in more detail later) of 20 mV/km amplitude will appear as a deviation of only 0.2 mV amplitude across a 10 m measuring dipole; a virtually undetectable value. However, this deviation is included in the error value discussed above and, depending on the period of the variation, will be added either as point-to-point random error or as overall accumulated error.

Therefore, even though time variations are less visible across a gradient array, their effect is the same as for configurations using larger electrode separations. The disadvantage of the gradient configuration in this regard is that it is much more difficult to recognize and correct the effect of time-varying noise.

Cumulative errors can be reduced by interconnecting survey lines at numerous tie-in points and distributing tie-in loop closure errors among all the readings around the loop. However, this arbitrary procedure may reduce the amplitude of real anomalies as well as errors. Also, it does not account for errors within a loop, which can be considerable. For these reasons, the use of the gradient configuration probably should be restricted to situations where operational difficulties such as vulnerability to wire damage or rough terrain prevail.

The fixed-base (or total field) configuration uses a stationary base electrode, a reel carrying the greatest practicable length of connecting wire, and a moving measuring electrode (Corry 1985, Corwin and Ticken 1988). This technique allows the maximum number of readings to be made directly from a single base electrode and thus minimizes accumulation of reading errors. When operational considerations require the use of more than one base station, establishment of multiple tie points helps to minimize tie-in errors between the base stations.

A major advantage of this configuration compared with the gradient configuration is the lower level of cumulative error. Although each reading is subject to the three error components mentioned above, these errors do not accumulate as for the gradient configuration. Also, as described in more detail later, it is relatively easy to estimate the magnitude of electrode polarization and time-varying errors and to remove these errors from the readings. Thus the reproducibility of data obtained using the fixed-base configuration generally is considerably better than that for the gradient configuration, and the probability of mapping spurious "anomalies" is lower. Therefore unless the terrain is especially difficult or the probability of wire damage is high, the fixed-base configuration generally is preferable to the gradient configuration.

A variation on the fixed-base or gradient technique is to survey the measurement stations and dig and water electrode holes a few hours to a day in advance of the actual measurements (Semenov 1974). Watering of electrode holes to reduce contact resistance or equalize soil moisture content between stations generally is not necessary. However, if the holes are watered time must be allowed for mobile water to diffuse out of the holes (Corwin and Hoover 1979).

A multi-electrode configuration is similar to a long-term SP monitoring network (Koester et al. 1984) in that an electrode is installed at each measuring station and all these electrodes are connected to a base station terminal through a multiconductor cable. Measurements are made by sequentially connecting each electrode through the meter to the base station electrode, or by using a multichannel data acquisition system. If a multiconductor cable is not used, measurements can be made using a gradient or fixed-base procedure.

An advantage of this configuration is the ease of making repeat measurements to check for time variations, and of applying data processing techniques such as stacking or filtering for removal of the effects of such variations. As for the fixed-base array, cumulative error is minimized by the use of a single survey base station. Also, electrode drift can be monitored and readings made after values stabilize. As discussed later, there is some question as to whether initial electrode readings or values recorded after drift has ended better represent potentials related to sources of interest.

Disadvantages of the multi-electrode configuration include initial equipment costs that are considerably higher than those for the configurations described above, relatively inflexible arrangement and spacing of measurement stations, and the difficulty of measuring initial and final electrode polarization values for correction of polarization errors (discussed in more detail later).

Once the configuration for a given survey is chosen, it is necessary to select survey line orientation and spacing and to determine the spacing of measurement stations along the survey lines. As for any other geophysical technique, if elongated anomalies are expected the survey line orientation should be perpendicular to the anticipated anomaly orientation. Because signal-to-noise ratios for engineering SP surveys often are low, it usually is preferable to conduct closely-spaced measurements along widely-spaced survey lines rather than the converse (Semenov 1974).

Measurement station spacing depends on the anticipated anomaly wavelength. As for other geophysical potential fields, anomaly wavelength depends on the configuration, size, and depth of burial of the source of the anomaly. A number of algorithms for modeling and interpretation of SP anomalies generated by simple geometric sources have been published (Corwin and Ticken 1988). Examples include point sources (Stern 1945), horizontal line sources (Rao et al. 1970), spheres (Petrowsky 1928), horizontal cylinders (Bhattacharya and Roy 1981), and dipolar sheets (Fitterman 1979). Modeling for more complex sources can be done using techniques described by Sill (1983).

As there is no universal zero potential reference level for SP measurements, selection of the zero potential point for a given data set is arbitrary. Usually, a station remote from expected or observed anomalous activity is assumed to be at zero potential. Locating the survey base station in such a quiet area and assuming the base station potential to be zero facilitates computation and may improve data reproducibility (Corry 1985). However, locating base stations centrally within the survey area reduces required wire lengths, and running survey lines outward from anomalous areas makes it easier to determine when anomalous activity has ended and sufficient background has been measured.

In some cases it is desirable to obtain SP data in water--covered areas such as the upstream face of a dam or the floor of a reservoir. Using submersible non-polarizing electrodes, data can be obtained successfully in either fresh or salt water with either gradient of fixed-base configurations (Corwin 1976). Because of the continuous contact between the electrodes and the water, continuous data profiles can be recorded on a strip chart recorder or other data acquisition system. This allows very rapid coverage as well as infinite lateral resolution. Although offshore SP signal amplitudes are reduced by the relatively low resistivity of the water (especially in salt water), noise and error levels also are lower, and signal-to-noise ratios for offshore SP data usually are equal to or greater than those for onshore data.

3 Equipment

Equipment required to obtain SP measurements using the gradient configuration includes electrodes, connecting wire, and a measuring meter. For the fixed-base configuration, additional equipment includes a much greater length of connecting wire (several hundred to several thousand meters) and a reel to hold the wire. As discussed above, the multi-electrode array requires additional electrodes (equal to the expected number of measurement stations), and possibly a multiconductor cable and data acquisition system.

Although electrodes of stainless steel (Parasnis 1966) and copper-clad steel (Koester et al. 1984) have been used for SP field measurements, so-called nonpolarizing electrodes have been found to give much more reproducible data (Parasnis 1966; Corwin and Ticken 1988). Such electrodes consist of a metal element immersed in a solution of a salt of the metal, with a porous junction forming the boundary between the solution and the soil (Ives and Janz 1961). Although such electrodes are not truly nonpolarizing and are more accurately called liquid junction electrodes, the nonpolarizing label is commonly employed and will be used for this discussion.

High-frequency electrode noise levels in the microvolt range are an important consideration for other geophysical applications such as magnetotelluric measurements (Petiau and Dupis 1980). For SP measurements, however, electrode response to variation of environmental parameters such as temperature and soil moisture content and chemistry is of greater interest. Laboratory and field measurements of such responses have been performed for a variety of nonpolarizing electrode types, including copper--copper sulfate, silver-silver chloride, lead-lead chloride, cadmium-cadmium chloride, and zinc-zinc sulfate (Ewing 1939; Semenov 1974; Morrison et al. 1979a).

Results of these measurements indicate that although these different electrode types respond in varying degrees to changes in the environmental parameters listed above, (for example, see Fig. 1), under most conditions these differing responses do not seem to significantly affect the error and noise level of field SP measurements (Morrison et al. 1979b).

For most applications, commercially available copper-copper sulfate electrodes, sold for use in pipeline corrosion surveys, have been found to give SP field data of acceptable quality if drift and polarization effects are monitored and corrected (these procedures are discussed later). Commercially available silver--silver chloride electrodes, sold as reference electrodes for pH and other electrochemical measurements, are similarly acceptable for offshore SP measurements. To minimize leakage, electrolyte solutions can be gelled if desired (Semenov 1974).

The response of copper-copper sulfate and silver-silver chloride electrodes to environmental variations is well documented (Ewing 1939; Ives and Janz 1961; Corwin and Conti 1973; Morrison et al. 1979a, 1979b; Petiau and Dupis 1980). Because the liquid junction construction of nonpolarizing electrodes tends to suppress response to electrochemical variations, temperature and soil moisture content differences tend to produce the largest electrode effects.

Figure 1 shows the response of a number of electrode types to changes in soil moisture content. For copper-copper sulfate and silver-silver chloride electrodes, this response is of the order of about +0.3 to +1 mV per percent moisture content increase,

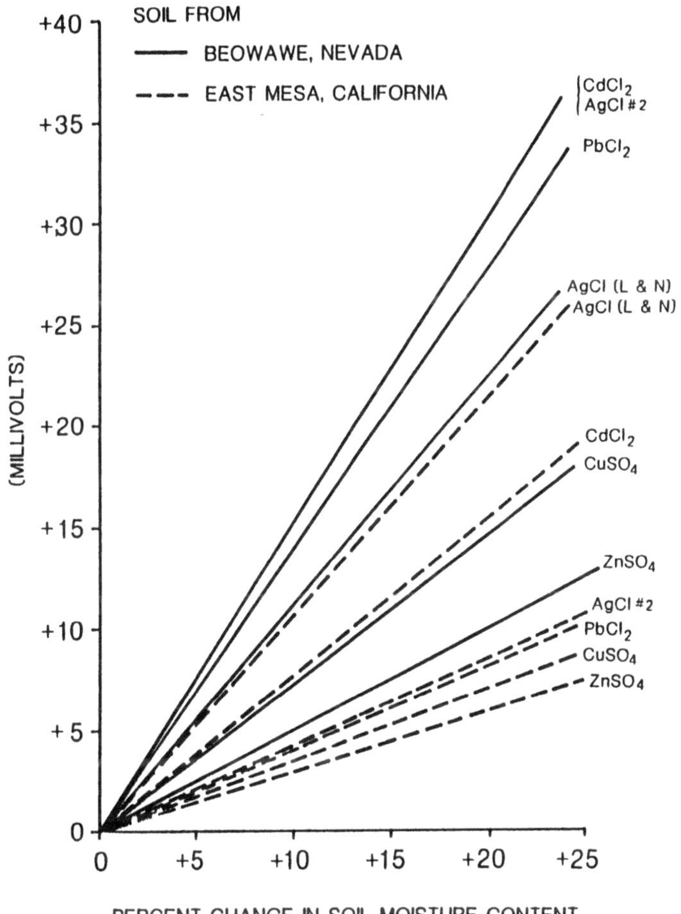

Fig. 1. Effect of soil moisture content

depending on soil type and electrode construction. Other measurements (Morrison et al. 1979b) have shown a maximum response of about 70 mV for a copper-copper sulfate electrode pair connected between saturated and very dry desert clay soil.

Thus soil moisture variations can represent a significant noise source when signal levels are a few tens of mV. Careful field notes documenting observed soil moisture levels can be helpful in distinguishing between anomalies caused by sources of interest and those related to soil moisture variations. This is especially important in seepage investigations, where positive SP readings caused by wet soil must be separated from those caused by the upward movement of subsurface water.

The temperature response of copper-copper sulfate and silver-

-silver chloride electrodes given in the preceding references are of the order of +0.5 to +1 mV per °C. Note that this refers to the temperature of the electrolyte rather than that of the soil, so the effect of a change in soil temperature will not be seen until the electrolyte temperature begins to change. As temperature changes are the major cause of electrode potential drift, and as temperature polarization values can reach levels of 10 to 20 mV under severe conditions, care should be taken to minimize electrode temperature changes.

Important considerations for the voltmeter used for SP field measurements include resolution, range, input impedance, interference rejection, and suitability for field use. Resolution of 1 mV is sufficient for SP field measurements, and a range of +/- 10 VDC will cover even very large anomalies generated by DC current grounds. As noise in the 10 to 100 Hz range is common in developed areas, inclusion of low-pass filtering in the voltmeter is necessary.

Electrode contact resistance ranges from a few hundred ohms in water or very conductive soil to several megohms in snow, frozen soil, or very dry or rocky soil. In most areas, contact resistance will be of the order of a few kohms to a few tens of kohms. Thus a voltmeter input impedance of ten megohms generally will be sufficient, but several hundred megohms or more may be required for high-resistance conditions. It is very desirable to measure and record electrode contact resistance at each station. This ensures that there are no breaks in the connecting wire, that contact resistance is low enough to avoid loading the measuring circuit, and that ground contact conditions are relatively uniform from station to station.

Fortunately, most of the requirements listed above are met by inexpensive, commercially available digital multimeters (DMM's). Ground contact resistance values exceeding about 100 kohm may require use of a high-impedance instrument such as a portable pH meter with a voltage display output or a battery-operated electrometer.

Because most of the resistance in the measuring circuit is in the electrode-to-soil contact, the resistance and gage of the connecting wire usually are not important. As a section of exposed wire in contact with wet soil can generate irreproducible error potentials of hundreds of millivolts, maintenance of insulation integrity is critical. To avoid ground loops it is important to insulate the wire conductor from the body of the reel on which the wire is carried.

A telluric monitor is used to record time-varying potentials in the earth that could be mistaken in the survey data for spatial variations. Instrumentation for such a monitor consists of a recorder (usually a battery-operated strip chart recorder), electrodes, and connecting wire. For optimum definition of telluric current directions, it is desirable to use an orthogonal electrode array and a two-channel recorder. The comments above regarding electrodes and instrument specifications also apply to the telluric recorder. Deployment of the monitor and use of the record for data correction are discussed later.

4 Measurement Procedures

Selection and implementation of appropriate survey field procedures are critical for maintenance of SP data quality. Appropriate field procedures are those which eliminate or minimize errors related to the data acquisition process, and which provide methods for recognition of errors and their removal, to as great a degree as possible, from the field data. Because little has been published regarding SP field procedures, they are discussed in some detail in the following sections.

The apparent "geologic noise" level of SP data at station spacings from a few cm to tens or hundreds of meters ranges from almost zero to tens or hundreds of mV. This "geologic noise" includes measurement errors caused by electrode polarization and drift, changing soil contact conditions, and time-varying potentials as well as noise related to changing soil conditions and other natural and artificial sources. Much of this "noise" can be eliminated or reduced, leaving only variations that are unavoidable or uncorrectable to be filtered or smoothed.

4.1 Electrode Polarization and Drift

Electrode polarization and drift are a major component of SP measurement error. For a survey conducted from a single base station, the first field reading will be in error by the initial polarization potential between the base and measuring electrodes and the last field reading will be in error by the final polarization value. Therefore measurement of these polarization values in a bath of electrolyte solution immediately before installation and after removal of the base electrode will allow subtraction of these polarization errors from the measured values, with corrections for intermediate measurements obtained by interpolation (Corwin and Ticken 1988).

For situations in which the base electrode is in the soil for long periods of time (more than about 1 h), intermediate drift correction values can be obtained by periodically reading the potential between the measuring electrode and an auxiliary ("portable reference") electrode carried in a container of electrolyte solution. By assuming that the potential between the base and auxiliary electrodes remains constant, drift of the measuring electrode with respect to the base electrode can be determined and removed. These procedures apply specifically to the fixed-base configuration, but the general concepts are equally applicable to other configurations.

Although this assumption is not always valid (sometimes the auxiliary electrode drifts significantly with respect to the base electrode), the procedures above have been found empirically to improve SP data reproducibility. Even in cases where the corrections are incomplete (for example, where the potential at the base station has changed due to drying of the soil or other factors) these procedures at least allow recognition of drift and polarization errors and estimation of their magnitude.

There is some controversy as to whether readings should be made quickly, before significant drift has occurred, or after the readings have stabilized. In most cases, readings become essentially constant (drift of less than a few tenths of a mV per minute) within a few seconds after electrodes are installed in the soil, and observation time will be determined by the need to detect time variations due to tellurics or artificial sources (discussed later).

In some cases, however, readings are observed to drift significantly (more than about 1 mV per minute) for several minutes or even hours. In such cases the question is whether the initial or the final reading represents the desired value. If the source itself is changing with time (for example, a variation in the rate of underground seepage flow), then both readings will of course be of interest. However, if the drift is due to electrode response to changing soil conditions such as moisture content, temperature, or pore fluid chemistry, the "correct" reading should be that obtained before the measuring electrode is significantly polarized by reaction to the new soil parameters. Further research is needed to determine the origin and effects of long-term drift.

The process of measuring contact resistance forces current to flow through the electrodes, resulting in electrode polarization. Therefore the contact resistance measurement should be made as quickly as possible to minimize this polarization. Because the exact value of the resistance is not important, the resistance should be measured only for a second or two even though the resistance reading will not completely stabilize in this time.

Occasionally an unusual condition is seen when measured contact resistance values become very high or infinite even though soil conditions do not appear to change, SP readings are stable, and no evidence of measurement circuit loading is seen. In many cases this condition is associated with crossing of a fault or vertical contact. Although the SP readings obtained under these conditions appear to be valid, more study of this phenomenon is needed.

4.2 Electrode-to-Soil Contact

Whenever an electrode is removed from and replaced in the same location, the SP reading will almost always change. For moist, conductive, compact soils this change may be only a few mV, and generally is less than 5 or 10 mV for most soils. However, for dry, resistive, loose soil it may amount to several tens of mV or more. This uncertainty represents an important component of irreproducible reading error.

As discussed above, this error may accumulate rapidly for measurements made using the gradient configuration and will accumulate to a lesser degree through tie-in points with the fixed-base configuration. Even when the values do not accumulate, they contribute to the noise level of the data. Therefore it is important to attempt to minimize this error by digging electrode holes deeply enough to contact moist soil below the surface layer and by maintaining constant contact conditions.

Because the temperature and moisture content of soil exposed to air will change, electrode holes should be refilled and flagged if the station is to be reoccupied or used as a tie point. Remeasurement then can be made under conditions as close as possible to original, even days or weeks later.

A technique for statistical reduction of the effects of both contact potentials and soil property variations is to use multiple electrode holes at each station (Sill and Johng 1979). The holes are placed in some consistent geometric pattern around the station location, within a radius of a few meters or less. The recorded value then is the average of the measurements, and the deviation provides an estimate of the noise and error level. The added time and cost of the additional measurements must be compared with the expected improvement in data quality. However, whenever measured values change suddenly along a profile additional nearby measurements should be made to verify the new value and to determine the spatial wavelength of the variation.

Often, both signal amplitudes and levels of noise and error are seen to be strongly related to near-surface soil resistivity. Thus signal-to-noise levels and the ratio of tie-in error to signal amplitude generally are roughly similar for widely varying soil conditions. The main exceptions to this are found in developed areas, where soil disturbance and artificial noise sources tend to reduce data quality levels relative to those seen in undeveloped areas.

5 Time-Varying Effects

In addition to the electrode drift effects discussed above, time variations of SP readings may be caused by changing site conditions or by electric fields generated by natural or artificial sources (Ernstson and Scherer 1986). It is important that these variations be recognized and removed from the data to avoid reading errors and misinterpretation of time variations as spatial anomalies. As mentioned above, for engineering surveys the magnitude of the source of an anomaly of interest also may change with time (e.g., the rate of subsurface seepage flow).

Time variations may be divided into those which would be expected to be recognized during the course of an individual measurement and those which occur too slowly to be so recognized.

Because telluric activity (discussed below) exhibits a spectral peak around periods of about 20 to 30 s, a similar reading time often is used to help detect such activity. Thus time variations may be arbitrarily divided into those of less than 30 - s period, which would be detected during the course of a measurement (if of large enough amplitude), and those of more than about 30 - s period which usually must be detected by other methods to recognize them and to differentiate them from electrode drift.

Changing site conditions can generate SP variations over periods ranging from minutes to months or longer. Thus some of these variations will be significant over the course of a survey lasting several hours or days, while others will be seen only if an area is resurveyed. Such variations can be generated by changes in soil properties due to temperature variations, rainfall, or construction activity; changes in topographic effects (discussed later) due to rainfall; and changes in corrosion fields (discussed later) due to changing soil conditions.

Because the magnitude of streaming potentials generated by subsurface water flow is related to both pore water resistivity and to the resistivity of the surrounding medium, changes in these parameters will affect observed SP values. The depth of the water table, the degree of soil saturation, or the temperature or ionic composition of the pore water all may be affected by rainfall or by the elevation of water behind a dam or other impoundment structure. Thus potentials related to subsurface water flow may change following rainfall or changes in impoundment levels even if the flow rates remain constant. Resistivity measurements can be helpful in detecting such effects (Corwin and Ticken 1988).

Significant time variations of SP readings also may be caused by vertical movement of near-surface vadose water. Such movement may be related to evapotranspiration by vegetation (Ernstson and Scherer 1986) or response to solar heating of near-surface soil (Semenov 1974). Under some conditions such variations may reach several tens or hundreds of mV and may be a significant source of noise and error. These variations often are characterized by a strong vertical dependence of SP values within the measurement hole.

The second category of time-varying noise is that generated by earth currents. These currents may be of natural origin (tellurics) or may have an artificial source (stray currents).

Natural telluric current variations have periods ranging from milliseconds to hours (Keller and Frischknecht 1966; Kaufman and Keller 1981). For SP surveys, the most significant periods are those in the 20-30 s range and those of about 0.1 s, which sometimes appear as high-frequency noise on DMM potential readings. Amplitudes of potential variations generated by telluric currents usually are of the order of several mV/km, so they usually do not exceed a few mV for surveys conducted over areas of less than 1 km extent.

In some areas of high resistivity, or during magnetic storms, telluric variations may be tens or even hundreds of mV/km. Under such conditions it is difficult to obtain usable data unless the correction procedures described below are employed. Lightning strikes generate voltage spikes of very high amplitude and very short duration. These spikes usually are easily recognized and generally are not a major source of error.

Time-varying stray currents are generated by grounded electrical machinery and are very common in the developed areas in which engineering SP surveys usually are conducted. Stray currents generated by corrosion processes also may change with time, but these changes are of much longer period than those associated with electrical machinery. High-frequency (50 or 60 Hz) noise may be generated by overhead or buried powerlines, but such noise usually is well suppressed by filters in the DC measuring circuits of most DMM's. In severe cases, additional low-pass filtering may be needed to suppress reading fluctuations caused by such high-frequency noise.

Lower-frequency stray currents may have periods ranging from a few tenths of a second to hours, days, or longer; and may have amplitudes of hundreds or even thousands of mV/km (Fröhlich 1971; Hoogervorst 1975). An example of noise generated by stray currents is shown in Fig. 2. The source of the noise was the electrically-operated San Francisco Bay Area Rapid Transit System, California, and the measurements were made across an 8.5 m dipole oriented perpendicular to the tracks and about 1 km to the east. Even at this distance, the observed noise of about 5 mV extrapolates to a level of about 600 mV/km.

Obviously, it is very difficult to obtain usable SP data in the presence of such noise. One method of avoiding stray current noise is indicated by the lower portion of Fig. 2. As for many sources of industrial noise, this transit system does not operate during evening hours and weekends, so SP readings sometimes may be made during these periods.

If measurements must be made during periods of significant natural or artificial time-varying noise, techniques to reduce the effects of such noise may be used. One such method is to use a multi-electrode array as discussed previously. The data from such an array may be stacked or filtered until satisfactory reproducibility is obtained. Care must be taken to differentiate between telluric fields, which increase in amplitude with increasing electrode separation, and fields from artificial sources, which decrease away from the source.

For fixed-base surveys, a variation on a technique described by Fröhlich (1971) for reduction of time-varying noise on DC electrical resistivity measurements may be used. A stationary electrode dipole of length comparable to that of the entire survey line is installed along or parallel to the survey line,

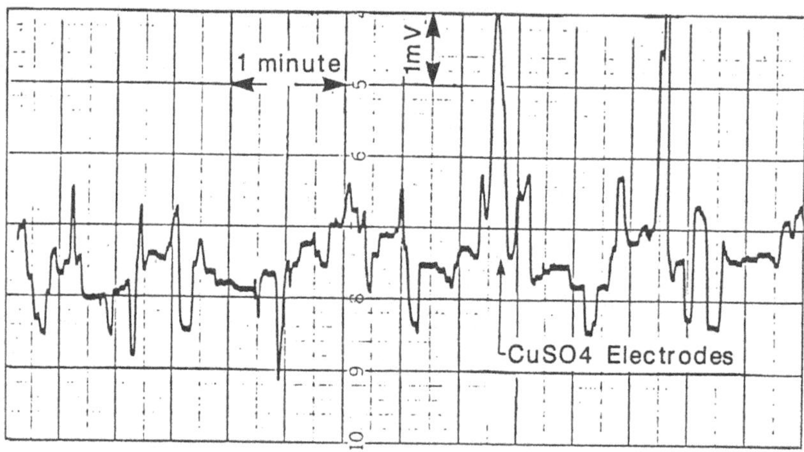

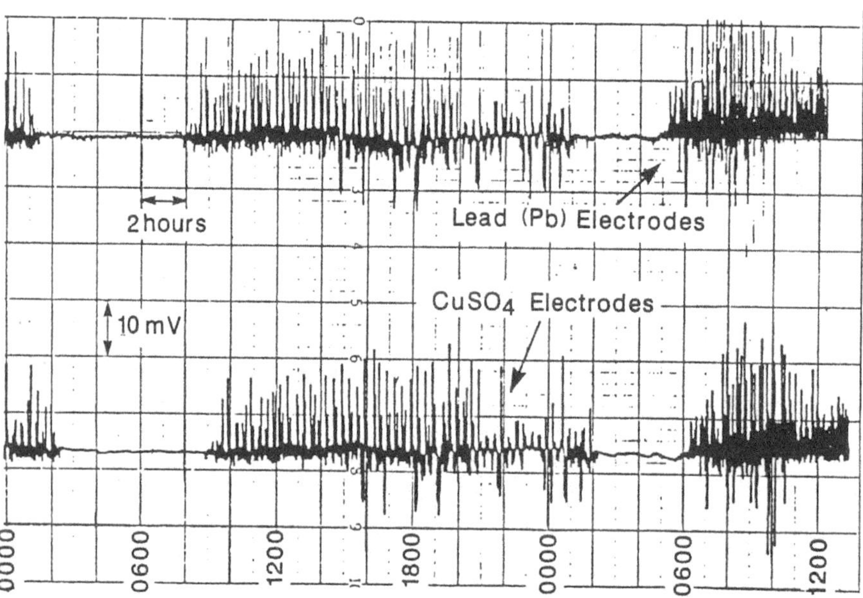

Fig. 2. Potentials generated by stray currents

and the signal from this dipole is brought to each measurement station by appropriate connecting wires. This signal is recorded for a period of a few hours to a day to estimate an approximate zero potential level for the variations.

Once this level is determined, SP readings at each measuring station are made at the instant that a zero potential reading is seen on the monitoring dipole. If the noise does not cross the zero potential level during the measurement interval, SP readings can be made at a number of different noise potential values and the readings can be linearly extrapolated back to the zero level (Fig. 3). Because this procedure does not account for resistivity

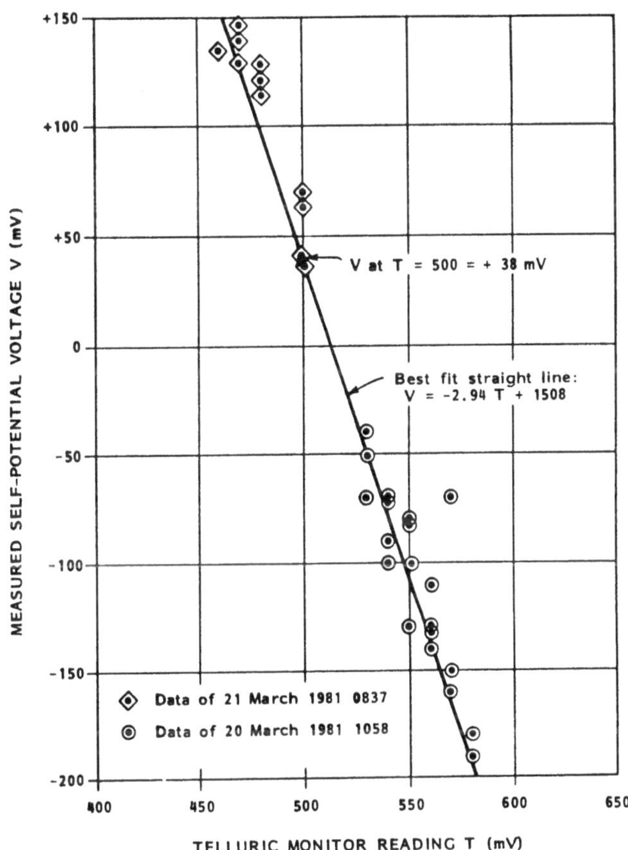

Fig. 3. SP readings vs telluric noise potential

variations along the survey line, and because the true DC potential across the monitoring dipole is difficult to determine, corrections will not be exact. Nevertheless, the procedure can result in very significant noise reduction and allow data to be taken under otherwise impossible conditions.

Figure 4a shows SP variations recorded across a 155-m dipole at a site in northern Canada, and Fig. 4b shows an SP profile obtained in the presence of these very large variations using the technique described above (the example of Fig. 3 also is from this site). Further examples of noise reduction using a similar technique are given by Fröhlich (1971).

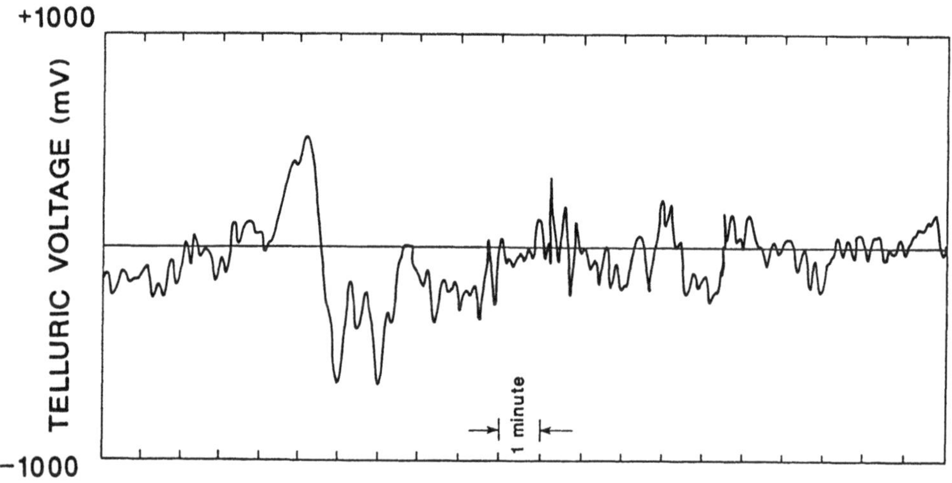

Fig. 4a. Telluric noise, northern Canada
Electrode separation is 155 m

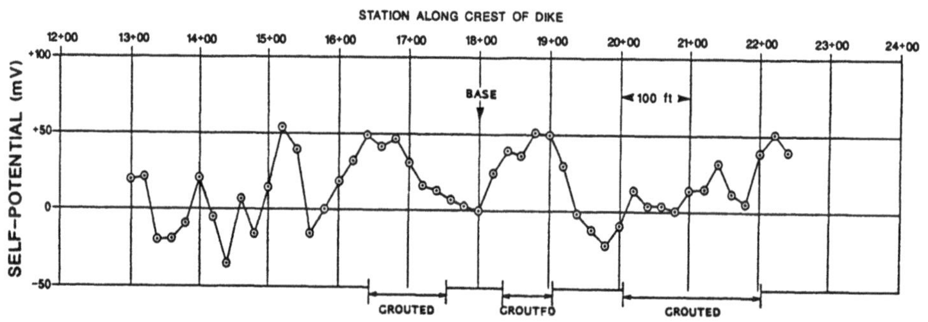

Fig. 4b. SP data along an impoundment dike, taken in the presence of noise shown above. Note negative SP readings in nongrouted areas.

The routine use of a telluric monitor to record time-varying potentials in the survey area is recommended. Without such a record it is difficult to detect long-period variations, and potentials generated by such variations can be mistaken for spatial anomalies. Even if the recorded data are not used for quantitative corrections, periods of significant noise can be recognized and values measured during these periods can be rechecked. This is facilitated by recording the time of each field SP measurement.

6 Other Sources of Noise

In addition to the electrode and time-varying effects discussed above, SP data are subject to a number of noise sources that generally are stable or change relatively slowly with time. These include topographic effects, grounds of electrical machinery, corrosion of buried metal, corrosion protection systems, electrochemical potentials, unwanted streaming potentials, distorting effects of terrain or lateral resistivity variations, conductive mineral deposits, and geothermal activity. Recognition of such noise sources, and removal of their effects from the field data if possible, will improve data quality and assist in the separation of desired signals from unwanted noise.

Topographic potentials generally tend to become more negative with increasing elevation (the so-called negative summit phenomenon) and are thought to be caused by the downslope movement of subsurface water. These potentials are not seen consistently, but when present may be of large amplitude, up to a few mV/meter of elevation (Poldini 1938, 1939; Zablocki 1976; Corwin and Hoover 1979; Nayak 1981). Topographic effects usually are largest in areas having volcanic geology, porous near-surface soil or rocks, large elevation changes, and high precipitation producing an abundant supply of fresh near-surface groundwater. In some cases topographic potential gradients are easily recognized and are consistent enough to permit their removal from field data, but these gradients may vary within a survey area and may change following rainfall.

Grounds of direct-current electrical machinery may generate relatively constant potential fields in the earth if the machinery is in operation during the entire period of the field survey. As these fields may be of very large amplitude, it is important to record the location and operation of such machinery in the survey field notes. These grounds often are equivalent to line or point sources of current, so their fields sometimes can be calculated and removed from the measured data.

Corrosion of buried metal such as well casings, pipelines, debris, and reinforcing rods in concrete can generate large potential fields. Vertical well casings often exhibit negative SP anomalies around the top of the casing (Fig. 5), generated by an oxidation-reduction mechanism similar to that for conductive mineral deposits (Sato and Mooney 1960). Reservoir outlet tunnels

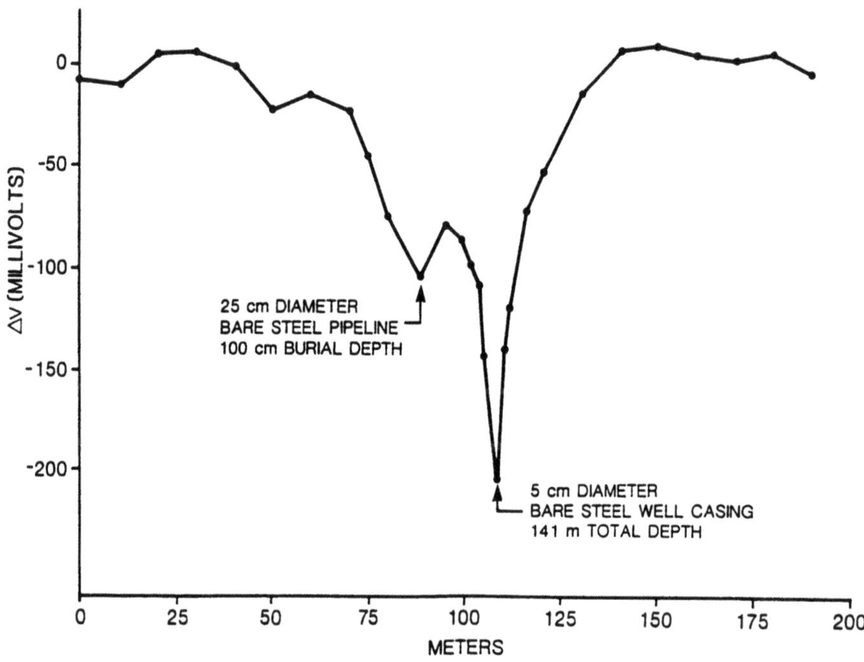

Fig. 5. Potentials generated by buried metal sources

or spillways of reinforced concrete often show a negative anomaly at the onshore end of the structure and a positive anomaly over the submerged offshore end, generated by a similar mechanism. Buried unprotected pipelines and grounded elevated pipelines are characterized by alternating positive and negative potentials (Uhlig 1963). By design, active and passive corrosion protection systems generate very large potentials (Uhlig 1963).

Because such buried metal sources are very common in the developed areas where engineering SP surveys are conducted, it is important to record their presence in the field survey data. Inspection of site plans and the use of magnetic and electromagnetic techniques can help to disclose buried metal sources that are not visible from the surface. In some cases corrosion fields can be modeled as generated by point or line current sources and, to some extent, removed from the data.

Electrochemical potentials can be generated across boundaries separating formations with differing pore water composition

(Heiland 1940; Nourbehecht 1963; Semenov 1974; Sill 1982). Such potentials sometimes are observed when crossing faults or contacts. In some cases these potentials may be the desired signal (e.g., for geologic mapping or contaminant detection); otherwise, they contribute to geologic noise.

Changes in vegetation patterns sometimes correlate with SP variations. In some cases this is caused by corresponding changes in soil properties such as moisture content or pore water chemistry, but SP variations also may be caused directly by the vegetation. Examples include bioelectric effects (Scott 1962) and streaming potentials generated by near-surface water flow related to plant evapotranspiration.

Other streaming potentials that are not the desired survey target also can contribute to noise. Examples include the topographic effects discussed above; movement of subsurface water along faults, fractures, or stratigraphic boundaries; and downward diffusion of surface water. For seepage investigations, it sometimes is difficult to separate streaming potentials generated by these sources from anomalies of interest related to the seepage flow.

Lateral resistivity variations and topography will distort SP fields generated by subsurface sources. If resistivity data are available, the effects of lateral resistivity variations can be calculated using DC potential field theory. Terrain effects can be estimated qualitatively (Kunetz 1966) or modeled mathematically (Xu et al. 1988).

Finally, SP anomalies related to conductive mineral deposits (Sato and Mooney 1960) or geothermal activity (Corwin and Hoover 1979) may constitute noise for engineering SP investigations. If surveys are conducted in areas of known mining or geothermal activity (e.g., seepage investigations for tailings ponds or heap leach sites) the possibility of such anomalies should be considered.

7 Conclusions

SP surveys performed in support of engineering investigations often are subject to levels of error and noise that are high with respect to the amplitude of anomalies generated by sources of interest. Even with such problems, the increasing use of the SP method for engineering applications indicates that usable SP data can be obtained under difficult conditions. To maximize data quality under such conditions, it is necessary to select appropriate survey configurations, equipment, and field procedures, and to recognize and correct sources of error and noise.

SP data acquisition and processing procedures have not been standardized as they have for other geophysical methods, and the recommendations in this chapter represent only a beginning toward the development of optimal procedures. Additional contributions that reflect the experience of other investigators in this field are needed to continue this development.

References

Bhattacharya B B, Roy N (1981) A note on the use of a nomogram for self-potential anomalies. Geophys Prosp **29**: 102-107

Corry C E (1985) Spontaneous polarization associated with porphyry sulfide mineralization. Geophysics **50**: 1020-1034

Corwin R F (1976) Offshore use of the self-potential method. Geophys Prosp **24**: 79-90

Corwin R F, Conti U (1973) A rugged silver-silver chloride electrode for field use. Rev Sci Instrum **44**: 708-711

Corwin R F, Hoover D B (1979) The self-potential method in geothermal exploration. Geophysics **44**: 226-245

Corwin R F, Ticken E J (1988) Development of self-potential interpretation techniques for seepage detection. Final Rep Contr DACW39-86-C-0059. US Army Corps Engrs Waterways Exp St, Vicksburg, MS

Ernstson K, Scherer H U (1986) Self-potential variations with time and their relation to hydrogeologic and meteorological parameters. Geophysics **51**: 1967-1977

Ewing S (1939) The copper-copper sulfate half-cell for measuring potentials in the earth. Tech Sec Am Gas Assoc Distrib Conf 1939

Fitterman D V (1979) Calculations of self-potential anomalies near vertical contacts. Geophysics **44**: 195-205

Fröhlich R K (1971) The influence of industrial stray currents on the measurement of earth potentials and their elimination. Geophys Prosp **19**: 118-132

Heiland C A (1940) Geophysical exploration. Prentice-Hall, New York

Hoogervorst G H T C (1975) Fundamental noise affecting signal-to-noise ratio of resistivity surveys. Geophys Prosp **23**: 380-390

Ives D J G, Janz G J (1961) Reference electrodes. Academic Press, New York, London

Kaufman A A, Keller G V (1981) The magnetotelluric sounding method. Elsevier, New York

Keller G V, Frischknecht F C (1966) Electrical methods in geophysical prospecting. Pergamon, New York

Koester J P, Butler D K, Cooper S S, and Llopis J L (1984) Geophysical investigations in support of Clearwater Dam comprehensive seepage analysis. US Army Eng Waterways Exp St Misc Pap GL-84-3

Kunetz G (1966) Principles of direct current resistivity prospecting. Geopub Assoc (Berlin)

Morrison H F, Corwin R F, Harding R, and de Moully G (1979a) Interpretation of self-potential data from geothermal areas. Semi-Ann Tech Prog Rep April 30, USGS Contr 14-08-0001-16546. Univ Cal, Berkeley

Morrison H F, Corwin R F, Harding R, and de Moully G (1979b) Interpretation of self-potential data from geothermal areas. Semi-Ann Tech Prog Rep Sept 30, USGS Contr 14-08-0001-16546. Univ Cal, Berkeley

Nayak P N (1981) Electromechanical potential in surveys for sulphides. Geoexploration **18**: 311-320

Nourbehecht B (1963) Irreversible thermodynamic effects in inhomogeneous media and their applications in certain geoelectric problems. PhD Thesis, Mass Inst Technol, Cambridge

Parasnis D S (1966) Mining geophysics. Elsevier, New York

Petiau G, Dupis A (1980) Noise, temperature coefficient, and long-time stability of electrodes for telluric observations. Geophys Prosp **28**: 792-804

Petrowsky A (1928) The problem of a hidden polarized sphere. Philos Mag **5**: 334-353; 914-933

Poldini E (1938) Geophysical exploration by spontaneous polarization methods. Min Mag **59**: 278-282; 347-352

Poldini E (1939) Geophysical exploration by spontaneous polarization methods. Min Mag **60**: 22-27; 90-94

Rao B S R, Marthy I V R, and Reddy S J (1970) Interpretation of self-potential anomalies of some geometric bodies. Pageoph (Pure and Applied Geophysics) **78**: 66-77

Sato M, Mooney H M (1960) The electrochemical mechanism of sulfide self-potentials. Geophysics **25**: 226-249

Scott B I H (1962) Electricity in plants. Sci Am **207**: 107-117

Semenov A S (1974) Electrical prospecting with the natural electric field method. Nedra, Leningrad (in Russian)

Sill W R (1982) Diffusion coupled (electrochemical) self-

-potential effects in geothermal areas. DOE/DGE Rep DOE/ID/12079-73, Dep Geol Geophys, Univ Utah

Sill W R (1983) Self-potential modeling from primary flows. Geophysics **48**: 76-86

Sill W R, Johng D S (1979) Self-potential survey, Roosevelt Hot Springs, Utah. DOE Rep IDO/78-1701.a.2.3, Dep Geol Geophys, Univ Utah

Stern W (1945) Relation between spontaneous polarization curves and depth, size, and dip of ore bodies. Trans AIMME **164**: 189-196

Telford W M, Geldart L P, Sheriff R E, and Keys, D A (1976) Applied geophysics. Cambridge Univ Press, New York

Uhlig H H (1963) Corrosion and corrosion control. Wiley, New York

Xu S, Gao Z, and Zhao S (1988) An integral formulation for three-dimensional terrain modeling for resistivity surveys. Geophysics **53**: 546-552

Zablocki C J (1976) Mapping thermal anomalies on an active volcano by the self-potential method, Kilauea, Hawaii. Proc 2nd Symp Development and use of geothermal resources, San Francisco, CA, US Gov Print Off, Washington, DC, vol 2, pp 1299-1309

NUMERICAL MODELING OF SELF-POTENTIAL ANOMALIES DUE TO LEAKY DAMS: MODEL AND FIELD EXAMPLES

M.J. Wilt[1] and R.F. Corwin[2]

Abstract

Substantial self-potential anomalies are known to be associated with zones of discharge in leaky dams. The mechanism for generating these SP anomalies. however, is not well understood. In this paper we apply a two-dimensional computer-code to calculate self-potential anomalies for a model of a leaky dam and then apply the code to some field data for a dam site in the United States. Fluid flow and electrical current flow are coupled processes; that is, there are small currents associated with fluid flow processes and small fluid flows associated with large electrical currents. The processes are connected via cross-coupling terms in Darcy's law and Ohm's law. The coupled equations have been solved numerically for two-dimensional geometry using the finite difference technique. A computer code developed by Sill (1983) is used in this study.

The two-dimensional schematic model for a leaky dam features a leakage zone, a seepage area and an earthen dam structure with physical properties similar to the field case considered below. We consider examples where the depth of the leak, the variation of anomalies with flow path, and the effect of a change in rock type within the dam are examined. In general, the SP profiles show a negative anomaly over the area where the leak is occuring and a positive anomaly over the seep. The relative magnitudes of these anomalies and the shapes of the profiles depend on the location of the sources, the geometry of the flow, and the distribution of physical parameters. Varying the source

[1] Engineering Geosciences, University of California, Berkeley, California, USA

[2] 419 Seaview Dr., El Cerrito, California, USA

Lecture Notes in Earth Sciences, Vol. 27
G.-P. Merkler et al. (Eds.)
Detection of Subsurface Flow Phenomena

(leak) depth mainly affects the negative SP anomaly. Shallow leaks produce larger magnitude and more abrupt anomalies, while deeper leaks produce broader and lower amplitude anomalies. Variations in flow paths affect the observed voltages in complex ways. Where the fluid circulates above the leak and then downwards in the seepage area the anomalies seem to be somewhat enhanced. Where the fluid initially flows downwards into the bedrock the anomalies are smaller, especially near the source. A change in coupling coefficient within the dam structure, caused by a fault for example, mainly results in a scaling of the anomalies over the source and seepage in proportion to the magnitude of the change.

SP surveys over the Beaver dam site in Arkansas have similar characteristics to the model studies, they show positive anomalies where seepage is occuring and negative anomalies over leaks. The leakage through the dam seems to be controlled by an east-west graben fault that is associated with a string of negative SP anomalies. Although field data appear considerably more complex than the two-dimensional models considered above we attempt to fit an SP profile that connects the leakage and seepage area using the 2-d code. Model parameters were obtained from previous field surveys and "educated guesses" when no other information was available. A surprisingly good fit was obtained using a very simple model. The only complexity that the model features is a change in SP cross-coupling coefficient associated with the graben fault. The modeling suggests that the SP anomalies are dominantly controlled by the positions of the sources and sinks and not by any complex flow processes.

1 Introduction

It is well known that substantial self-potential (SP) anomalies are associated with regions of discharge in leaky dams (Butler 1988). The general mechanism ascribed to these anomalies is a streaming potential caused by rapidly flowing groundwater through the porous rock of the dam fill. Although numerical and physical models have confirmed that this is the dominant mechanism (Hötzl and Merkler 1988), the specific relationship between observed SP anomalies and the geology and flow characteristics of leaky dams is still poorly understood.

In this paper we examine the relationship between observed SP anomalies and the position and flow paths of leaks. We first develop some fundamental theory that illustrates the connection between SP and fluid flow processes. Then using a two-dimensional computer code we calculate some SP profiles for a schematic earth dam. Finally, we apply two-dimensional modeling to some field data for a leaky dam site in the United States.

2 General Principles

The basic principle relating self-potential and fluid flow processes is that the flow of electrical current and fluid are coupled. That is, there are electrical potentials ϕ related to fluid flow processes and hydraulic potentials P associated with electrical current flow. The mathematical relationship between the potentials is given Eqs. (1) and (2) (Nourbehecht 1963):

$$Q = C_{11}\nabla P + C_{12}\nabla\phi \quad (1)$$

$$J = C_{21}\nabla P + C_{22}\nabla\phi \quad (2)$$

where Q and J are fluid flow and current flow, respectively, and
C_{11} = K, the hydraulic conductivity;
C_{22} = σ, the electrical conductivity and
C_{12}, C_{21} are the cross-coupling coefficients.

The cross-coupling coefficients in Eqs. (1) and (2) are typically much smaller than the primary flow coefficients. Fluid flow processes do not generate very large electrical currents and imposition of electrical potential does not generate large fluid flows (although this method has been used for dewatering low permeability material). Note that if we neglect the cross-coupling terms in Eqs. (1) and (2), the equations decouple into the more familiar Darcy's law and Ohm's law.

In the absence of external electrical current we note that the current is divergenceless:

$$\nabla \cdot J = \frac{dq}{dt} = 0$$

where q is electrical charge. Applying this relation to Eq. (2) and doing some algebra we obtain:

$$-\nabla \cdot (\sigma \nabla \phi) = \nabla C_{11} \cdot \nabla P + C_{21} \nabla^2 P \tag{3}$$

This is the classical DC conduction equation with electrical potential and conductivity on the left-hand side and current sources on the right-hand side. Equation (3) shows that there are SP current sources wherever (1) the cross-coupling coefficient C_{21} changes in the direction of flow and (2) wherever there are fluid sources ($\nabla^2 P \neq 0$).

In the case of a leaky dam Eq. (3) shows that there are SP anomalies associated with the leak area and the surface discharge zone since these are source areas. There are also SP anomalies associated with changes in rock and fluid type (i.e., where the cross-coupling coefficients change).

In Eq. (4) we examine the cross-coupling coefficient C_{12}:

$$C_{21} = \zeta \frac{\varepsilon}{4\pi\eta} \tag{4}$$

where ζ is the zeta potential, ε is the fluid dielectric, and η is the fluid viscosity. For a dam site the most significant of these parameters is the zeta potential which is a function of both fluid and rock type. The zeta potential, in general, increases with porosity and salinity. For unconsolidated dam-site material the cross-coupling coefficient is expected to be fairly large (Ishido and Mizutani 1981).

Equation (3) has been solved numerically for three-dimensional sources and two-dimensional geometry by Sill (1983). Sill developed a finite difference code that calculates the potentials within and on the surface of a two-dimensional half-space. In this solution models for the permeability, cross-coupling coefficients and electrical resistivity are required. The current sources in Eq. (3) are calculated by first solving the fluid flow problem and then, knowing the cross-coupling coefficients, calculating the position and strength of current sources using Eqs. (1) and (2). Once the current is known Eq. (3) becomes a DC resistivity problem and the solution is obtained with the same finite difference routines used to solve the fluid flow problem. For a reasonable sized mesh Sill's code (program SPXCPL) runs in about 1 min. on a VAX-11780.

3 Model Studies

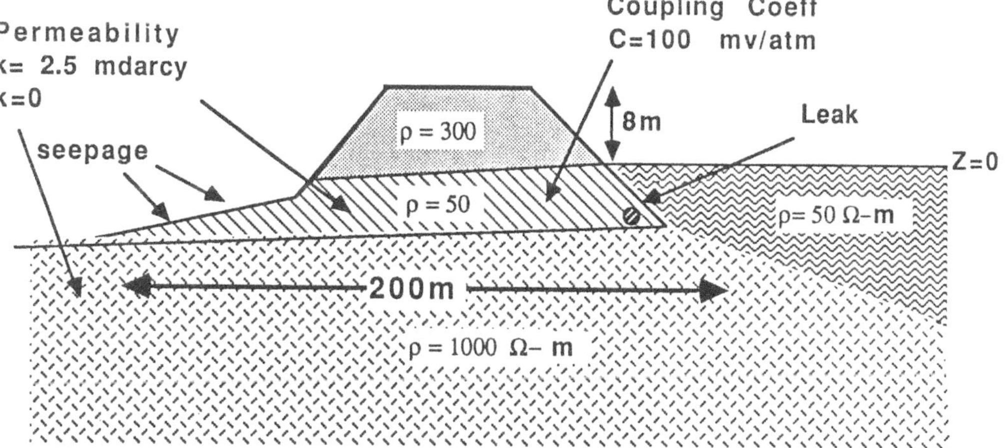

Fig. 1: Schematic two-dimensional model of an earthen dam

Figure 1 is a schematic of a leaky earth dam used in our model studies. Our dam features a leakage area, a downstream seep, and a dam structure where physical properties are assigned corresponding to different cases. Physical property values assigned to the rock and fluid in this model are based on field observations for a dam site in the USA. In our model studies we consider the effects of variations in source (leak) depth, and flow path. We also examine a case where the dam is made of different materials, i.e., there is a fault within the dam structure.

In Fig. 2 and 3 we show the results of varying the source depth. We consider three depths, 8, 10, and 12 m; the magnitude of the leak is the same in all three cases (100 l/s). The SP profiles shown in Fig. 3 are similar in style to most of the calculated results in this study. They display a negative anomaly over the region where water is flowing into the dam and a positive over the area where the seepage is occurring.

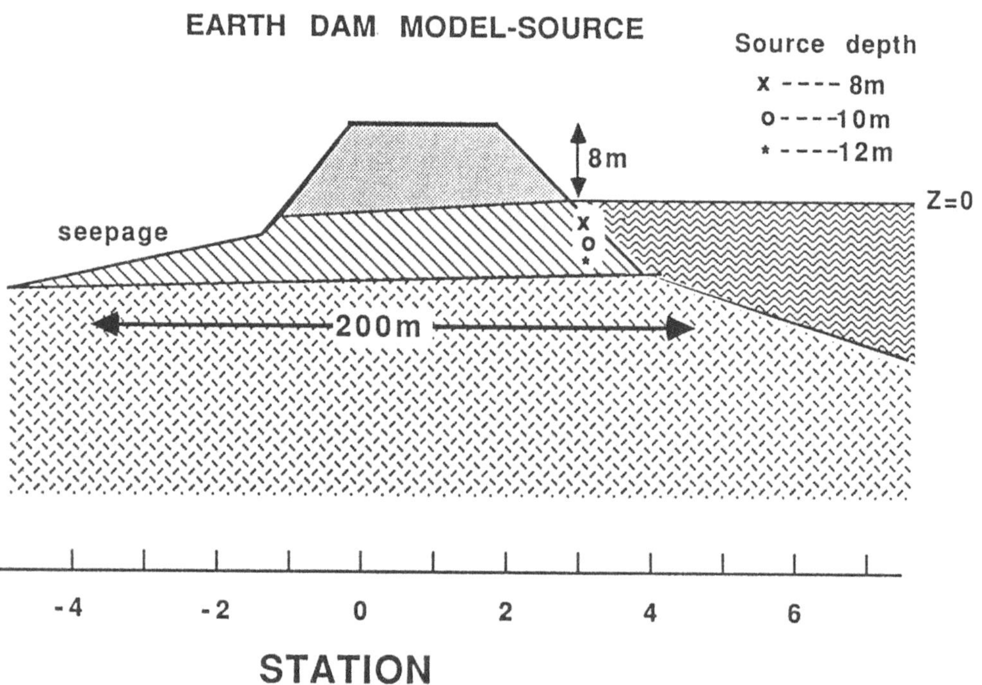

Fig. 2: Model showing the location and depth of sources

Figure 3 shows that the main effect of source depth variation is to change the magnitude and shape of the negative anomaly over the leak. The anomalies for the shallow sources are larger and more abrupt; the deeper sources produce broader, lower amplitude anomalies.

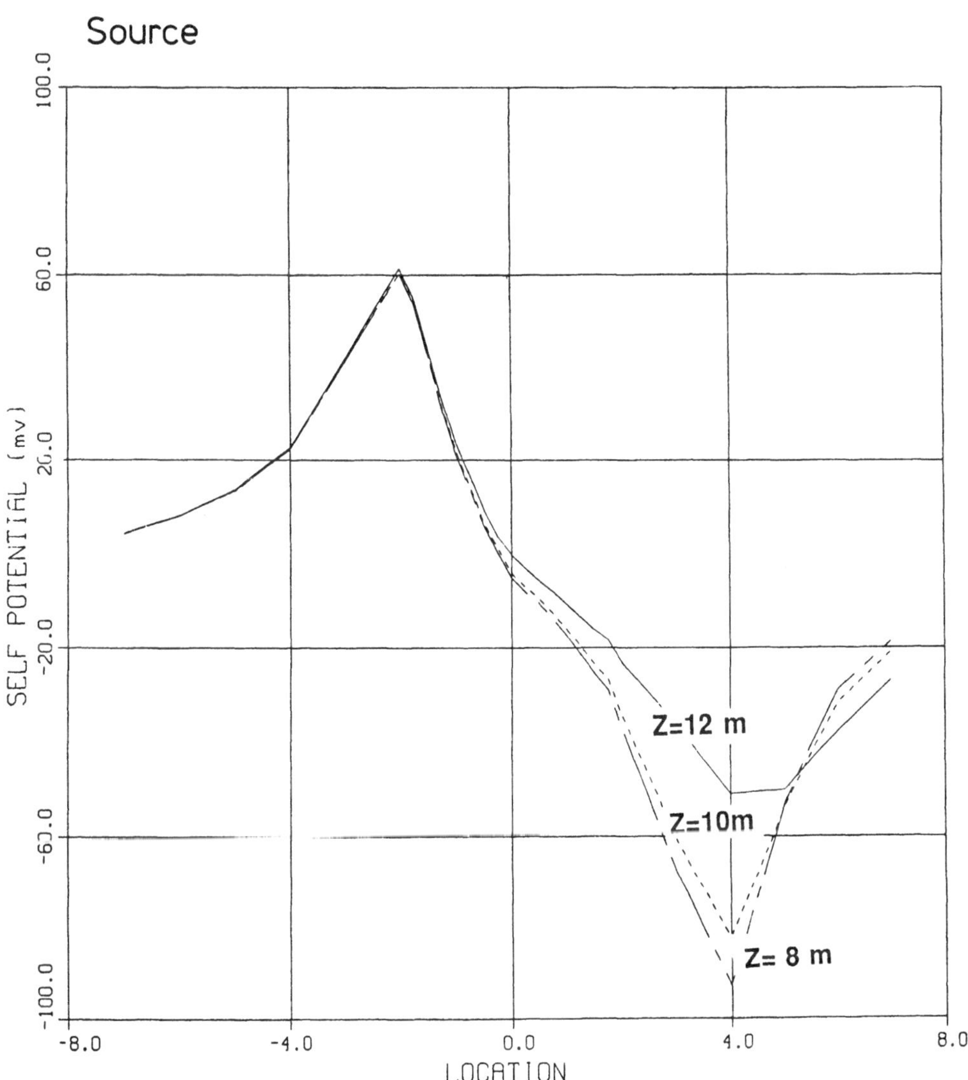

Fig. 3: SP profiles over the source models in Fig. 2

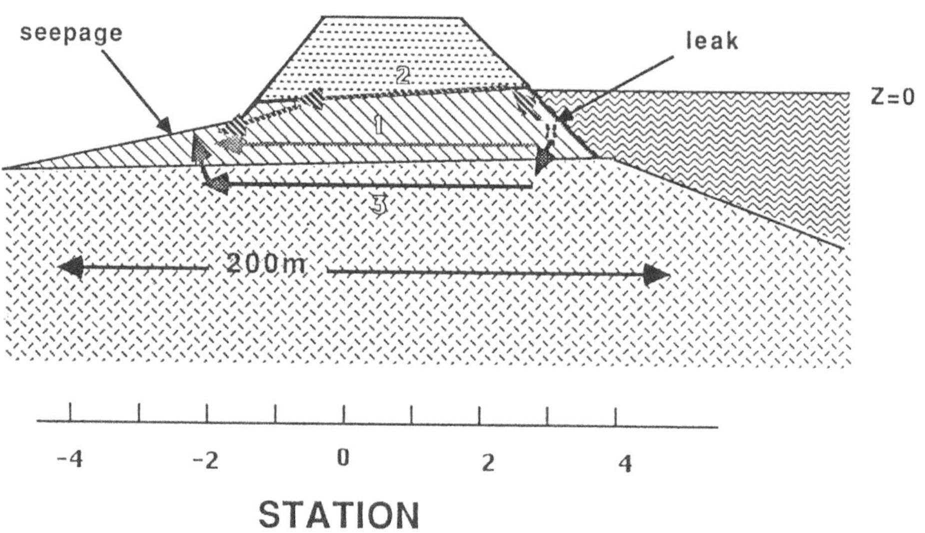

Fig. 4: Model showing different high-permeability leakage pathways

In Figs. 4 and 5 we examine the effect of leakage pathway on the SP profiles. In Eq. (3) we can see that the SP voltage is dependent on the pressure gradient so that the geometry and permeability of high permeability pathways should have a significant affect on the observed SP. We consider three high permeability pathways for fluids to flow from the source to the seep. In the first case the pathway directly connects the source and the seep areas, in the second case fluid initially moves upwards toward the top of the dam and then laterally before descending into the seepage area. For the third case the fluid moves downwards and then laterally before moving upwards and out of the dam (Fig. 4).

We immediately notice from the plots given in Fig. 5 that the anomaly is substantially smaller than for the uniform, lower permeability model in Fig. 3. This is due to the lower pressure difference between source and seep area because of the high permeability pathway. Eq. (3) shows that this pressure difference is the driving force in generating the SP anomaly. The results in Fig. 5 show that the pathway has a significant effect on the shape of the self-potential anomaly, particularly near the seepage area. Where the fluid moves directly across the dam (case 1) the anomaly is more or less symmetrical, where

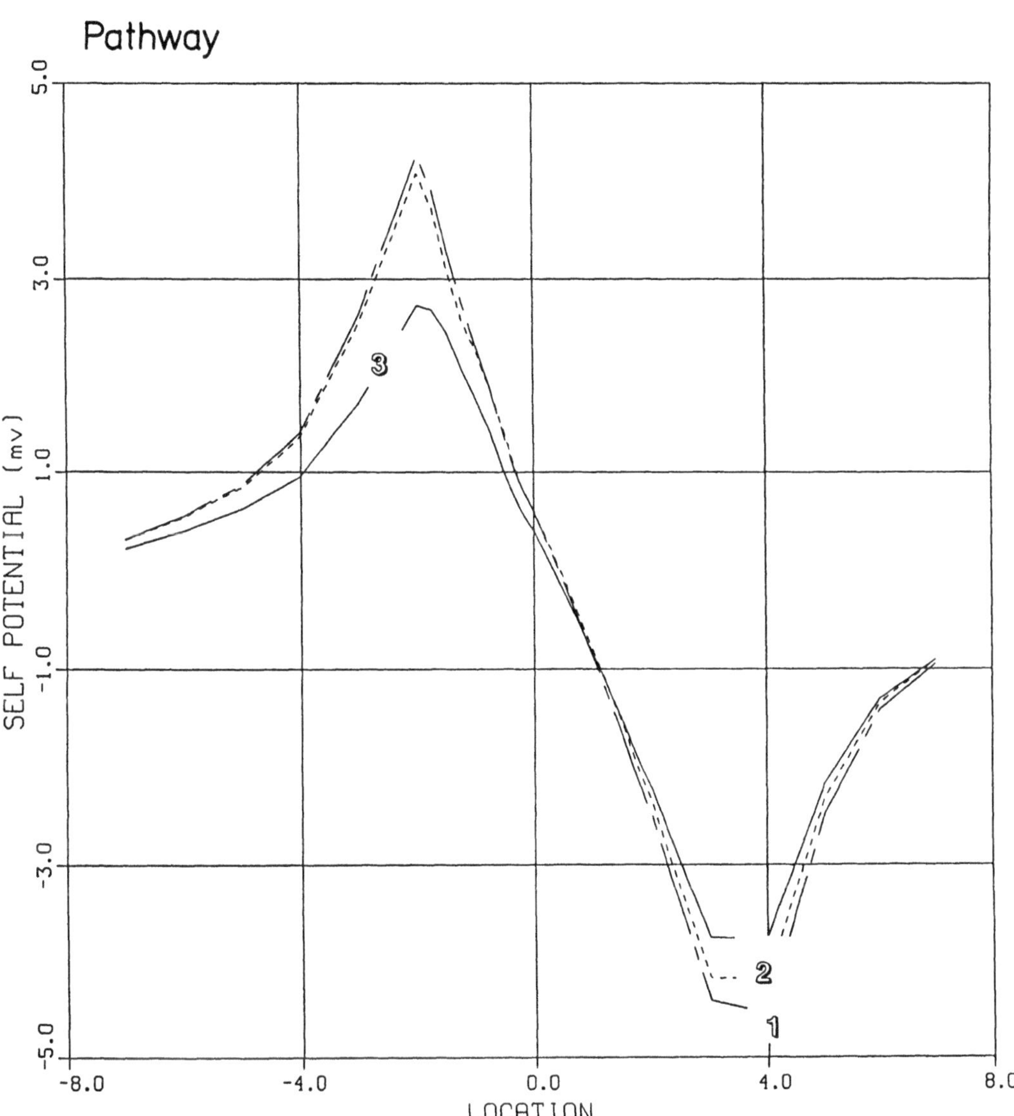

Fig. 5: SP profiles over pathway models in Fig. 4

the fluid first moves upwards (case 2) the anomaly is slightly larger. Where the fluid first moves downwards the anomaly is slightly smaller. These results show that the observed SP is only slightly affected by the configuration of leaky flow paths.

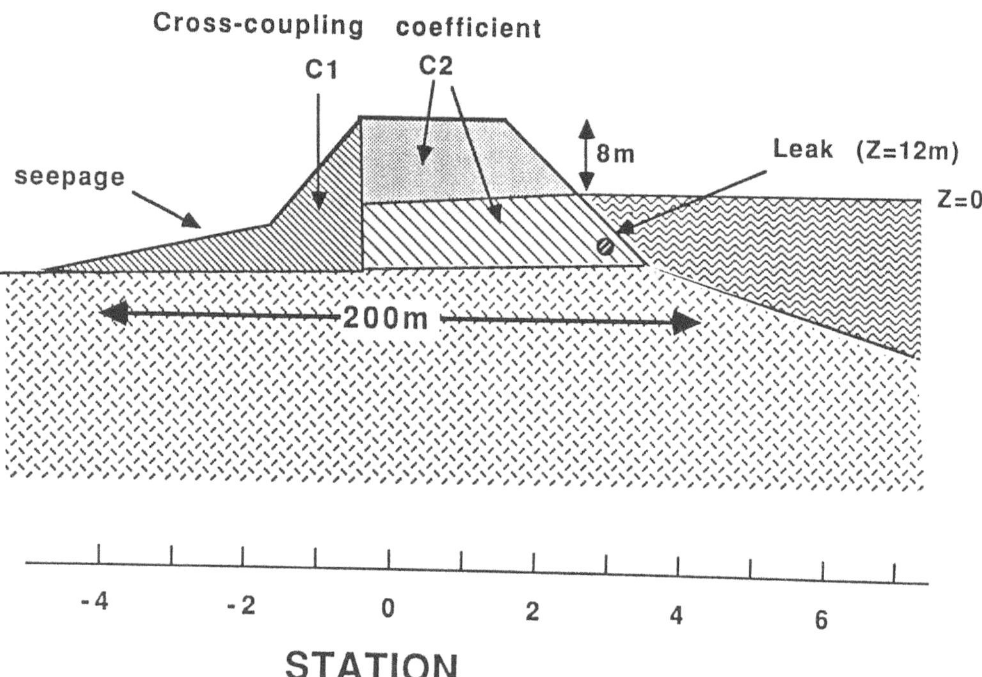

Fig. 6: Model showing a change in the cross-coupling coefficient within the dam

The fault model is illustrated in Fig. 6. In our model, the fault has obliquely crossed the dam structure at the position of the cross-section. At the position of the cross-section we consider that the dam is constructed of materials with different cross-coupling coefficients. This might be the case if dam fill material has been imported and added to existing rock. It might also be the case if there is a fault within the host rock of the dam. Although changes in cross-coupling coefficients are often accompanied by changes in resistivity and permeability in the interest of simplicity we are not considering these effects.

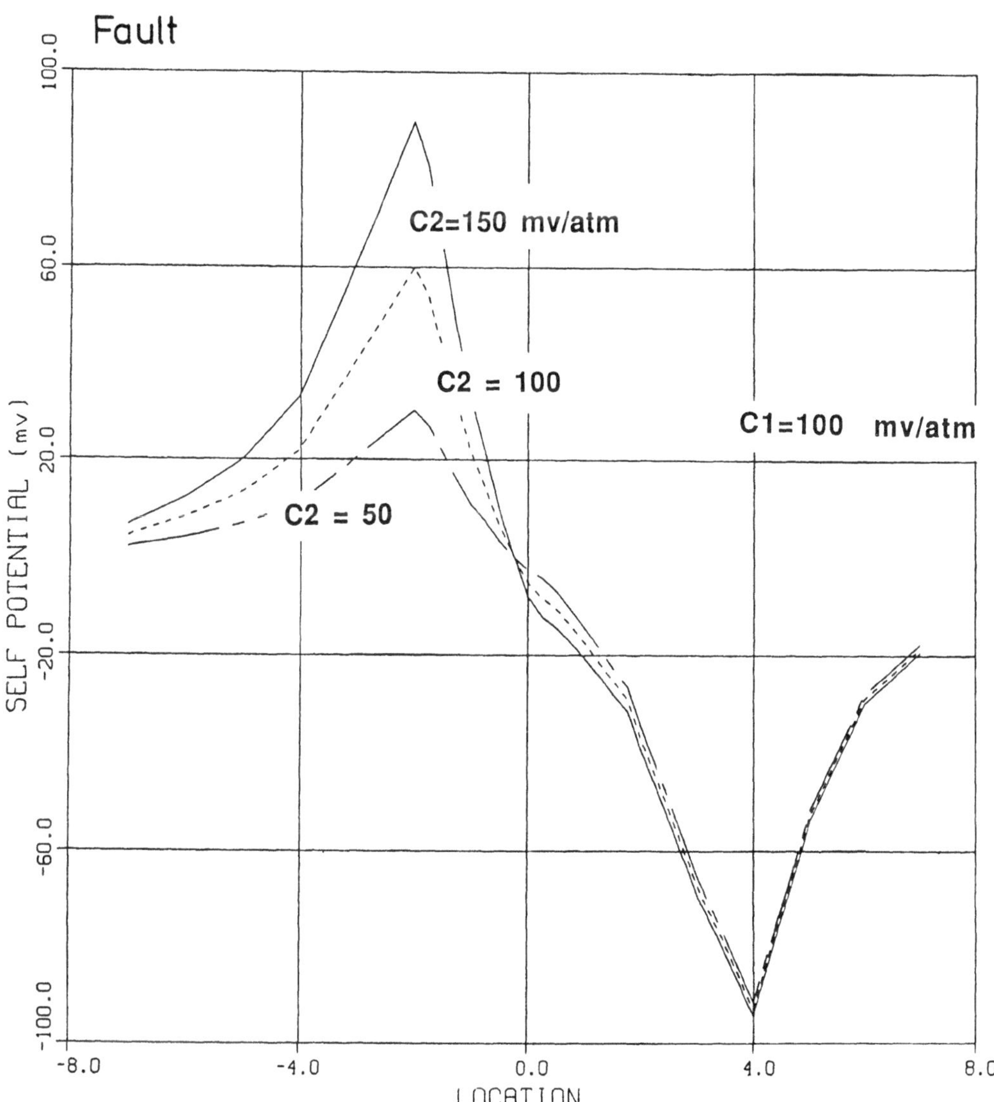

Fig. 7: SP profiles over the fault model

Variations in cross-coupling coefficients affect the observed SP in two ways. The first is that the anomaly over a pressure source or sink is scaled by the magnitude of the coupling coefficient [Eq. (3)]. The second is that abrupt changes in the cross-coupling coefficients in the flow direction results in the generation of secondary sources. In Fig. 6 we consider a fault where the coupling coefficient on one side is half as large (case 1), the same size (case 2), and twice as large (case 3) as the other side. The resulting SP anomalies are shown in Fig. 7.

The results show that the dominant effect of the cross-coupling coefficient change is the scaling of the anomalies referred to above. Where the cross-coupling coefficient contrast is larger, the resulting SP anomaly is correspondingly larger.

One conclusion that can be drawn from the above models is that the SP anomalies related to fluid flow processes can be very complex. Even in this simple two-dimensional schematic model almost endless variations are possible. The other conclusion is that the dominant forces controlling the size and shape of the SP anomalies are the locations of leakage (inlet) and seepage (outlet) zones.

4 Field example

In Fig. 8 we show a self-potential anomaly map for the Beaver dam site in the state of Arkansas in the southern part of the USA. This is an earthen dam through which significant seepage occurs in the region labeled "new wet area". The seepage in the southern part of the survey area is thought to be controlled by an east-west trending fault zone that obliquely crosses the dam structure. The fault is thought to form the southern boundary of a graben structure and is associated with a string of negative SP anomalies (Butler 1988).

Although the map is far more complex than any of the models previously considered it does have some of the same general characteristics. That is, there are negative anomalies where fluid is leaking into the dam (near A) and positive anomalies over the surface discharge areas ("new

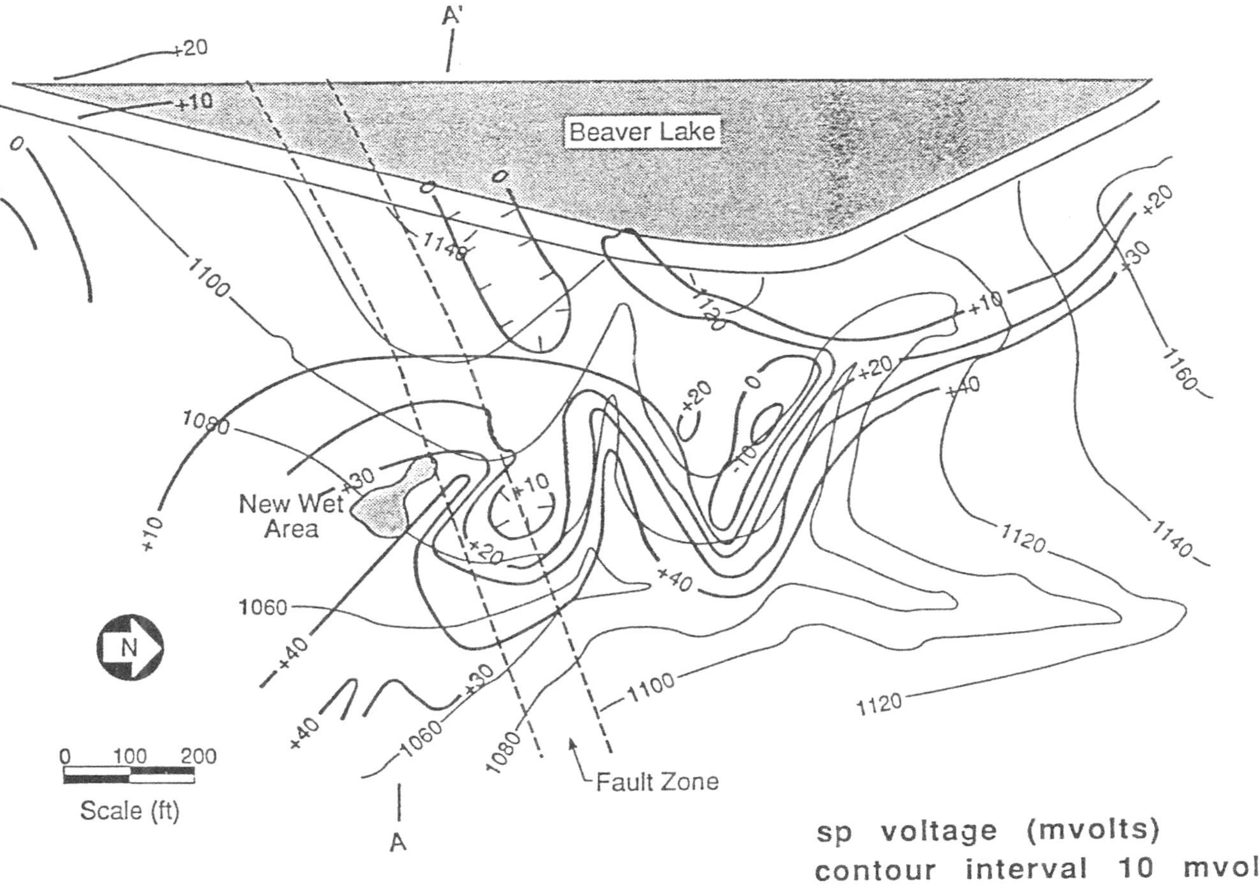

Fig. 8: Observed SP data at the Beaver dam site in Arkansas, USA

wet area"). The alignment of the anomalies with the known fault zone strongly suggests that the fault provides a low impedance leakage pathway for fluids. The map also shows several other anomalies, some of which are related to other leakages and some are due to topography which can have a significant effect (Corwin 1988).

We selected a single profile from the map (A-A') that connects source and seep areas and fits the observed data to calculated data for a two-dimensional geometry. The permeability and resistivity distribution for the model shown in Fig. 9 were obtained from field measurements (Butler 1988).

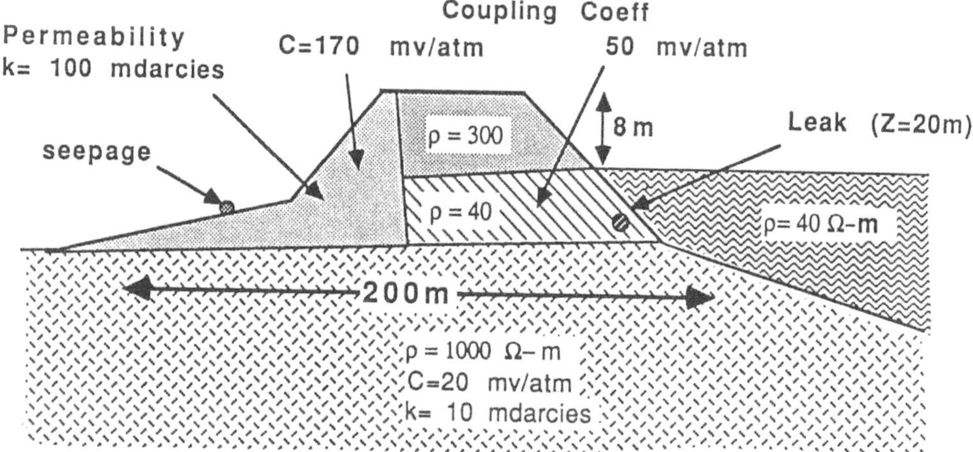

Fig. 9: Model used to fit observed Beaver dam data for profile A-A' in Fig. 8

Values given for the cross-coupling coefficients are "educated guesses" based on general relationships given by Ishido and Mizutani (1981). Figure 9 is a very simple model for the dam. The model features single source and seep locations and a single couple-coefficient contrast corresponding to where the graben fault crosses the profile line. The figure shows that the magnitude of the leak is about 50 l/s and about half of this amount is discharging in the new wet area.

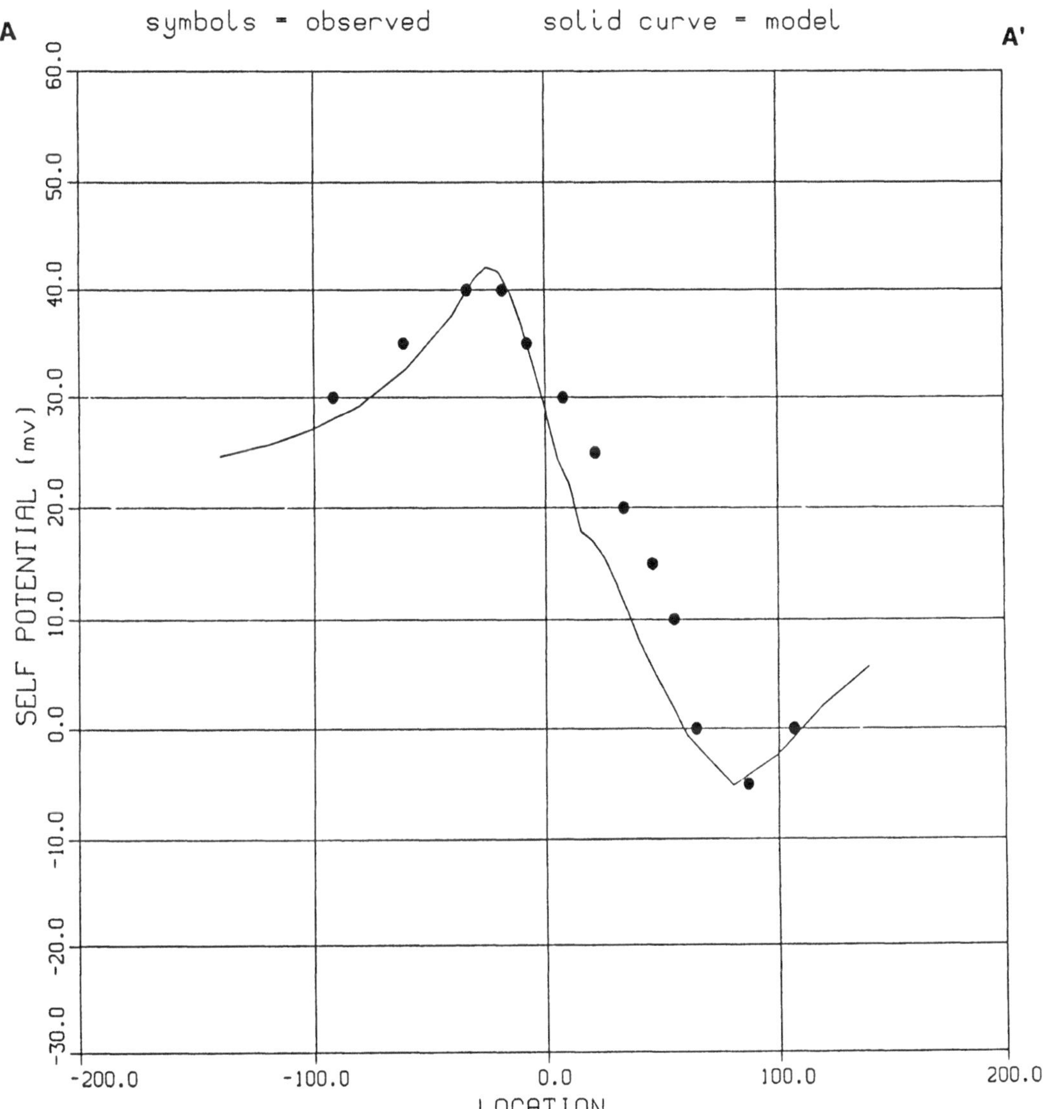

Fig. 10: Comparison of calculated to observed SP results for the model in Fig. 9

The fit between calculated and observed SP is shown in Fig. 10. A very good match was achieved even though the assumed model is considerably simpler than the known geology. This suggests that the SP anomalies are dominantly controlled by the positions of the sources and seeps and not by complex flow geometry or rock-type variations. We note, however, that this is not a unique model and the parameters are not very well constrained. The model is consistent with the known geology and hydrology but it does not account for three-dimensional features.

5 Conclusions

This short paper has demonstrated that numerical modeling can be a powerfull tool in understanding SP anomalies over leaky dams. The modeling can provide physical insight to the factors controlling the observed SP anomalies and can provide means for locating leaks for remedial action. The use of a two-dimensional code clearly has limitations in three-dimensional cases and for many dam sites the code may be inappropriate. In cases approximating two dimensions substantial information must be known about the physical properties of the dam before numerical modeling can be applied at a particular site. If this is the case, then two-dimensional modeling can be rewarding.

References

Butler D. (1988): Geophysical Methodology for Subsurface Fluid Flow Detection Mapping and Monitoring: An Overview and Selected Case History. Proc. Detection of Subsurface Flow Phenomena by Self-Potential/Geoelectrical and Thermometrical Methods. Int. Symp. Karlsruhe, FRG, March 14-18, 1988

Corwin R.F. (1988): Data Acquisition, Reduction and Reproducibility for Engineering Self-Potential Surveys. Proc. Detection of Subsurface Flow Phenomena by Self-Potential/Geoelectrical and Thermometrical Methods. Int. Symp. Karlsruhe, FRG, March 14-18, 1988

Hötzl H., Merkler G.P. (1988): Model Experiments in a Channel. Empirical Correlations between Streaming Potentials and Hydraulic Fields. Proc. Detection of Subsurface Flow Phenomena by Self-Potential/Geoelectrical and Thermometrical Methods. Int. Symp. Karlsruhe, FRG, March 14-18, 1988

Ishido T., Mizutani H. (1981): Experimental and Theoretical Basis of Electrokinetic Phenomena in Rock-Water Systems and its Application to Geophysics. Geophys. Res. 86. B3: 1763-1775

Nourbehecht B. (1963): Irreversible Thermodynamic Effects in Inhomogeneous Media and their Application to Certain Geoelectric Problems. Ph D Thesis, Mass. Inst. Technol., Cambridge, Mass.

Sill W.R. (1983): Self-potential Modeling from Primary Flows. Geophysics 48. 1: 76-86

MATHEMATICAL MODELS OF SELF-POTENTIAL FIELDS (GEOELECTRICAL OR GEOTHERMAL) FOR DETECTION OF SUBSURFACE FLOW PHENOMENA

S.G. Kostyanev[1]

Abstract: This chapter offers some mathematical models for determination of self-potential fields (geoelectrical or geothermal) in media with an arbitrary distribution of conductivity. Furthermore, heterogeneity occurs in this media.

1 Introduction

To solve the inverse problem of geoelectrical (geothermal) investigation in order to detect some subsurface flow phenomena, it is necessary to determine the electrical (thermal) conductivity of the media on the basis of the self-potential measured at the earth's surface. In developing methods for solving the inverse problem it is assumed that self-potential is measured at the horizontal relief of the earth's surface. However, the potential observed in rugged topography is considerably deformed due to the effect of the irregular terrain relief. Such deformation effects should not be considered in the interpretation of the problem. The existing electrical (geothermal) investigation methods cannot effectively eliminate the deformation exerted upon the potential. In the first part of this chapter a new method of eliminating the rugged topography effect upon potential fields for the detection of subsurface flow phenomena is discussed.

Geothermal observation can be used for detecting heterogeneities of media. The investigation comprises heterogeneities of a hydrogeological nature. The above mentioned heterogeneities cause the disturbances of the deep heat flow field on the earth's surface. By using the observation and analysis of the thermal regime in the surface layer, we can distinguish an anomalous component from the thermal field studies;

[1] Higher Institute of Mining and Geology, Department of Mathematics, Sofia-1156, Bulgaria

Lecture Notes in Earth Sciences, Vol. 27
G.-P. Merkler et al. (Eds.)
Detection of Subsurface Flow Phenomena

this provides information on the subject under investigation. The basic interpretation of the measured anomalous thermal field consists in comparing the surface heat flow distribution with that calculated for some models. This procedure requires effective methods of calculation for direct problems of stationary thermal field distributions in heterogeneous media. In the second part of this chapter a mathematical model of the geothermal field in gradient media is discussed.

2 Formulation and solution of the problem concerning the elimination of the topography effect on potential fields

Suppose that there is homogeneity V_0 with electrical (thermal) conductivity K ; its boundary is the earth's surface S_0, partially smooth (Fig. 1). Let V_0 contain regions V_i (i=1, 2,..., N) (hydrogeological sites), respectively, with conductivity K_i (i=1, 2, ..., N). In the general case the conductivities are not constant magnitudes. Suppose

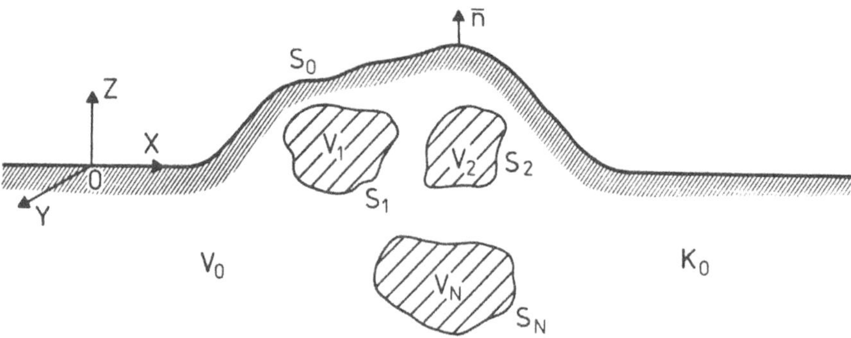

Fig. 1: Assumed situation for eliminating the topography effect on potential fields

that S_i (i=1, 2,..., N) are also partially smooth surfaces and represent the boundaries to regions V_i. Suppose that potential fields are generated by regions V_i. Let us present the observed (summary) potential U (M) of the surface S_0 as consisting of components whose sources are located below S_0, as well as components whose emergence is due to the uneven surface S , i.e., the rugged topography of the earth's surface (Kostyanev 1977, 1985; Nedelkov 1978).

Suppose that potential function U(M) satisfies the equation:

$$\Delta U(M) = \frac{d^2U}{dx^2} + \frac{d^2U}{dy^2} + \frac{d^2U}{dz^2} = 0, \qquad (1)$$

where

$$M = (x, y, z), \; M\varepsilon(V_0 + S_0 - \sum_i V_i).$$

Besides, for the geoelectrical field:

$$U(M) = A(M), \text{ for } M\varepsilon S_0 \qquad (2)$$

is a known, measured magnitude, and

$$\frac{dU}{d\bar{n}} = 0; \text{ for } M\varepsilon S_0. \qquad (3)$$

For the geothermal field in the general case we have:

$$\frac{dU}{d\bar{n}} + k^2U = f, \text{ for } M\varepsilon S_0. \qquad (4)$$

where $\bar{n}$ is normal to the surface S_0 and f is a known function.

Applying Green's formula against function U(M) and Green's function $G(M, M_0)$, the anomalous geoelectrical field in an arbitrary point $M_0\varepsilon S_0$, generated solely by internal nonhomogeneities V_i of the media, is represented as follows:

$$W(M_0) = 2\pi U(M_0) + \iint_{S_0} U(M) \frac{dG}{dn} dS_M, \; M_0\varepsilon S_0. \qquad (5)$$

Applying Green's formula against function U(M) and function $Q(M, M_0)$, analogous to Green's function for the geoelectrical problem, the anomalous geothermal field in an arbitrary point $M_0\varepsilon S$, generated solely by internal nonhomogeneities V_i of the media, is represented as follows:

$$\tilde{W}(M_0) = 2\pi U(M_0) - \iint_{S_0} f\cdot Q\cdot dS, \; M_0\varepsilon S_0. \qquad (6)$$

3 Formulation and Solution of some Problems concerning Geothermal Fields in Gradient Media

Suppose that there is a heterogeneity V (hydrogeological site) with a constant heat transfer coefficient in the bed $0<z<H$ with an arbitrary distribution of the heat transfer coefficient $K(z)$ (Fig. 2). It is assumed that the V is located in a layer with constant $K(z) = K$, i.e.,

$$K(z) = \begin{cases} K_1(z), & 0<z<H \\ K_0, & h_1<z<(h_1+h_2) \\ K_2(z), & (h_1+h_2)<z<H \end{cases} \qquad K(M) = \begin{cases} K(z) & \text{at } M \notin V_0 \\ K_T & \text{at } M \in V_0 \end{cases}$$

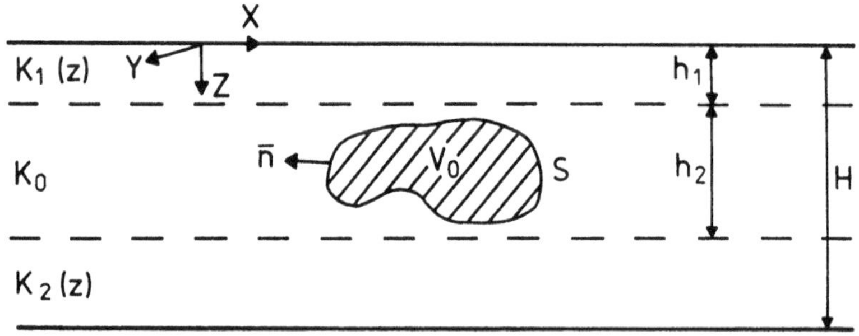

Fig. 2: Assumed situation for solving the problem concerning geothermal fields in gradient media.

The temperature field in the layer $T(x, y, z)$ can be found, as well as the heat flow at the earth's surface, if constant heat flow is given at the lower surface at the layer ($z = H$) and constant temperature at the upper surface ($z = 0$). The mathematical problem is reduced to finding the function $T(x, y, z)$, satisfying the equation:

$$\operatorname{div}(K(M)\operatorname{grad} T) = 0 \qquad (7)$$

within the layer.
With conjugation conditions on the surface of disruption:

$$[T] = 0; \; [K \frac{dT}{dn}] = 0, \qquad (8)$$

where $\vec{n}$ is the normal to the disruption surface, and square brackets are used to indicate the difference betwen the boundary values at both sides of the disruption surface $K(M)$. The boundary conditions must also be fulfilled:

$$T(x, y, z) = \text{const}, \text{ for } z = 0; \frac{dT}{dz} = P_0 = \text{const.}, \text{ for } z = H \quad (9)$$

with the infinity condition:

$$\frac{dT}{dr} \to 0 \text{ at } r = \sqrt{x^2 + y^2} \to \infty. \quad (10)$$

The given problem is solved by using the integral equation method.

The general temperature field $T(x, y, z)$ is subdivided into a normal field $T_n(z)$ (field without heterogeneities) and an anomalous field $T_a(x, y, z)$, resulting from the disturbing effect of a heterogeneous region. Thus,

$$T(x, y, z) = T_n(z) + T_a(x, y, z). \quad (11)$$

<u>Calculation of the Normal Field.</u> The normal field is obtained by the solution of the problem:

$$\frac{d}{dz}\left(K \frac{dT_n}{dz}\right) = 0$$
$$T_n(z) = 0; \; K \frac{dT_n}{dz} = 0$$
$$T_n = T_0, \text{ for } z = 0; \; \frac{dT_n}{dz} = P_0, \text{ for } z = H. \quad (12)$$

The solution of this problem can be written in the following analytical form:

$$T_n(z) = T_0 + P_0 K(H) \int_z^0 \frac{dz}{K(z)} \quad (13)$$

<u>Calculation of the Anomalous Field.</u> The following problem is given for the anomalous field:

$$\text{div}\,(K \text{ grad } T_a(x, y, z) = 0; \; 0<z<H; \; -\infty < x, y < \infty \quad (14)$$

At the disruption $K(z)$ conjugation conditions are satisfied:

$$[T_a] = 0; \; \left[K(z) \frac{dT_a}{dz}\right] = 0. \quad (15)$$

At the surface of heterogeneity S, the following conditions are satisfied:

$$[T_a] = 0; \; K_0\left(\frac{dT_a}{dn}\right)_{out} - K_T\left(\frac{dT_a}{dn}\right)_{in} = (K_T - K_0) \left.\frac{dT_n}{dn}\right|_S , \quad (16)$$

where $(\frac{dT}{dn}a\text{-})_{out}$ and $(\frac{dT}{dn}a\text{-})_{in}$ are boundary values of the normal derivative from the outer and inner side of the surface S. The following homogeneous boundary conditions are satisfied at the layer surfaces:

$$T_a(x, y, z) = 0, \text{ for } z = 0; \frac{dT}{dz}a = 0, \text{ for } z = H \tag{17}$$

and for infinity, the condition can be expressed:

$$\frac{dT}{dr}a\text{-} \rightarrow 0 \quad \text{at} \quad r = \sqrt{x^2 + y^2} \rightarrow \infty \quad . \tag{18}$$

The boundary value problem (14-18) thus obtained for the anomaous field can be easily reduced to an integral equation. Suppose $G(M, M_0)$ is the Green's function for the boundary value problem:

$$\begin{cases} \text{div}(K(z) \text{ grad } G(M, M_0)) = - \delta(r_{MM_0}) \\ [G] = 0, \quad [K(z) \frac{dG}{dz}] = 0 \\ G = 0, \text{ for } z = 0, \quad \frac{dG}{dz} = 0, \text{ for } z = H, \\ \frac{dG}{dr} \rightarrow 0 \quad \text{at } r \rightarrow \infty \end{cases} \tag{19}$$

Then the anomalous temperature field can be represented as a simple layer potential:

$$T_a(M) = \oiint_S R(M_0)G(M, M_0)dS_{M_0} . \tag{20}$$

The field represented in this way satisfies all the conditions of problem (14-18), except for those of a normal derivative disruption on the surface S. Substituting expression (20) into this condition yields the following integral equation:

$$\frac{K_0+K_T}{2} R(M)+(K_0+K_T) \oiint_S R(M_0)\frac{dG}{dn_M} dS_{M_0} = (K_T-K_0) \frac{dT}{dn_M}n\text{-} \tag{21}$$

Having solved the integral equation, according to Eq. (20), we define the field within the layer. For solving Eq. (21) it is necessary to know the Green's function $G(M, M_0)$ for which we have designed a method of calculation (Kostyanev 1982).

4. Conclusion

The results we obtained permit (if not completely, yet very simply) the elimination of the influence of the relief upon the geoelectrical (geothermal) potential measured at the earth's surface. Such a simple calculation permits obtaining the distribution of the potential without false anomalies. As a whole we can, to a considerable extent, avoid mistakes in solving an inverse geoelectrical (geothermal) problem.

The designed mathematical models of geothermal fields in gradient media with heterogeneity permit the solution of inverse geothermal problems for detection of subsurface flow phenomena.

References

Kostyanev, S. (1977): Integral method of eliminating the relief effect in some geoelectrical investigations.- Acad. Bulg. Sci. Tome 30, 4: 523.

Kostyanev, S. (1982): Mathematical modeling of geothermal processes in gradient media.- Bulg. Acad. Sci. 8, 2.

Kostyanev, S. (1985): On the potential of the geoelectrical fields in rugged topography.- Int. Meet. Potential fields in rugged topography; Lausanne, Switz.

Nedelkov, I. (1978): Improper problems of potential theory and their application in geophysics. Sofia, Bulgaria.

Streaming Potential in Nature

M. SCHUCH

Bayer. Landesanstalt für Bodenkultur und Pflanzenbau,
Menzinger Straße 54, D-8000 München 19, FRG

Abstract

For the first time, QUINCKE found in 1859 the phenomenon of electric streaming potential. Twenty years later HELMHOLTZ published a mathematical expression for the streaming potential.
In the following years a number of scientists studied the phenomenon. BIKERMAN (1932) showed that each electric streaming potential causes an electric current in the contrary direction. SWARTZENDRUBER postulated in 1967 that this electric field tries to stop the streaming potential as a result of the energy balance.
It seems that streaming potentials are very general in nature. In applied geophysics streaming potential was used in the past in auger hole prospecting. But steaming potentials were determined in each soil type with sufficient capillarity (SCHUCH and WANKE 1968a), as also in peat bog (SCHUCH and WANKE 1967) and in living plants, e. g. spruce (SCHUCH and WANKE 1968b). It is not easy to measure streaming potentials. Besides streaming water or soil solution, temperature differences or concentration differences etc. cause electric potential differences. Unpolarized electrodes must be used. A special device with calomel electrodes shows good results.

1 Introduction

Between different materials exists a difference of electric potential. The causes of this difference may be various, e. g. physical differences like different temperature, different pressure or chemical differences. There exists also a difference of electric potential between a solid body and a solution, two different solutions or different concentrated solutions.
If the origin of the electric potential is not exactly determined, one uses in applied geophysics the expression "self-potential, SP" in French "polarisation spontané, PS" or in German "Eigenpotential".

Lecture Notes in Earth Sciences, Vol. 27
G.-P. Merkler et al. (Eds.)
Detection of Subsurface Flow Phenomena

2 Some Historic Remarks on the Theory of Streaming Potentials

Beside this kind of SP another phenomenon causes very often in mineral soil (SCHUCH and WANKE 1968a *), peat soil (SCHUCH 1963; SCHUCH and WANKE 1967,* 1969 a,b *) or plants (SCHUCH and WANKE 1968b *), [2] a difference of electric potential, the so-called electric streaming potential.

More than 100 years ago QUINCKE (1859) discovered the following phenomenon: water forced through a porous medium develops an electric potential. Between water or a solution and a capillary there exists a double layer (after HELMHOLTZ 1879). By movement of the water one can observe at both ends of the capillary an electric streaming potential. The value of this potential is determined by an equation after HELMHOLTZ:

$$U = (\varepsilon \zeta \ / \ 4 \pi \sigma \eta \ ). \ \Delta P$$

Where ε = dielectric constant;

ζ = ζ - potential;

ΔP = pressure difference;

σ = electric conductivity;

η = viscosity.

By varying the length l of the capillary or the sample and the distance of both electrodes one must consider the pressure gradient $\Delta P/l$ and the distance of electrodes a:

$$U = (\varepsilon \zeta \ / \ 4 \pi \sigma \eta \ ). \ (\Delta P/l \ ) \ a \ .$$

Besides QUINCKE and HELMHOLTZ, BIKERMAN (1932), CLARK (1877), LAMB (1888), SAXEN (1892), GOUY (1910), CHAPMAN (1913), STERN (1924) and NEALE (1946) studied this phenomenon.

---

2) Investigations denoted by * were supported by the Deutsche Forschungsgemeinschaft during the "Internationalen Hydrologischen Dekade".

3 Observing the Phenomenon PS

The first and main problem which occurs during the observation of PS is represented by the electrodes. In an electrolyte system one connot use metallic or polarized electrodes, because the galvanic potential between metallic electrodes and the electrolyte is much greater than the electric streaming potential one will observe. In some special cases metallic electrodes are used, if the galvanic potential on both electrodes is identical but with opposite polarity. In this case the galvanic potential between the metallic electrode and electrolyte is eliminated. This case seldom occurs when studying the phenomenon PS.

In general, unpolarized electrodes are used: copper Cu in concentrated copper vitriol ($Cu\ SO_4$) or zinc in sulphate of zinc ($ZnSO_4$). The polarization of such an electrode is zero. But there are other complications: between the concentrated solution and the measuring object exists a concentration potential because the cation Cu^{2+} and the anion SO_4'' have different movements (SCHUCH 1963).

By using the calomel electrode this error is eliminated because the cation K^+ and the anion Cl' of the potassium chloride bridge have nearly the same movement (e. g. MILAZZO 1952). The calomel electrode is used by measuring the pH value. It has a very constant potential compared to the normal hydrogen electrode. By determination of pH one observes mostly the electric potential of a glass membrane compared to a calomel reference electrode. Observing PS one uses two calomel electrodes (SCHUCH 1963, 1985). If there is no electromotoric force in an electrolyte system, e. g. if both electrodes are placed in water, no difference in electric potential will be found. It is useful to make this test before every observation of PS. If the two electrodes are brought into contact, for example with different soil, different peat, or soil with water movement (evapotranspiration), one can measure or register the PS exactly.

Since electronic instruments are generally used in the laboratory, there are no problems with the input resistance. It must be very high ($< 10^{12}\Omega$)

and the current, by measurement or registration, very low (10^{-8} A, SCHUCH 1963). One can use any suitable instrument for observing pH.

4 Some Laboratory Experiments and Observations on a Spruce

By pressing water through a plastic tube filled with mineral soil or peat an electric potential difference at the ends of the tube could be observed. The end of the pipe where water exits always has a positive potential.

We took an undisturbed peat sample, a monolith of about 125 l and put the sample in a plastic box with a dense cover. We then pressed water through the sample from one side to the other and found an electric streaming potential in the monolith. By changing the direction of water flow, the plus and the minus signs were alternated (SCHUCH and WANKE 1969a).

We put in the same plastic box a peat monolith with a grass cover. While heating the grass cover slightly we recorded an increase of the electric potential in the vertical direction. The positive sign shows us that there was water movement to the surface. If heating was ceased the electric potential returned to zero after some time.
If we interpret this electric potential as a function of evapotranspiration, we should find electric streaming potentials even in the trunk of a spruce.

We put electrodes in small drill holes at a vertical distance of 0.2 m in the trunk of a spruce and recorded the electric potential over several days. We found a daily variation of electric potential.
We could show that the electric field power in the trunk is a function of radiation balance (SCHUCH and WANKE 1968b). It was very interesting that on sunny days during the maximum of radiation balance a short time after midday, a depression in the electric field power was observed. If the transpiration of a plant increases too much, the cells could be destroyed. Botanists speak of a midday depression.

5 Some Field Experiments

In bog soils, pseudogley, sandy soils and moraines we placed the end of one electrode on the surface and the end of the other electrode at a depth between 0.2 and 2 m beneath the surface (SCHUCH and WANKE 1967, 1968a, b, 1969a, b). The electrograph shows in each case a variation of the electric potential during days with fine weather. On cloudy days there was less variation. If the fields were covered with snow, no variation could be expected. At a vertical distance of the electrodes of about 0.3 m we found the greatest variations of the electric potential differences. WANKE (1957) also found sometimes on fine days a midday depression, the same effect explained in the previons section.
These experiments demonstrate other interesting phenomena. The electrograph shows after a dry period during strong rainfall in pseudogley a water movement from subsoil to topsoil. By drying, pseudogley is lacerated by many rents. During heavy rainfall, water flows down these rents causing only a small streaming potential. When water increases in the capillaries, a large streaming potential is obtained. The recorded electric potential, the difference of both, shows water movement to the surface.

To study the soil moisture movement in a drain field we installed 24 electrodes beside a drain pipe and recorded on 12 lines the electric potential (SCHUCH 1978). It was possible to determine the direction and the amount of water flow as a function of precipitation and flow from the drain pipe.

6 Results

Water movement in soil or in the trunk of a tree produces an electric streaming potential. Thus, it is possible to record exactly with three-component arrangements the direction and amount of water flow and its variations as a function of time and soil-physical, hydrological and climatic parameters. A calibration of the method is, however, necessary.

7 Noise Effects

A major problem of recording electric streaming potentials is represented by the electrodes. Many noise effects could be eliminated by using calomel electrodes in special devices. For instance electric diffusion and concentration potentials created by the contact of electrode/soil could be eliminated. By placing both electrodes close together an electric potential between both electrodes due to temperature differences can be prevented. The circuit from electrode to soil in this arrangement passes over a potassium chloride bridge in a plastic hose. In each case the electrodes should be protected from sun radiation.

Some noise effects, however, could not be eliminated, for example, electric concentration potentials between different soil layers (SCHUCH 1963). But usually these electric potential differences are constant over a long time period. They have a fossil character.

Constant subsurface water flow produces a permanent electric potential. If there are other electromotoric forces in the subsoil, one cannot distinguish between the different effects. But by observing variations of water flow, e. g. the daily water flow in soils as a result of evapotranspiration, it is easy to eliminate the constant noise effects.

The potential differences we recorded by soil water movement in the laboratory and in the field are only some mV. Noise effects are numerous. Thus, we looked for other methods to confirm our results. If the daily variation of electric potential between two vertical points in soil is caused by water movement from subsoil to topsoil, we should find a higher water content in the topsoil during a day with sunshine. But it is impossible to determine the water content exactly. However, swelling and shrinking of soils by a change in water content is well known. Thus, we recorded the swelling of bog soil and found a good correlation between the daily swelling and shrinking of the surface and the results of our electrograph (SCHUCH 1963; VIDAL and SCHUCH 1966).

8 Conclusions

Environmental pollution represents a great danger to mankind. Ecological hazards must be recognized and prevented. New methods to observe detrimental effects, especially in the field, are necessary. Streaming potential in nature could not only be applied by studying water balance, but also by testing the nutrient loss in fields or nitrogen dynamics in soils.

References

BIKERMAN J (1932) Ionentheorie der Elektroosmose, der Strömungsströme und der Oberflächenleitfähigkeit. Z phys Chem A 163: 378 - 394

CHAPMAN D.L. (1913) A contribution to the theory of electrocapillarity. Philos Mag 25 (6): 475 - 481

CLARK J.W. (1877) Über die beim Durchströmen von Wasser und Capillarröhren erzeugte elektromotorische Kraft. Ann Phys 2 (3): 335 - 346

GOUY G (1910) Sur la constitution de la charge êlectrique â la surface d'un êlectrolyte. J Phys 9 (4): 457

HELMHOLTZ H (1979) Studien über elektrische Grenzschichten. Ann Phys 7 (3): 337 - 382

LAMB (1988) Philos Mag 25 (5): 52

MILAZZO G (1952) Elektrochemie, theoretische Grundlagen und Anwendungen. Springer Wien

NEALE S.M. (1946) Electrical double layer, the electrokinetic potential and the streaming current. Trans Karaday Soc 42: 473 - 478

QUINCKE G (1859) Über eine neue Art elektrischer Ströme. Ann Phys 107 (2): 1 - 47

SAXEN U (1892) Über die Reciprocität der electrischen Endomose und der Strömungsströme. Ann Phys 47 (3): 46 - 68

SCHUCH M (1963) Beobachtung von Eigenpotentialen an Torflagerstätten mit verschiedenartigen Elektroden. Z Geophys 29 (4): 175 - 169

SCHUCH M (1978) Regulation of water regime of heavy soils by drainage, subsoiling and liming and water movement in this soil. Proc Int Drainage Worksh Wageningen, Neth

SCHUCH M (1985) Natürliche, elektrische Spannungen im Boden, im Torf und an der menschlichen Haut beim Moorbad. TELMA 15: 245 - 250

SCHUCH M, WANKE R (1967) Strömungsspannungen in einigen Torf- und Sandproben. Z Geophys 33 (2): 94 - 109

SCHUCH M, WANKE R (1968a) Die zeitliche Variation elektrischer Strömungsspannungen auf kurzen vertikalen Meßstücken in Mineralböden. Z Geophys 34: 599 - 610

SCHUCH M, WANKE R (1968b) Die zeitlichen Variationen der elektrischen Strömungsspannung in einem Fichtenstamm, verursacht durch die tägliche Änderung des Saftstromes. Z Oecol Plant 3: 169 - 176

SCHUCH M, WANKE R (1969a) Ein neues Verfahren zur Beobachtung der Wasserbewegung in Böden. Dtsch Gewässerkdl Mitt, Sonderh: 37 - 40

SCHUCH M, WANKE R (1969b) Die zeitliche Variation der elektrischen Strömungsspannung auf kurzen Meßstrecken im Torfboden als Folge kapillarer Wasserbewegungen. Z Pflanzenernähr Bodenkd 2, 122: 112 - 128

STERN O (1924) Zur Theorie der elektrischen Doppelschicht. Z Elektrochem 30: 508 - 516

SWARTZENDRUBER D (1962) Non-Darcy flow behaviour in liquid saturated porous media. Soil Sci 93: 22 - 29

SWARTZENDRUBER D (1967) Non-Darcian Movement of soil water. Proccedings II of Internat Soil Water Symposium Prag: 207 - 221

VIDAL H, SCHUCH M (1966) Eine hochempfindliche Schlauchwaage mit photoelektrischer Registriereinrichtung. Z Felsmech Ing Geol IV/2: 154 - 159

WANKE R (1967) Die Variation der elektrischen Strömungsspannungen auf kurzen Meßstrecken in Torf- und Mineralböden. Diss Naturwiss Fak Ludwig-Maximilians-Univ, München

Self-Potential Surveys on Waste Dumps
Theory and Practice

M.Weigel

Prakla - Seismos AG. Bucholzer Str. 100
D-3000 Hannover 51

Measuring self-potential differences is one of the oldest geophysical prospection methods. For conventional measurements one electrode functions as a base electrode, whereas a second is used as a moving electrode. The latter is moved fixed distances along a previously surveyed line. The difference in voltage between the two electrodes is the potential difference. This type of survey is time-consuming and noise potentials which arise are not recognized, for example potentials that occur when the electrode holes are struck, periodic variations of the potential field resulting from telluric currents, stray currents produced by industry or variations due to the weather.

To achieve quicker and more accurate surveying, PRAKLA-SEISMOS developed a survey arrangement in which instead of just one base electrode (which under certain conditions may be located in a particularly "noisy" area, i.e. unfavourably located) and one moving electrode, two or more base electrodes and a number of survey electrodes (presently up to 216) are deployed (Fig 1).

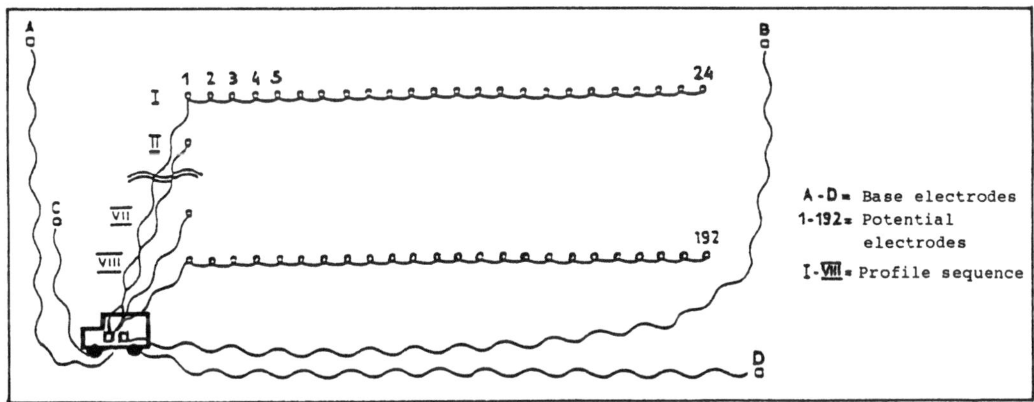

Fig. 1. Survey configuration

Computer controlled recording is made at time intervals of about 5 to 10 min and in doing so the entire survey configuration is recorded virtually simultaneously. It is recommended to record over a long period, e.g. several hours, so that the behaviour of the electrodes can be recognized.

Lecture Notes in Earth Sciences, Vol. 27
G.-P. Merkler et al. (Eds.)
Detection of Subsurface Flow Phenomena

Moreover, it is then also possible to eliminate noise potentials, which was not the case with the previously applied "base electrode/ moving electrode" configuration.

All the base and survey electrodes are laid out in 1 day and connected up to the survey truck using special 24 trace cable. The electrodes adjust themselves overnight to the surrounding rock with the potential field around the electrodes "relaxing" to a certain extent. Fig. 2 shows a recording of this type of relaxation.

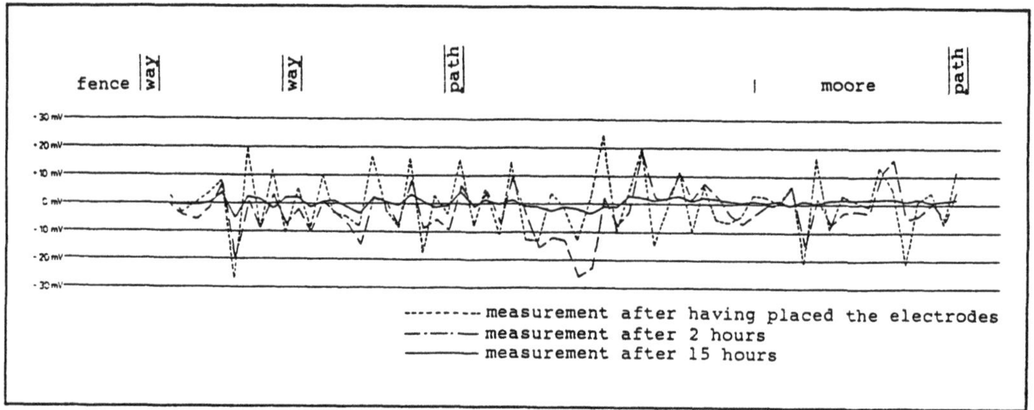

Fig. 2. Potential field measured at different times along a survey line

The curve can be seen to be very erratic immediately after the electrodes were set. However, after just 2 h the curve quietened down and after 15 h (overnight) was relatively smooth.

A scanner and desk-top computer are used for data recording. Potential differences with respect to the various base electrodes can be measured as often as required. Subsequently the recorded data can be printed in the field on the thermal printer or displayed on the screen. The recorded data are subjected to different computer processes before being interpreted.

In recent years we have tested the method on different objects.

The limits of propagation of a gas injected into a man-made gas cavity are depicted in Figs. 3,4.

[V]
mv1
mv 2
mv=mean value
Water
Transition Zone
Gas
Electrode
[V]2

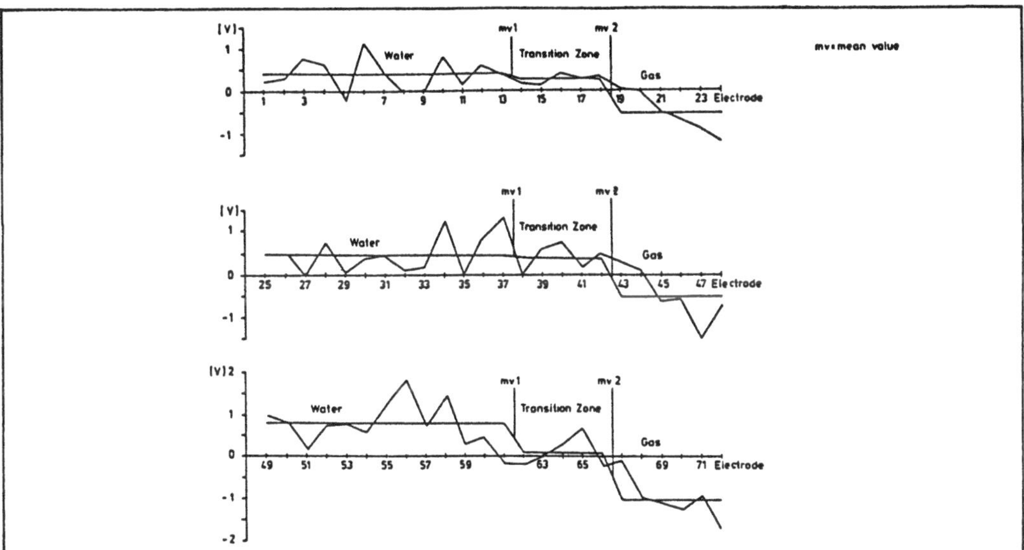

Fig. 3. Self-potential measurement

[mV]
profile 3
Storage
profile 4
profile 5
Self potential over the gas reservoir
weighted running average (three-point)
Legend
Electrode-positions (Separation 50m)
Gas-Water-contact
Pore-storage (Gas) in south Germany during the injection phase

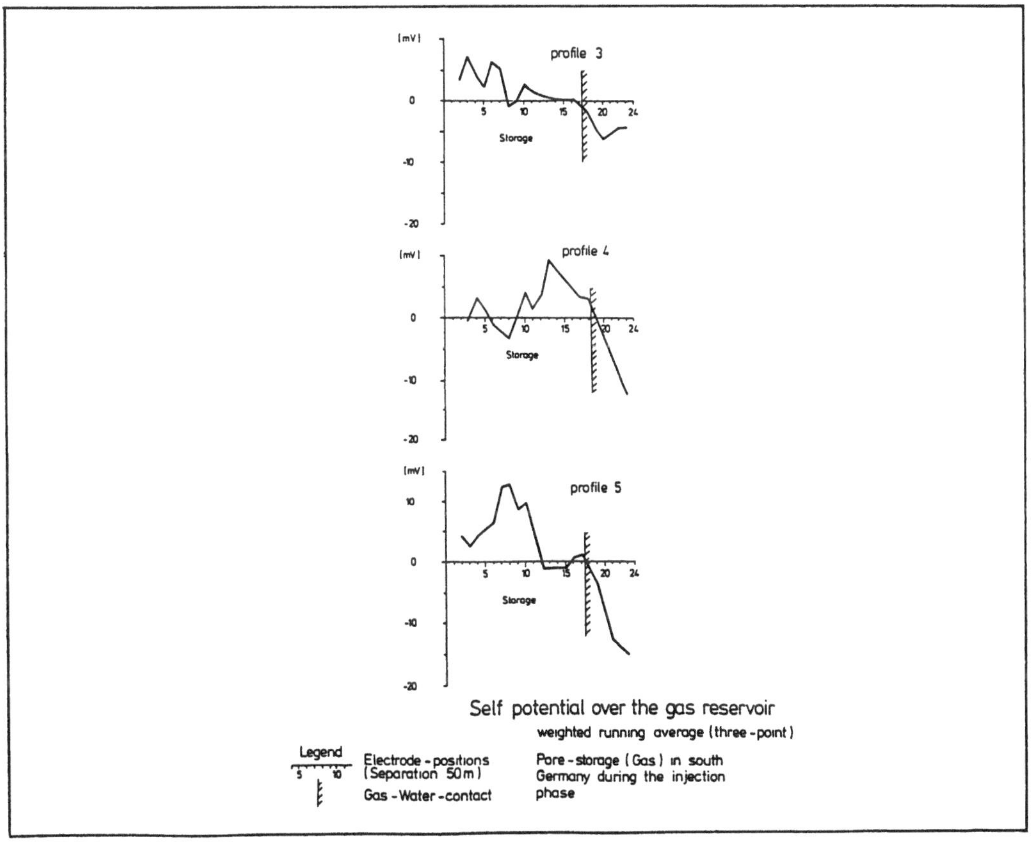

Fig. 4. Self-potential measurement over the gas reservoir

Measurements over several waste dumps have also been carried out. Several observation wells are available at a waste disposal site which is located in a former limestone quarry.

From the wells it is known that contaminated water exists in a certain area, and although contamination is not very great (CSB about 839) groundwater pollution cannot be ruled out. In order to determine the precise limits of contamination a number of self-potential lines (Fig. 5) were laid out and recorded over several hours. The survey results were subjected to a series of computer processing steps. In Fig. 6 the average values for the individual electrodes and the average values over specific ranges are shown graphically. The contaminated area is distinguished by lower self-potential values, whereas the limits of this area are marked by distinct downward points on the curve.

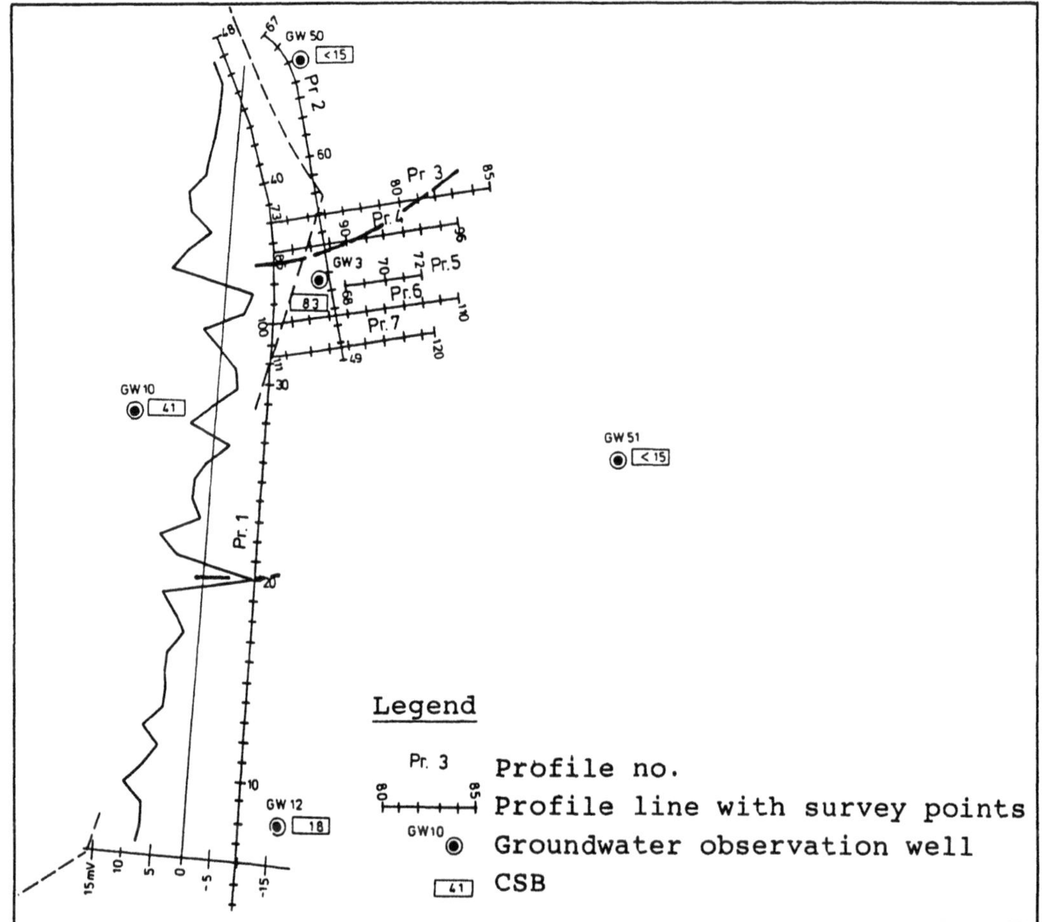

Fig. 5. Disposal site, self-potential measurement

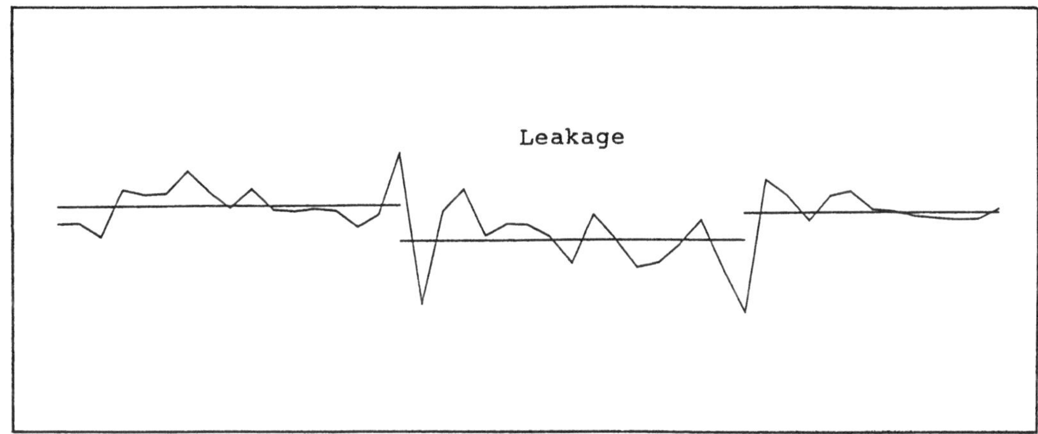

Fig. 6. Disposal site, averaged values

At a different waste disposal site the existance of contaminated water was suspected. Self-potential measurements were consequently used to localize the leakage point (Fig. 7), which was confirmed by drilling.

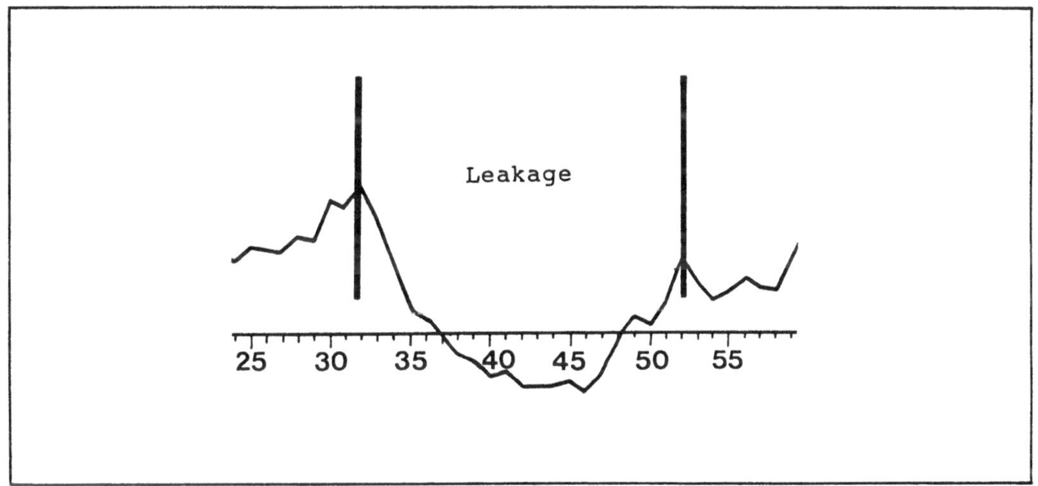

Fig. 7. Disposal site, simple survey

In yet another area fruit trees adjacent to a waste dump were damaged by gases originating from the ground. A subsequent gas analysis confirmed the suspicion that the gases in the garden and those in the waste dump had the same composition. We believe our surveys have found the path these gases have taken. It must be noted here that the groundwater in this area had sunk by about 15 to 20 m and consequently only the water of imbibition remained. Nevertheless, several peaks were determined which I feel could represent rock fractures that permit the gases to pass more freely.

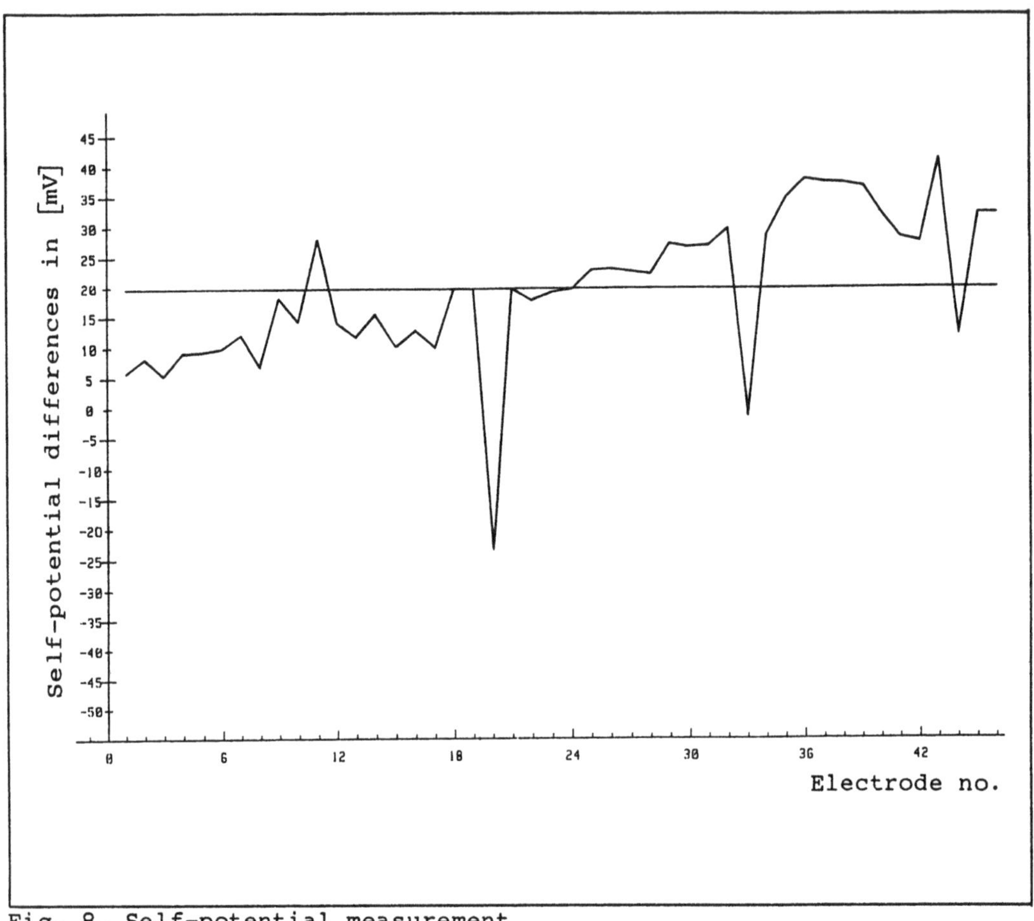

Fig. 8. Self-potential measurement

In the next example (Fig. 8) the extent of contamination by dissolved salts from a potash dump had to be determined. The survey was accompanied by miserable weather with unexpected snow and it was feared that the self-potential field would be greatly distorted. Notwithstanding, within 5 days 600 self-potential electrodes had been laid out with an electrode spacing of 10 m along the 14 survey profiles; the distance between profiles being 50 and 100 m with a profile length of generally 470 m. The self-potential values of the configuration were recorded a total of 24 times at 5-min intervals; the value for each electrode being averaged from two readings. As potential differences between the individual electrodes and the base electrodes were measured, a total of 28 800 survey values were recorded.

A desk-top computer processed the results, which were plotted by a plotter. The individual measurements were displayed in plots of six profiles each.

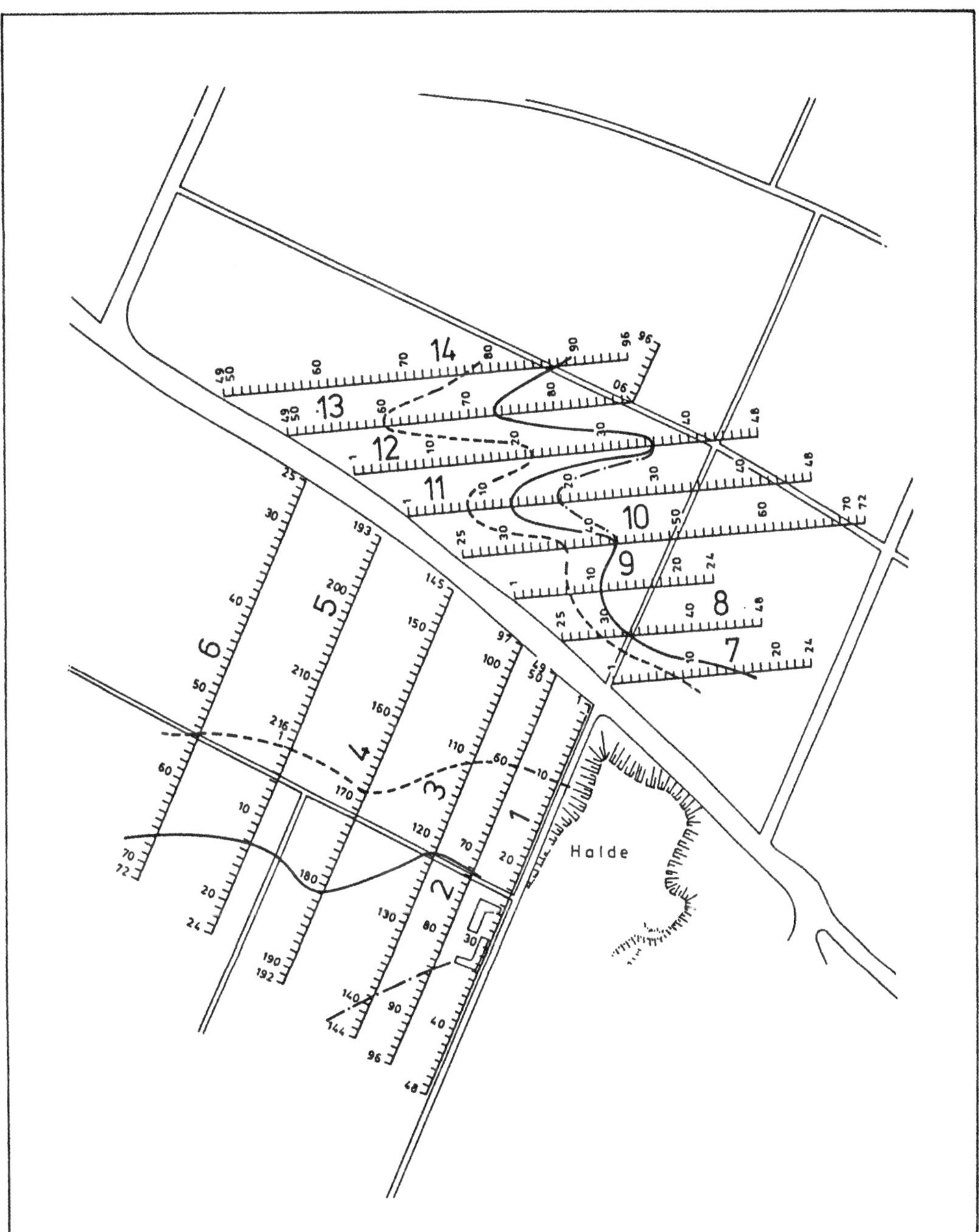

Fig. 9. Salt disposal site

Profile 1 (Fig. 9) is located next to the dump. Several years ago the buildings stood between electrode points 25 and 33 but they have since been pulled down. The curve (Fig. 10) itself exhibits distinct anomalies which are caused by the faulted subsurface layering.

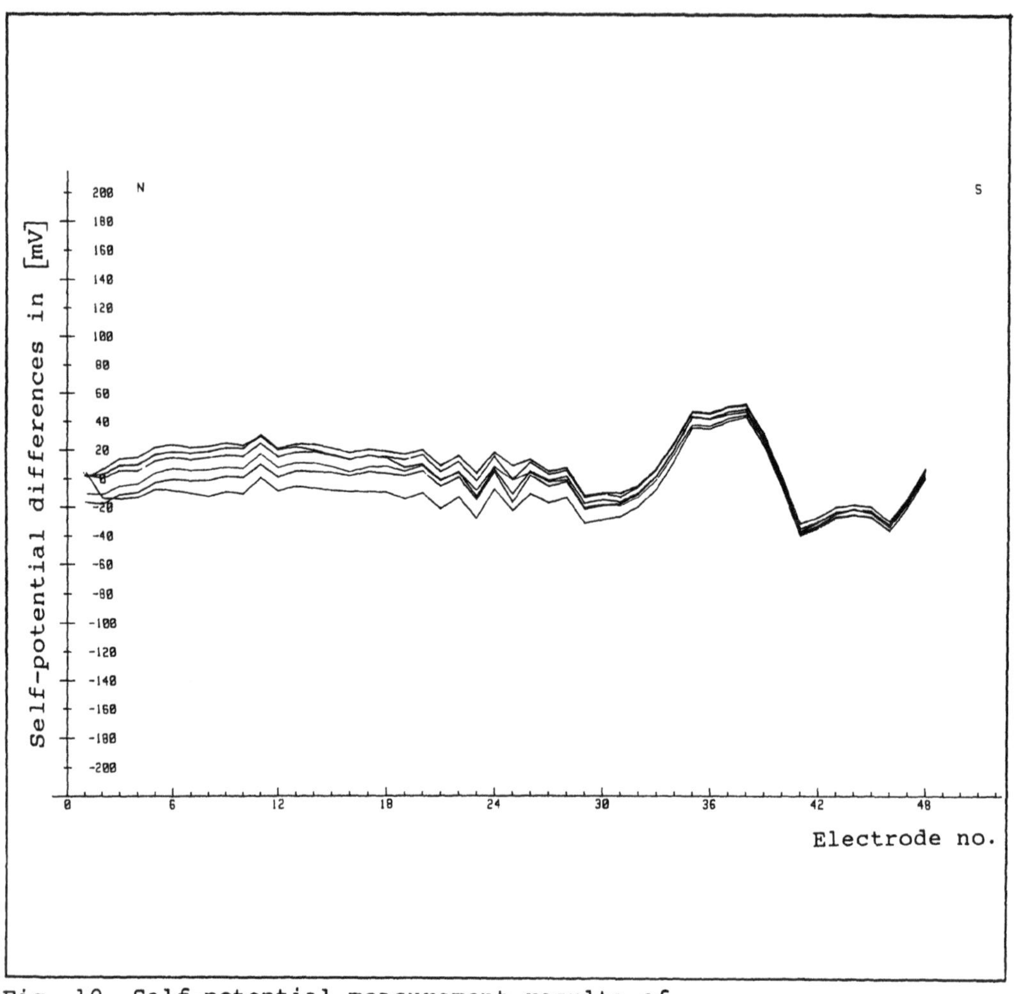

Fig. 10. Self-potential measurement results of the salt disposal site

When studying Figs. 11 and 12, one could ask "what would a self-potential profile look like if it were recorded with the conventional moving electrode configuration."

Recording here was made at 5-min intervals and within a short time the self-potential field had altered by more than 100 mV. The cause of these variations is unknown. The greatest variation occurred in the middle of the field approximately 120 m away from a little used road. Although a telephone line was located next to the road, it could hardly have caused such an effect 120 m away. It was only after the entire recording had been averaged that an interpretation could be made.

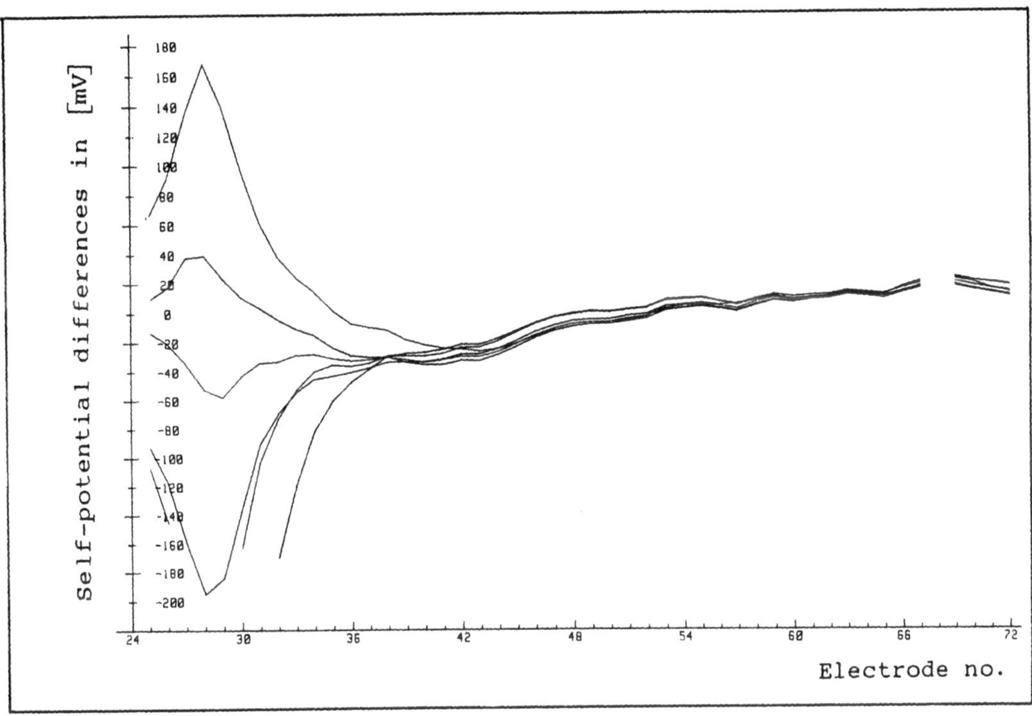

Fig. 11. Self-potential measurement

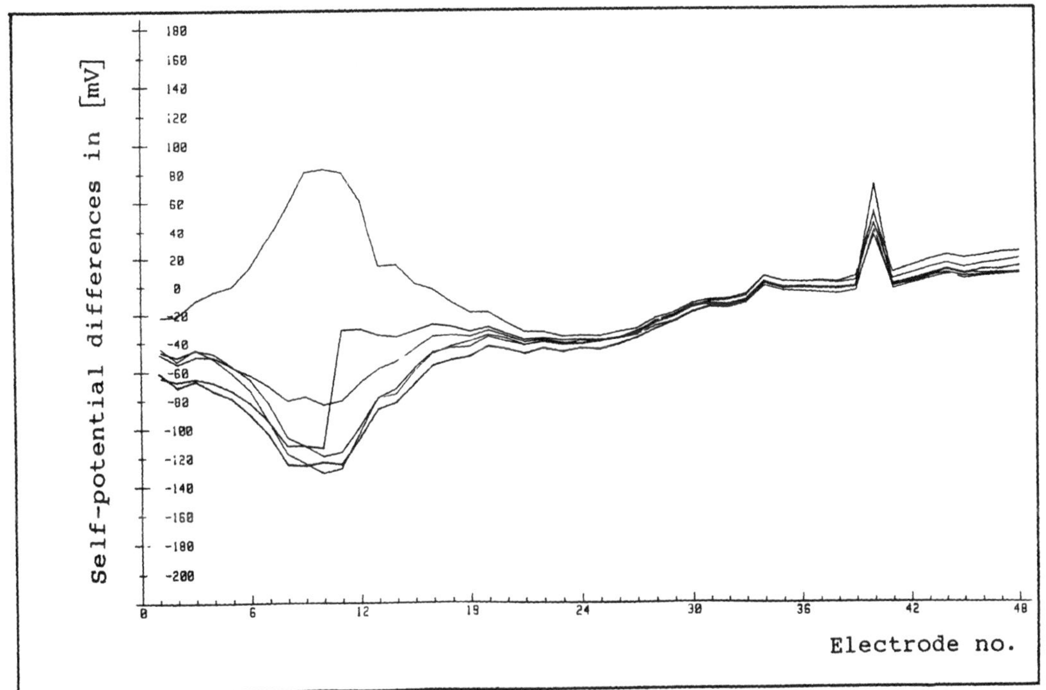

Fig. 12. Self-potential measurement

This is also a reason why long recording times are recommended.

Another important observation can be made in this area, that is, technical disturbances such as power lines cause positive peaks, whereas real anomalies exhibit negative peaks.

The measured values can generally be reproduced, the exceptions being certain individual cases such as previously mentioned. Slight differences in the reproductions can be traced back to large-scale variations of the self-potential field. Such variations affect all the electrodes and cause a shift in the level. The measured values of the different recordings generally run parallel to each other.

Looking at the interpretation as a whole the following can be seen:

The anomalies diverge, as expected, towards the north-west, that is in the direction of groundwater flow (Fig. 9). The display of the averaged values indicates that the values decrease towards the axis of the survey area, that is in the vicinity of the assumed higher salt concentration (Fig. 13). Nevertheless, at present no definite statement can be made on extension in depth and on the concentration of salt.

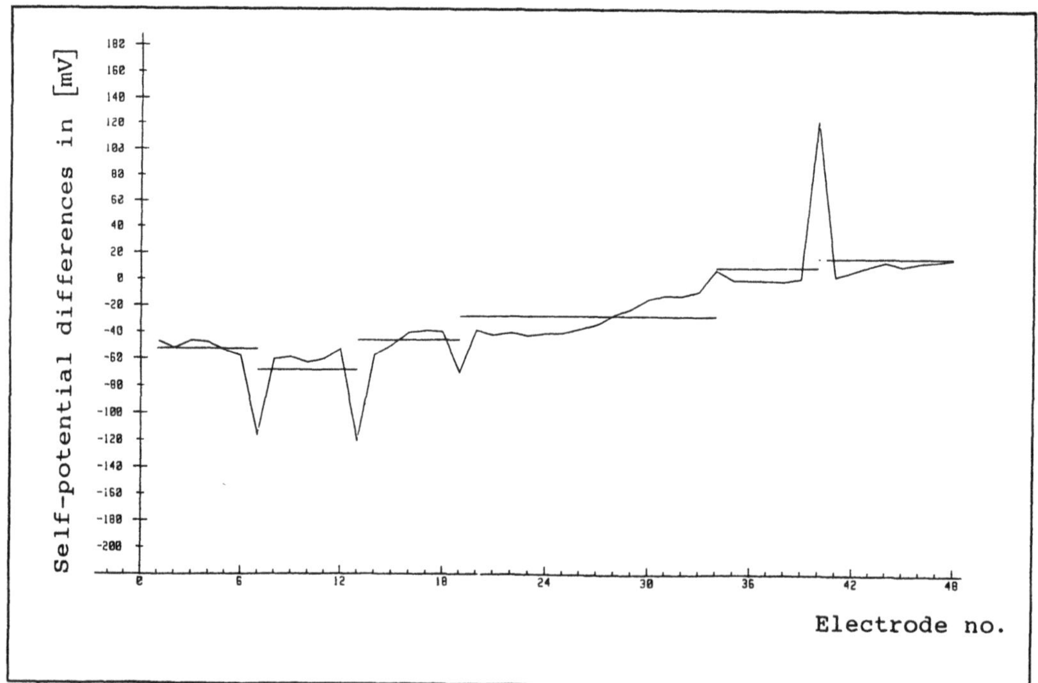

Fig. 13. Self-potential measurement

Test measurements were carried out together with the Institute for Geology and Paleontology of the University of Tübingen to determine the propagation paths and speed of gas injected into the near-surface layers. Testing was made using gas fertilization methods in which methane was injected into the ground with the gas paths being investigated by means of large numbers of potential electrodes (refer to the dissertations of STARKE 1986 and MARX 1988). Other tests were concerned with impermeability checks at dams, dykes and dumps in which methane or carbon dioxide was pumped into the ground. STARKE detected inhomogeneities in dams in these tests.

Measurements such as these can be used in the future for a wide variety of problem areas.

In summary, the deposition of waste, whether household rubbish or industrial waste, is generally made by dumping at waste disposal sites at the surface. These sites are frequently disused clay, gravel or rock quarries. As a result groundwater pollution often cannot be ruled out.

Nowadays the seals of waste disposal sites are generally checked by observing the groundwater in wells in which possible contamination can be determined. The accuracy of the checks depends on the number of wells, however, it must be noted that such wells are relatively expensive to construct.

Previous self-potential measurements using a base and a moving electrode were not particularly convincing because the moving electrode, which was set up at various sites, falsified the measurements.

Now we work with up to 216 self-potential electrodes and with several base electrodes. The self-potential differences are preferentially determined 2 h after the electrodes have been set up with measurements being made over a continuous period. The configuration of potential electrodes can be either along profiles or in a grid.

Application of electrodes firmly fixed in the ground enables, owing to the large number used, the self-potential values to be determined considerably quicker than when using the base electrode/moving electrode configuration.

As measurements are performed simultaneously at different locations it is also possible to recognize and eliminate other noise signals, and for this reason it is recommended to maintain long measurement periods.

References:

MARX, E., (1988) Überprüfung der Durchlässigkeit tertiärer Westerwaldtone im Einbau als Deponieabdichtung mit der CO_2-Injektions- und Eigenpotentialmethode.
Tübingen, Dissertation. Not published.

STARKE, U.W., (1986) Messungen der Strömung injizierter Gase mit der Eigenpotentialmethode.
Tübingen, Dissertation. Not published.

Self-Potential Generation by Subsurface Water Flow Through Electrokinetic Coupling

T. Ishido [1]

[1] Geological Survey of Japan
Higashi 1-1-3, Tsukuba, 305 Japan

Abstract

Electrokinetic phenomena associated with source-free ground water flow can produce electric potential anomalies at the earth's surface. Sources of conduction current (required for the appearance of electric potential at the surface) are located at the air-earth interface for terrain-related self-potentials (SP). Induced current sources for SP generation by hydrothermal circulation are thought to be located at underground thermal interfaces which act as boundaries between regions of differing streaming potential coefficients. Field survey results obtained in Japanese geothermal areas can be explained by these electrokinetic mechanisms.

1 Introduction

A self-potential (SP) survey is conducted by mapping the natural time-invariant electric field at the surface of the earth. In recent years, the SP method has attracted increasing interest in geothermal prospecting and engineering geophysics. Among the various mechanisms which can cause SP, it is essential to consider electrokinetic (streaming) potentials when making hydrologic investigations (e.g., Ogilvy et al. 1969; Zohdy et al. 1973; Zablocki 1976; Anderson and Johnson 1976; Mizutani et al. 1976; Corwin and Morrison 1977; Corwin and Hoover 1979).

In general, electrokinetic effects are described on the basis of irreversible thermodynamics (de Groot and Mazur 1962). There are, however, several difficulties involved in quantitative interpretation of electrokinetic effects in the earth. First, in situ values of the cross-coupling coefficients (zeta potential and/or streaming potential coefficient) are hard to estimate. This difficulty has been partially alleviated by recent experimental studies of the zeta potential and streaming potential coefficient for crustal rock-water systems (Ishido and Mizutani 1981; Ishido et al. 1983). An experimental study by Morgan (1988) has provided additional evidence concerning this question.

Quantitative SP interpretation is also difficult because of the complicated character of SP generation by subsurface electrokinetic sources. Theoretical studies by Nourbehecht (1963), Fitterman (1978), and Sill (1983) have helped to explain these processes. These studies provided useful theoretical models which form an analytical bridge between the basic electrokinetic source mechanisms and observed SP anomalies. This chapter describes quantitative modeling of SP generation by subsurface water flow, focusing on source-free flow which is very important in groundwater and hydrothermal circulation problems.

Lecture Notes in Earth Sciences, Vol. 27
G.-P. Merkler et al. (Eds.)
Detection of Subsurface Flow Phenomena

2 Mechanisms of Self-potential Generation by Subsurface Water Flow

The flow of a fluid through a porous medium may generate an electrical potential gradient (called the electrokinetic or streaming potential) along the flow path by the interaction of the moving pore fluid with the electrical double layer at the pore surface. This process is known as electrokinetic coupling. The general relations between the electric current density **I** and fluid volume flux **J** (on the one hand), and the electric potential gradient $\nabla \phi$ and pore pressure gradient $\nabla \xi$ forces (on the other) are

$$\mathbf{I} = - L_{ee}\nabla \phi - L_{ev}\nabla \xi \quad ; \tag{1}$$

$$\mathbf{J} = - L_{ve}\nabla \phi - L_{vv}\nabla \xi \quad ; \tag{2}$$

where the L_{ab} are phenomenological coefficients. The first term on the right-hand side in Eq. (1) represents Ohm's law and the second term in Eq. (2) represents Darcy's law. The terms with coefficients L_{ev} and L_{ve} represent the electrokinetic effect; $L_{ev} = L_{ve}$ according to Onsagar's reciprocal relations (de Groot and Mazur 1962).

In the absence of current sources, $\nabla \cdot \mathbf{I} = 0$, so from (1):

$$\nabla \cdot \mathbf{I}_{cond} = \nabla L_{ev} \cdot \nabla \xi + L_{ev}\nabla^2 \xi \quad , \tag{3}$$

where

$$\mathbf{I}_{cond} = - L_{ee}\nabla \phi \quad . \tag{4}$$

As pointed out by Sill (1983), sources (nonzero divergence) of conduction current (current driven by the gradient of the electric potential) will appear wherever there are gradients of the cross-coupling coefficient L_{ev} parallel to the direction of fluid flow (flow perpendicular to boundaries) or wherever there are external or induced sources of fluid. Three types of conduction current sources are depicted schematically in Fig. 1. Type I sources result from fluid flow perpendicular to a boundary between regions with differing values of L_{ev}. Type II sources can appear at the air-earth interface, which acts as a boundary between regions of differing L_{ev} (since L_{ev} on the air side is always zero). Type III sources are associated with nonzero divergence (sources or sinks) of fluid flow.

The presence of a source does not in itself guarantee that an electric potential will be present on the surface, however. In a homogeneous half-space with nonzero L_{ev}, even if fluid flow caused by a fluid source at depth intersects the surface, no electric potential will appear at the surface if the surface pressure is constant (Fitterman 1978). This is because the induced electrical source at the air-earth interface (type II) exactly cancels the effects of the type III source at depth (Sill 1983). On the other hand, if instead a condition of zero vertical pressure gradient is imposed at the surface (i.e. no type II source), an electric potential will be generated at the surface by the underlying type III source (Sill 1983).

If the fluid flow is confined to depth and vanishes at the earth's surface, some inhomogeneity in L_{ev} (and hence a type I source) is required for the appearance of an electric potential anomaly on the surface. On the basis of a total (pseudo) potential approach,

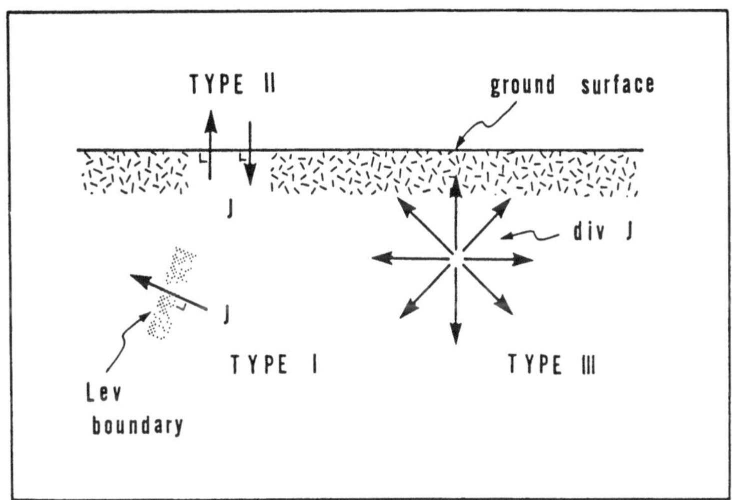

Fig. 1. Three types of conduction current sources important for electrokinetic coupling in subsurface water flow problems

Nourbehecht (1963) and Fitterman (1978) proposed that the following conditions are necessary and sufficient for the appearance of a surface potential. First, there must be a boundary separating regions of differing L_{ev} (or alternatively of differing streaming potential coefficient, as discussed below); second, there must be a nonzero component of pressure gradient parallel to this boundary. These conditions imply that no surface electric potential will appear if the L_{ev} boundary is a closed surface and coincides with one of the equipotential surfaces of the fluid flow pattern.

It should be noted that this discussion is applicable to other voltage cross-coupling manifestations such as thermoelectric or electrochemical effects as well as to electrokinetic phenomena. If the cross -coupling coefficient depends only on the value of the primary driving potential (pressure, temperature, concentration of chemical species, etc.), the cross-coupling coefficient boundary will always coincide with an equipotential surface of the primary driving force and no electric potential anomaly will appear on the surface. This situation does not occur in practice in the electrokinetic case, however, because L_{ev} depends primarily upon such parameters as temperature, pore water chemistry, and/or the hydraulic radius of the pores and cracks rather than upon the primary driving potential (pressure) (Ishido and Mizutani 1981).

Substituting (2) into (1) yields:

$$\mathbf{I} = - L_{ee}(1 - L_{ev}L_{ve}/L_{ee}L_{vv})\nabla \phi + (L_{ev}/L_{vv})\mathbf{J} , \tag{5}$$

where the quantity $L_{ev}L_{ve}/L_{ee}L_{vv}$ is $O(10^{-5})$ for typical geologic situations and may be safely neglected. Thus, Eq. (5) becomes:

$$\mathbf{I} = - L_{ee}\nabla \phi + (L_{ev}/L_{vv})\mathbf{J} . \tag{6}$$

This reformulation of Eq. (1) with $\nabla \xi$ replaced by $\mathbf{J}$ is useful for fluid-flow problems in which the flow cannot be deduced from the gradient of a pressure, as in thermally driven fluid convection. (We will discuss self-potential generation by hydrothermal circulation in the next section.) Using Eq. (6), Eq. (3) becomes:

$$\nabla \cdot \mathbf{I}_{cond} = - \nabla (L_{ev}/L_{vv}) \cdot \mathbf{J} - (L_{ev}/L_{vv}) \nabla \cdot \mathbf{J} \ . \tag{7}$$

Now consider fluid flow problems in which the divergence of flow is zero ($\nabla \cdot \mathbf{J} = 0$):

$$\nabla \cdot \mathbf{I}_{cond} = - \nabla (L_{ev}/L_{vv}) \cdot \mathbf{J} \ . \tag{8}$$

In these cases, sources for the conduction current are, of course, limited to types I and II (shown in Fig. 1). If we divide a large (inhomogeneous) region of interest into a number of homogeneous subregions, we may obtain the following equations for each subregion. Since the right-hand side of Eq. (8) is zero for a homogeneous medium, using Eq. (4) we may write:

$$\nabla^2 \phi = 0 \ . \tag{9}$$

Continuity of normal current flow requires the following boundary condition:

$$\mathbf{n} \cdot \mathbf{I}_1 = \mathbf{n} \cdot \mathbf{I}_2 \ , \tag{10}$$

where $\mathbf{n}$ is the unit vector normal to the boundary.

Let us consider steady-state fluid flow caused by spatial variations in the elevation of the water table as an example of a problem in which $\nabla \cdot \mathbf{J} = 0$. In this case, Eqs. (9) and (10) are appropriate to describe self-potential generation. Consider a mountainous region (Fig. 2).

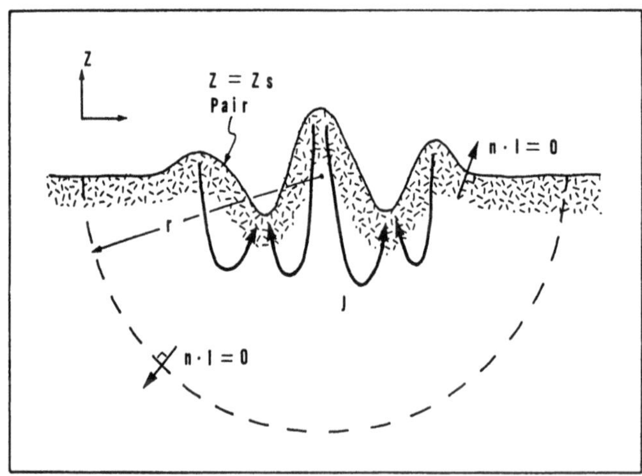

Fig. 2. Model for self-potential generation by groundwater flow caused by variations in the elevation of the water table

If the entire region is homogeneous, sources of conduction current will occur only at the surface (type II). The proper boundary conditions are: (1) the current normal to the ground surface must be zero, and (2) the current normal to a hemispherical boundary of radius r (see Fig. 2) should approach zero as $r \rightarrow \infty$.

Thus boundary condition (10) becomes:

$$\mathbf{n} \cdot \mathbf{I} = 0 . \tag{11}$$

This condition may be written as follows for a homogeneous medium:

$$\frac{\partial}{\partial n}(\phi + (L_{ev}/L_{ee})\xi) = 0 . \tag{12}$$

Using this boundary condition, a solution for the following Laplace equation is easily obtained:

$$\nabla^2(\phi + (L_{ev}/L_{ee})\xi) = 0 . \tag{13}$$

The above equation may be derived from Eq. (9) using the condition $\nabla \cdot \mathbf{J} = 0$ and the same approximation used to develop Eq. (6). The solution of Eq. (13) with boundary condition (12) is, using Green's theorem:

$$\phi + (L_{ev}/L_{ee})\xi = \text{constant} . \tag{14}$$

Substituting the relation $\xi = \rho gZ + p$ (elevation head + pressure; ρ is the density of the pore fluid, here assumed to be constant, and g is the acceleration of gravity) into (14) and applying p_{air} (= 1 bar) at the ground surface, we obtain:

$$\phi + (L_{ev}/L_{ee})\rho gZ_s = \text{constant} , \tag{15}$$

where Z_s is the elevation of the ground surface (or the air-pressure isobar). Equation (15) shows that the electric potential at the surface is linearly related to the elevation of the ground surface. As discussed in the next section, L_{ev} is positive in polarity (i.e. the streaming potential coefficient is negative) for typical geologic conditions; therefore, the electric potential generated by terrain-related fluid flow decreases as the ground surface elevation increases.

3 Self-Potential Associated with Hydrothermal Circulation

The generation of self-potential by thermally driven convection can be estimated using Eqs. (9) and (10), if we restrict our attention to single-phase (liquid) systems and adopt the Boussinesq approximation ($\nabla \mathbf{J} = 0$). Substituting expressions for the phenomenological coefficients L_{ab} developed by Ishido and Mizutani (1981) into Eq. (6), we obtain

$$\mathbf{I} = - \eta t^{-2}\sigma \{\nabla \phi + C(\mu /k)\mathbf{J}\} , \tag{16}$$

where I is the electric current density, $\mathbf{J}$ the volumetric fluid flux, and ϕ the electric potential. η, t, and k are the porosity, tortuosity, and permeability of the porous medium, respectively; μ is the dynamic viscosity of the pore fluid. $\sigma = \sigma_f + m^{-1}\sigma_s$ (σ_f is the pore fluid electrical conductivity, σ_s the surface electrical conductivity, and m is the hydraulic radius) and C ($= -L_{ev}/L_{ee} = \varepsilon \zeta / \sigma \mu$) is the streaming potential coefficient (ε is the dielectric constant of the pore water and ζ is the "zeta potential", the potential across the electrical double layer). If C is negative (positive), positive (negative) charge is carried by the fluid flow (**J**).

The streaming potential coefficient (C) is presumably distributed heterogeneously in the earth. Experimental results (Ishido and Mizutani 1981; Ishido et al. 1983) show that the principal parameters influencing the zeta potential (ζ) and/or the streaming potential coefficient (C) in silicate rock-water systems are the pH of the aqueous solution, the electrolyte concentration, and the temperature of the system. The ζ potential (and hence C) will be negative if the pH is greater than about 2, and will increase in magnitude with decreasing electrolyte concentration and/or increasing temperature. When small quantities ($\sim$1 ppm) of hydrolyzable metal ions such as Al^{3+} or Fe^{3+} are present in the solution, the ζ potential (and C) becomes positive at temperatures below about 100 ℃. The experimental results described above imply that C will be inhomogeneous in a hydrothermal convection cell in which there is a large temperature contrast. In addition to the above factors, the value of the hydraulic radius of the pores and cracks within the rocks influences C. If the hydraulic radius is smaller than a certain critical value determined mainly by the electrolyte concentration, C becomes smaller with decreasing hydraulic radius due to increasing effective pore fluid conductivity σ (Ishido and Mizutani 1981).

Using values of C estimated from the above-mentioned experiments, Ishido (1981) developed quantitative models of the electric potentials generated by hydrothermal circulation through electrokinetic coupling. Equation (9) was solved numerically subject to boundary condition (10). In one of the models (Fig. 3), a half-space below the surface was divided into two regions: one characterized by temperatures above 150℃ and the other by lower temperatures. The physical properties of the higher and lower temperature regions were assumed to be those appropriate for 200 ℃ and 100 ℃, respectively. Considering a typical crustal rock-water system containing water with pH = 7, 0.02 mol l^{-1} NaCl, and 10^{-5} mol l^{-1} Al^{3+}, Ishido (1981) estimated the appropriate values of C as -35 and 0 mV bar^{-1} for the higher and lower temperature regions, respectively (see also Ishido and Mizutani 1981). As shown in Fig. 3, positive electric potential appears around the portion of the thermal boundary intersected by outward fluid flow from the higher temperature region. The accumulation of positive charge at this portion of the boundary is caused by fluid flow carrying positive charge ($C < 0$) and no charge ($C = 0$) in the higher and the lower temperature regions, respectively. The opposite effect produces an accumulation of negative charge in the lower right corner of the diagram.

Ishido (1981) has shown that an observable self-potential anomaly (10-100 mV in magnitude) can appear at the surface if the following conditions are satisfied: (1) the circulating fluid is an aqueous solution of neutral pH (> 4) and moderate concentration of dissolved salt (< 0.1 mol l^{-1}), and (2) the fluid volume flow flux (**J**) is larger

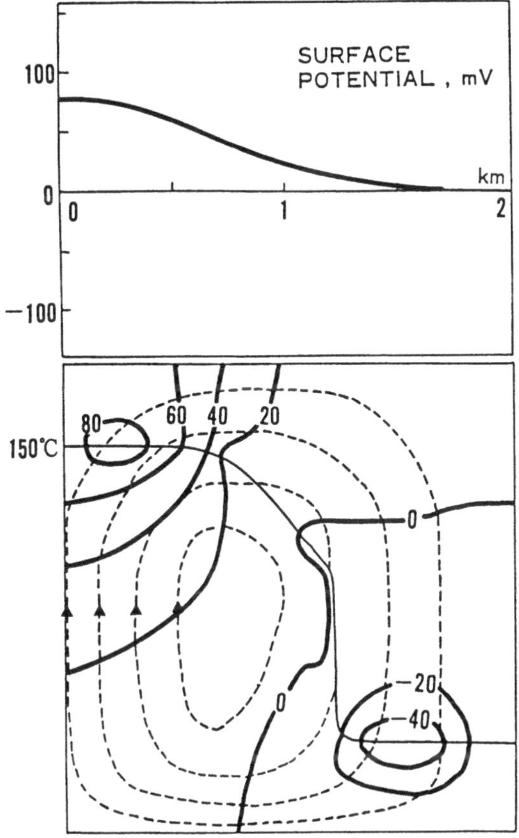

Fig. 3. The lower diagram shows the electric potential distribution (solid lines, in mV) generated by hydrothermal circulation (stream lines are shown as broken lines). The entire region is 2x2 km^2 and divided into two subregions by the 150 ℃ isotherm. Physical properties of the higher and lower temperature regions were assumed to be those at 200° and 100 ℃ , respectively. The upper diagram shows the SP distribution on the earth's surface (After Ishido 1981)

than 10^{-8} - 10^{-7} m s^{-1}. The polarity of the anomaly over a hot zone is always positive whether or not the fluid flow (with nonzero C) intersects the surface [the effects of type I source (shown in Fig. 1) dominate those of type II source] ; this is mainly because C is negative ($\zeta < 0$) and larger in magnitude under high temperature conditions.

4 Field Examples

Self-potential (SP) surveys of a number of geothermal areas in Japan have been carried out during the last decade. The Geological Survey of Japan (GSJ) performed SP measurements in six areas and the New Energy

Development Organization (NEDO) surveyed four additional fields. Each of these surveys covered an area of 50-100 km^2 with survey lines of about 100 km in total length and data sampling intervals of about 100 m. SP anomalies of various types were recorded, and obvious anomalies of positive polarity were found in seven different areas: the Kutcharo and Nigorikawa calderas in Hokkaido, the Sengan and Okuaizu geothermal areas in the northern part of Honshu, and the Hohi, Unzen and Kirishima geothermal areas in Kyushu. Most of these cases have provided additional support for correlations reported in the 1970's between positive anomalies and high temperature upflow zones (Zohdy et al. 1973; Zablocki 1976; Anderson and Johnson 1976). Streaming potential generated by hydrothermal circulation (described in the preceding section) is the most probable cause of these positive anomalies (Ishido et al. 1988).

Terrain-related self-potentials were observed in all of the fields surveyed. Figures 4 and 5 show SP profiles measured along two survey lines which traverse the central lava domes within the Kutcharo caldera

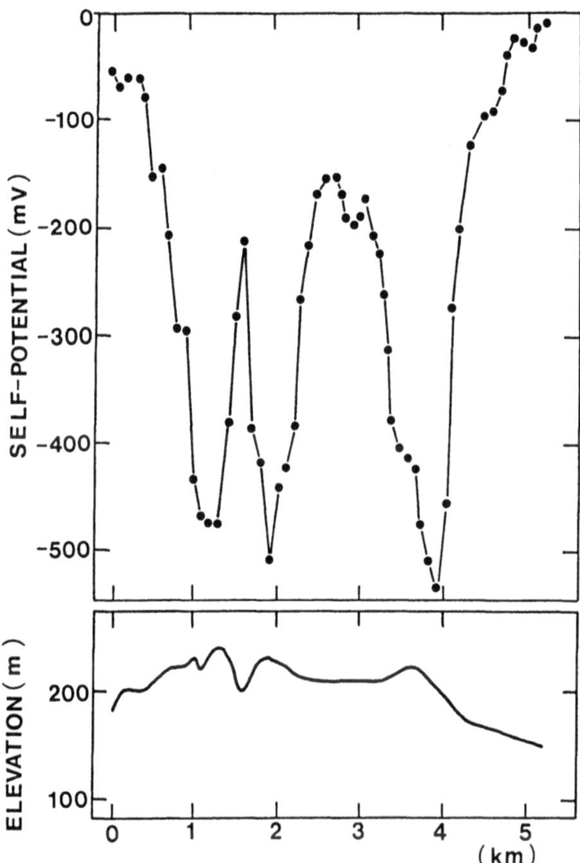

Fig. 4. Self-potential profile measured along a survey line which traverses a mountainous area within the Kutcharo Caldera, Hokkaido Island, Japan

(Hokkaido island). A strong correlation between the SP profile and ground surface elevation is evident in Fig. 4; SP decreases about 100 mV as elevation increases 10 m (100 mV/10 m). This is the largest value we have observed in Japanese fields; terrain-related SP usually ranges from 1 mV/10 m to 10 mV/10 m. The 100 mV/10 m value is, however, reasonable for this system since the groundwater is quite low in dissolved solids ($< 10^{-2}$ mol l^{-1}) so that the streaming potential coefficient C is around -100 mV bar^{-1} (e.g., Ishido et al. 1983). If the water table is shallow and closely follows the ground surface, an SP decrease of 100 mV would be expected to result from an elevation increase of 10 m using this value of C (= $-L_{ev}/L_{ee}$) in Eq. (15).

The SP profile shown in Fig. 5 was obtained from a survey line which passes through a fumarole area. Here, a positive SP anomaly associated with the fumarole area is superimposed upon terrain-related self-potentials. The positive polarity anomaly is about 100 mV in magnitude and is thought to be generated by the thermally driven upflows through electrokinetic coupling. We have also observed similar SP patterns in two other active fumarole areas within the Kutcharo caldera.

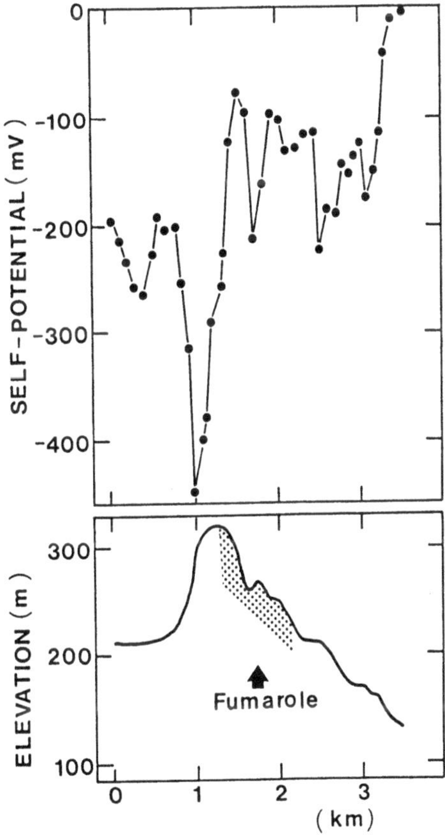

Fig. 5. Self-potential profile along a survey line passing through a fumarole area in the Kutcharo Caldera

5 Concluding Remarks

The observed correlation between positive SP anomalies and high temperature upflow zones is adequately explained by the proposed electrokinetic mechanism. Anomalies which result from natural hydrothermal circulation may or may not be observable depending upon the temperature, the chemical composition of the pore water, the flow rate, and the geometry of the hydrothermal convection cell. In order to delineate SP anomalies generated by natural thermally driven upflows, we must consider terrain-related SP (1 mV/10 m to 100 mV/10 m) in addition to other effects such as conductive mineral deposits and soil chemistry variations (Corwin and Hoover 1979). The effect of production-induced SP (Ishido et al. 1988) must also be considered in interpreting SP data from areas where production and/or injection of fluids is taking place for geothermal power generation.

Acknowledgments. I am grateful to K. Baba, K. Ogawa, M. Kawamura and K. Kimbara for their constant encouragement. H. Mizutani, T. Kikuchi and M. Sugihara are greatly acknowledged for helpful comments and discussions.

References

Anderson L A, Johnson G R (1976) Application of the self-potential method to geothermal exploration in Long Valley, California. J Geophys Res 81: 1527-1532

Corwin R F, Hoover D B (1979) The self-potential method in geothermal exploration. Geophysics 44: 226-245

Corwin R F, Morrison H F (1977) Self-potential variations preceding earthquakes in central California. Geophys Res Lett 4: 171-174

de Groot S R, Mazur P (1962) Non-equilibrium thermodynamics. North-Holland, Amsterdam, pp 405-452

Fitterman D V (1978) Electrokinetic and magnetic anomalies associated with dilatant regions in a layered earth. J Geophys Res 83: 5923-5928

Ishido T (1981) Streaming potential associated with hydrothermal convection in the crust: a possible mechanism of self-potential anomalies in geothermal areas. J Geotherm Res Soc Jpn 3: 87-100 (in Japanese with English Abstr)

Ishido T, Mizutani H (1981) Experimental and theoretical basis of electrokinetic phenomena in rock-water systems and its applications to geophysics. J Geophys Res 86: 1763-1775

Ishido T, Mizutani H, Baba K (1983) Streaming potential observations, using geothermal wells and in situ electrokinetic coupling coefficients under high temperature. Tectonophysics 91: 89-104

Ishido T, Kikuchi T, Sugihara M (1988) Mapping thermally driven upflows by the self-potential method. AGU Monogr (Proc) IUGG Symp U. 8 Hydrogeological regimes and their sub-surface thermal effects (in press)

Mizutani H, Ishido T, Yokokura T, Ohnishi S (1976) Electrokinetic phenomena associated with earthquakes. Geophys Res Lett 3: 365-368

Morgan D F (1988) Fundamentals of streaming potentials in geophysics.

In: Program Abstr Int Symp Detection of subsurface flow phenomena by self-potential/geoelectrical and thermical methods, Karlsruhe, p 32

Nourbehecht B (1963) Irreversible thermodynamic effects in inhomogeneous media and their applications in certain geoelectric problems. Ph D Thes. MIT, Massachusetts

Ogilvy A A, Ayed M A, Bogoslovsky V A (1969) Geophysical studies of water leakages from reservoirs. Geophys Prospect 17: 36-62

Sill W R (1983) Self-potential modeling from primary flows. Geophysics 48: 76-86

Zablocki C J (1976) Mapping thermal anomalies on an active volcano by the self-potential method, Kilauea, Hawaii. In: Proc 2nd UN Symp Development and use of geothermal resources, vol 2, pp 1299-1309

Zohdy A A R, Anderson L A, Muffler L J P (1973) Resistivity, self -potential, and induced-polarization surveys of a vapor-dominated geothermal system. Geophysics 38: 1130-1144

Fundamentals of Streaming Potentials in Geophysics: Laboratory Methods

Frank Dale Morgan[1]

Abstract

This brief article will discuss laboratory measurements of the streaming potential properties of rocks. Emphasis will be placed on: (1) those aspects of the experiment and associated theory that may give rise to problems, suggesting possible remedies, and (2) future areas/directions in which laboratory studies are needed.

1 Introduction

Streaming potentials are generated when subsurface fluids flow through rocks. The observed surface or borehole *self-potential* anomaly is a function of the streaming potential properties of the rock/fluid couple and also of the gross electrical conductivity structure in the vicinity of the anomaly. There is clearly the possibility of detection of subsurface flow patterns, especially those associated with recent fractures, by observing and analyzing self-potential anomalies.

In order to analyze the self-potential anomalies, appropriate modeling of the type initiated by Nourbehecht (1963) and recently extended significantly by Sill (1983), must be applied to the observed data. These models need accurate values of the self-potential properties of the particular rock/fluid combination under *in situ* conditions. Such measurements can be conducted in the laboratory. This presentation addresses such measurements.

[1]Geophysics Department, Texas A&M University, College Station, TX 77843 USA

Lecture Notes in Earth Sciences, Vol. 27
G.-P. Merkler et al. (Eds.)
Detection of Subsurface Flow Phenomena

The author has just completed an extensive article on streaming potentials (Morgan et al. 1988). Readers should consult this paper for extended references, current data and applications to earthquake prediction, geothermal exploration and rock electrical conductivity predictions. *The present chapter will complement the above paper* by stressing instead, problems that arise in laboratory measurements and making recommendations for their solutions. Furthermore, a number of areas of needed research will be discussed. This chapter should therefore be of most use to those contemplating or engaged in their own laboratory streaming potential measurements.

2 Why Study Streaming Potentials?

Streaming potentials are one of the main contributors to observed self-potential anomalies in geothermal areas, self-potential logs and as earthquake precursors. Furthermore, since streaming potentials are generated by fluid flow interacting with the rock surface, it is clear that fluid-filled subsurface fracture patterns may produce diagnostic self-potential signatures.

There are two main geophysical reasons for pursuing laboratory studies of the streaming potential properties of rocks. In order to model realistically the above mentioned self-potential anomalies, knowledge of the *streaming potential cross-coupling coefficient* (abbreviated, CC, for coupling coefficient), that is the voltage generated per unit pressure difference (mV/atm), must be known for the appropriate rocks. In other words, the laboratory measured CC is utilized in a direct way in models of subsurface flow.

The other reason for pursuing laboratory streaming potential studies, utilizes the CC data in a more indirect manner, to obtain a measure of the surface electrical potential of the rock relative to the bulk fluid. This surface potential is a critical parameter in most models of the electrochemical interface region. Currently, many rock physics models of such diverse phenomena as seismic attenuation and dispersion, electrical surface conductivity, dielectric spectra, induced polarization, permeability and nuclear magnetic resonance (NMR) responses are making use of the surface potential parameter. As I will discuss below, the CC data can be interpreted through the Helmholtz-Smoluchowski equation to give the zeta potential (ζ, mV). The ζ potential is an approximation to the true surface potential of the rock.

3 The Rock/Liquid Interface

In the present chapter there is little justification for an in-depth treatment of the solid/fluid interface, since many pertinent and excellent treatments already exist (see for example: Overbeek 1952; Bockris and Reddy 1970; Ishido and Mizutani 1981; Morgan et al. 1988).

Most solids in contact with a liquid become charged. Virtually all material of geological interest becomes negatively charged thereby resulting in an excess of positive ionic charge in the solution phase. The governing forces are both chemical and electrostatic in nature. Chemical adsorption forces are quite solid/solution dependent and act in close proximity to the surface. In contrast, most of the electrostatic forces obey Maxwell-Boltzman statistics, forming a Debye shielding region extending away from the surface. The fall-off of potential with distance from the rock surface is approximately exponential. The 'thickness' of the interface region, the *Debye length*, Δ, is given by:

$$\Delta = \left(\frac{\epsilon k T}{2 e^2 C} \right)^{\frac{1}{2}}, \tag{1}$$

where ϵ is the permittivity of the solution of ionic concentration C at temperature T. k is Boltzman's constant and e the electronic charge. For an NaCl solution, Δ is approximately 100, 30 and 10 Å at molarities of 10^{-3}, 10^{-2} and 10^{-1}, respectively.

4 The Helmoltz-Smoluchowski Equation

4.1 Applicability to Capillaries

The derivation of the classical Helmoltz-Smoluchowski (HS) equation is shown in many places (e.g. Overbeek 1952) and will not be repeated here. The flowing fluid in shearing off the excess charge from the electrochemical boundary layer produces a *convection* or drag current. In equilibrium this current is balanced by a simple Ohm's law *conduction current*, thereby establishing the streaming potential. The HS equation is given by:

$$\frac{\Delta V}{\Delta P} = \frac{-\zeta \epsilon}{\eta \sigma}, \tag{2}$$

where ΔV = potential difference across the sample;

ΔP = driving pressure difference across the sample;

ζ = zeta potential, defined as the potential on the plane closest to the surface on which fluid motion occurs;

ϵ = dielectric constant of the liquid;

η = viscosity of the liquid.

In many fields, such as chemistry, only the ζ potential is of any particular interest. However, as mentioned above, in geophysics the streaming potential cross-coupling coefficient, $\Delta V/\Delta P$, is of direct interest, since it is this CC that is responsible for streaming potential-induced, self-potential anomalies.

The conditions under which expression (2) applies are:

1. Flow is fully established;
2. The geometrical parameters of flow, such as the equivalent cross-sectional area and length, are identical for hydrodynamic (convection) and electrical (conduction) processes;
3. The radius of curvature of the capillary must be much greater than the thickness of the Debye layer and/or surface conductivity can be neglected.

Condition (1) can easily be satisfied by a "sufficient" reduction in the hydrodynamic flow rate. This problem has been investigated in great detail by earlier workers (Reichardt 1935; Bocquet et al. 1956; Boumans 1957; Rutgers et al. 1957; and Kurtz et al. 1976). The CC shows two plateaus. The "correct" CC is the value at the lower flow rate, where flow is fully established. Expressions can be derived which relate the Reynolds number to the established/unestablished flow regimes. However, it represents good practice to verify the condition of fully established flow experimentally. A word of caution is given here. A linear plot of ΔV vs ΔP is usually taken as evidence that Eq. (2) is upheld. Unfortunately, many workers have inadvertently chosen their lowest pressure so large that they completely jump over the fully established regime and obtain a CC of diminished magnitude, associated with unestablished flow.

Condition (2) is difficult to verify. However, for a straight capillary, it is not unreasonable to *assume* that the cross-sectional area and length are identical for hydrodynamic and electrical purposes, provided the electrochemical layer thickness or Debye length is small compared to the capillary radius. This has therefore led us naturally into condition (3).

Surface conductivity becomes important when the Debye length is a significant portion of the capillary radius. It has been shown that surface conductivity cannot be neglected if the Debye length is greater than 0.1 of the capillary radius (see Dukhin and Derjaguin 1974 for an excellent discussion of surface conductivity effects). Although the ratio of Debye length to capillary radius could be computed for a known electrochemical solution and capillary, it is again advisable to check the validity of (3) experimentally. There is a straightforward way of accomplishing this.

In geophysics, the *resistivity formation factor*, FF, is defined as the ratio of the resistivity of a rock sample over the resistivity of the saturating solution. Clearly, for a capillary the FF should be one. However, the FF will only be one for cases in which the Debye length is much smaller than the capillary radius. This occurs with high salinity solutions and large bore capillaries. For weak electrolytes and small bore capillaries the FF will be less than one due to the presence of surface conductivity.

Conditions (1) to (3) can therefore usually be satisfied, and Eq. (2) implemented. However, when surface conductivity σ_s is present, Eq. (2) can be modified (see for example Overbeek 1952) to read,

$$\left(\frac{\Delta V}{\Delta P}\right)_s = \frac{-\epsilon\zeta}{\eta\left(\sigma + \frac{2\sigma_s}{a}\right)}, \qquad (3)$$

where a is the average capillary radius. The CC with surface conductivity present, $(\Delta V/\Delta P)_s$, is simply the experimentally derived quantity. However, the ζ potential cannot be computed using Eq. (3), since σ_s is unknown. There has been much discussion in the literature about this point and possible ways around it (see Dukhin and Derjaguin 1974 for one of the best treatments). The simplest and perhaps most universally acceptable correction is to obtain a formation factor ratio F_0/F, where F_0 is the FF with a very high salinity solution and F the FF with the electrolyte under study. The ratio F_0/F will be greater than one if surface conductivity is present. The correction is applied by computing the ζ potential with Eq. (2) and then multiplying by the FF ratio F_0/F. This correction increases the magnitude of the ζ potential appropriately.

4.2 Applicability to Porous Media

Overbeek (1952) presents the best general derivation of the HS equation for porous media. The same general conditions, as stated previously for capillaries, apply. If these conditions are upheld then the same HS equation, Eqs. (2) and (3), are applicable to porous media or rocks. Unfortunately, because of the complexity of the microgeometry of porous media, it is much more difficult to ensure that these conditions exist because in porous media we are dealing with both highly curved surfaces and overlapping interface regions.

In the case of porous media with surface conductivity present, it is not clear whether or not an appropriate correction, such as the FF ratio F_0/F, can be applied to give corrected ζ potentials. It is again suggested that the reader consult the excellent reviews by Overbeek (1952) and Dukhin and Derjaguin (1974). We are currently investigating this problem for rocks. In the investigation, we are making measurements with varying grain size, applying the F_0/F correction, and then comparing the corrected ζ potentials. Preliminary results indicate that the FF correction does a reasonable job but at times the correction seems too small. In other words, in some cases the corrected ζ potential is still smaller in magnitude than the ζ potential obtained when surface conductivity was absent. Whenever possible, measurements of ζ potential should therefore be carried out without surface conductivity present. Unfortunately, for many low porosity rocks, this is often not possible.

5 Experimental Methodology

The streaming potential experiment seems simple at first glance but in practice it turns out to be quite the opposite. I now briefly discuss a number of points I consider essential for successful experimentation.

5.1 Determination of the CC Coefficient

Since the two electrodes seldom have identical potentials there is a finite rest potential for zero pressure difference. The CC should therefore be obtained not as $\Delta V/\Delta P$ but rather as a gradient $(\Delta V_1 - \Delta V_2)/(\Delta P_1 - \Delta P_2)$. Neglect of this point can often lead researchers to erroneous CC at low applied pressure differences.

5.2 Standard Electrodes

A substantial portion of the streaming potential literature has been devoted to electrode type and placement. Zucker (1959) has a good discussion of this problem. I favor some form of standard electrode. Since most experiments of geological interest are run with electrolytic solutions containing the chloride ion, and also because of their ruggedness and low cost, I use Ag-AgCl electrodes.

5.3 Electrode Placement

Electrode placement has probably represented the most serious experimental problem. It is strongly recommended that the electrodes be placed away from the flowing fluid.

5.4 Surface Conduction Effects

For measurements of the CC, the effects of surface conduction must be taken as is, that is, the experimentally measured CC needs no correction. However, if the "true" ζ potential is desired, surface conductivity must either be eliminated or accounted for as discussed previously.

5.5 Trapped Air

Trapped air produces a number of spurious effects. When the air bubbles are moving, the streaming potentials are enhanced as discussed below. Since bubbles do not move smoothly or continuously through porous media, the consequence is a very noisy streaming potential signal. Stagnant air bubbles display characteristic time-dependent behavior. The resulting streaming potential signals display rise/fall times of the order of one to a few minutes. Trapped air must therefore be completely removed from the experimental system.

5.6 Crushed Rock Samples

In some instances, if the experiment is performed on crushed rock samples, the conductivity of the rock particles may not be negligible because of trapped saline fluids and also surface conduction effects. Since the effect is in principal approximately the same as the gross surface conductivity effect, the analysis/corrections are similar to those discussed previously using the FF ratio.

5.7 Fluid Flow Rate

The necessity of ensuring fully established flow has been stressed above.

5.8 Voltmeter Input Impedance

In many situations, especially with weak electrolytes, the sample resistance can be extremely large. The experimenter must ensure that the input impedance of the voltmeter is much larger than the sample resistance.

5.9 System Check

The entire experimental system should be checked by performing experiments on some standard material such as glass (Rutgers and DeSmet 1945; Jednacak et al. 1974; and Broz and Epstein 1976).

6 Present-Day Data Base

I will not attempt, in this chapter, to reproduce or comment in any great detail on the present-day data base. There are a few excellent studies with pertinent streaming potential data (Berlin and Khabakov 1961; Ahmad 1964; Ishido and Mizutani 1981; Johnson 1983 and Morgan et al. 1988).

In general, the CC for a 1:1 electrolyte (e.g. NaCl) is approximately -4 mV/(atm Ω-m) and for a 2:1 electrolyte (e.g. $CaCl_2$) -2 mV/(atm Ω-m). Small quantities of higher valent cations such as Al^{+3} can reduce the magnitude of the negative CC substantially and in fact change the sign to positive. The CC values do vary, depending on the specific electrolyte/rock type but the quantities above are representative order of magnitude values.

The CC has been found to increase in magnitude with temperature (Ishido and Mizutani 1981) and also remain approximately constant with temperature (Morgan et al. 1988). The different behavior seems to be associated with the equilibration time at each temperature.

Recently, we have discovered the fact that streaming potentials are greatly enhanced with two-phase fluid flow (Morgan et al. 1988). To produce an enhancement, one of the phases must be relatively nonconducting and moving with the fluid flow.

Upon studying the above mentioned papers, it becomes clear that the data base is in fact quite sparse and that much more laboratory data is needed. The next section outlines a number of areas of streaming potential research we are investigating.

7 Future Research Directions

The Rock Physics Laboratory at Texas A&M University has embarked on a wide variety of laboratory streaming potential investigations. These investigations are now briefly discussed.

7.1 Fundamental Rock Properties

Many rock properties such as seismic attenuation, dielectric response, induced polarization, surface conductivity, and permeability depend on the electrochemical characteristics of the rock/solution interface. At the present time, many projects in this laboratory are aimed at a better understanding of such correlations. It is therefore necessary to obtain a large general data base of interface properties, particularly the ζ potential, in attempting to understand and model these diverse phenomena.

7.2 Streaming Potential Properties of Sedimentary Rocks

There have been few previous studies of the streaming potential properties of sedimentary rocks. The study incorporates various rock/electrolyte systems. The petroleum industry is showing a keen interest in this area.

7.3 Surface Conductivity Contribution to Total Conductivity

This is a fundamental aspect of the conductivity of rocks, particularly in low permeability rocks. The project tests the random network model derived by Madden (1976). Two things must be measured: (1) the zeta potential of the rock; (2) the conductivity of the rock with the same solution(s) in the pore space. Morgan et al. (1988) describes the appropriate methodology as applied to Westerly granite. The method will be applied to a suite of sedimentary rocks. It is also my intention to attempt to obtain a "pore-size distribution", from surface conductivity and other appropriate electrochemical interface properties.

7.4 Effects of Temperature on Streaming Potential Coupling Coefficients

A sparse data set exists mainly on quartz up to 100°C temperatures. The reports on whether the streaming potential coupling coefficient increases or decreases on a given mineral surface are not well documented or understood. Furthermore, the question of time equilibration, as discussed in Morgan et al. (1988), is a relevant and important problem to consider and also the accomplishment of measurements at plus 100° temperatures.

7.5 Effects of Two-Phase Flow on Streaming Potentials

Consider a capillary with a solution moving through it, across which the streaming potential is being measured. If a single bubble is inserted into the flow stream, the convection current will not be appreciably altered, however, depending on the bubble diameter, the capillary resistance will increase and hence the streaming potential cross-coupling coefficient is enhanced. As the number of bubbles increase, the convection current will eventually start decreasing. This enhanced streaming potential process may occur in the flashing regions of a geothermal reservoir and may also be of relevance to the phenomenon of earthquake lights and the detection of recent subsurface fractures.

7.6 Effects of Complex Chemistry on Streaming Potentials

Very small quantities of higher valent ions such as (Al^{3+}) can drastically affect the streaming potential properties. Usually the effect of strong (Al^{3+}) adsorption will cause a change in the sign of the streaming potential at higher concentrations. I plan to look, both theoretically and experimentally, at the relationship between the streaming potential and solutions with varying ratios of cations such as Na^+/Ca^{2+}, Na^+/Mg^{2+}, Na^+, Al^{3+}, etc.

7.7 Correlation Between Zeta Potentials and NMR Spectra

The decay times of the proton NMR responses of rocks are related to microscopic relaxation phenomena. Because of the high electric field ($\approx 10^7$ V/m) in the interface region, water molecules cannot relax as freely. In other words, they behave as though they were bound by the electric field to the surface. Consequently, at least two distinct relaxations are seen in rocks, one for ***bound*** water in the interface region and the other a function of the dipole moment of ***bulk*** water. The intention is to attempt to relate the NMR response of the *bound* water with the classical ideas of the double or triple electrochemical surface layer, by doing both NMR and streaming potential measurements on the same suite of rocks. The real situation in rocks is complicated by a distribution of pore sizes and by interacting triple layers. These studies will be mutually beneficial to both NMR and streaming potential investigations and may have broad implications in chemistry and rock physics.

7.8 Summary

The research areas described here will foster our understanding of the fundamental aspects of rock/solution interfaces. This will aid our modeling of such diverse rock properties as seismic wave attenuation and dielectric properties. Moreover, many of the results may have direct applications to such problems as geothermal exploration, earthquake lights and prediction, and the monitoring of subsurface fluid flows.

8 Conclusions

Morgan et al. (1988) should be read in close conjunction with the present chapter. Together they represent a current comprehensive treatment of laboratory measurements of streaming potentials. This presentation has emphasized the background theory and experimental methodology. It then proceeded to outline many areas of future research in this exciting area of rock physics.

Acknowledgements. I sincerely thank the sponsors, the Volkswagon Foundation (Hannover, FRG) and the US Army Research, Development and Standardization Group (UK), for the support to participate in the international symposium on the "Detection of Subsurface Flow Phenomena by Self-Potential and Thermometrical Methods", University of Karlsruhe, Germany, 14-18 March 1988. In particular, my deepest gratitude is extended to Profs. Hőtzl, Merkler and Wilhelm and also to the organizing secretary Ms. A. Kastner.

I also thank the National Science Foundation for their support of my streaming potential studies. The following people have all contributed to different aspects of the ideas presented in this paper, and I thank them: O. Agunloye, R. Corwin, C. Estrada, D. Fitterman, G. LaTorraca, D. Lesmes, A. Nur, S. Parks, S. Pride, E. Williams and in particular, the person from whom I have learned the most about streaming potential studies, T. R. Madden.

References

Ahmad MU (1964) A laboratory study of streaming potentials. Geophys Prospect 12:49-64

Berlin TS, Khabakov AV (1961) Differences in electrokinetic potentials of carbonate sedimentary rocks of different origin and composition. Geochemistry 3:217-230

Bockris JO'M, Reddy AKN (1970) Modern electrochemistry, vol 2. Plenum, New York

Bocquet PE, Sliepcevich CM, Bohr DF (1956) Effect of turbulence on the streaming potential. Ind Eng Chem 48:197-200 197-200.

Boumans AA (1957) Streaming currents in turbulent flows and metal capillaries (I-III). Physica 23:1007-1046

Broz Z, Epstein N (1976) Electrokinetic flow in fine cylindrical capillaries at high zeta-potentials. J Coll Interf Sci 56:605-612

Dukhin SS, Derjaguin BV (1974) Electrokinetic phenomena. In: Matijevic E (ed) Surface and colloid science, vol 7, chap 2. Wiley, New York, 356 p

Ishido T, Mizutani M (1981) Experimental and theoretical basis of electrokinetic phenomena in rock-water systems and its applications to geophysics. J Geophys Res 86:1763-1775

Jednacak J, Pravdic V, Haller W (1974) The electrokinetic potential of glasses in aqueous electrolyte solutions. J Coll Interf Sci 49:16-23

Johnson AR (1983) Rock property measurements and analysis of selected igneous, sedimentary and metamorphic rocks from world-wide localities. USGS Open File Rep 83-736

Kurtz RJ, Findl E, Kurtz AB, Stromo LC (1976) Turbulent flow streaming potentials in large base tubing. J Coll Interf Sci 57:28-39

Madden TR (1976) Random network and mixing laws. Geophysics 41:1104-1125

Morgan FD, Williams ER, Madden TR (1988) Streaming potential properties of Westerly granite with applications. J Geophys Res (in press)

Nourbehecht B (1963) Irreversible thermodynamic effects in inhomogeneous media and their applications in certain geoelectric problems. Ph D Thes, Mass Inst Tech, Cambridge

Overbeek JThG (1952) Irreversible systems. In: Kruyt HR (ed) Colloid science, vol 1. Elsevier, New York, 389 p

Reichardt H (1935) Elektrisches Strömungspontential bei turbulenter Strömung. Phys Chem A 174:15

Rutgers AJ, DeSmet M (1945) Research on electro-endosmosis. Trans Faraday Soc 41:758-771

Rutgers AJ, DeSmet M, deMyer G (1957) Influence of turbulence upon electrokinetic phenomena. Trans Faraday Soc 53:393-396

Sill WR (1983) Self-potential modeling from primary flows. Geophysics 48:76-86

Zucker ER (1959) A critical evaluation of streaming potential measurements. Ph D Thes, Columbia Univ, New York

COPPER-COPPER SULFATE ELECTRODES FOR SELF-POTENTIAL AND MAGNETOTELLURIC MEASUREMENTS

HELMUTH WINTER, EMIL AULBACH and JOHANNES STOLL

Institut für Meteorologie und Geophysik, Universität Frankfurt,
Feldbergstr. 47, D-6000 Frankfurt a.M., F.R.G.

Copper-copper sulfate electrodes have been designed for robust and easy handling in self-potential and magnetotelluric field campaigns. They are based on a type which was constructed and used for many years by the MT groups from the universities of Munich and Berlin. The set-up was essentially as follows (Fig.1):

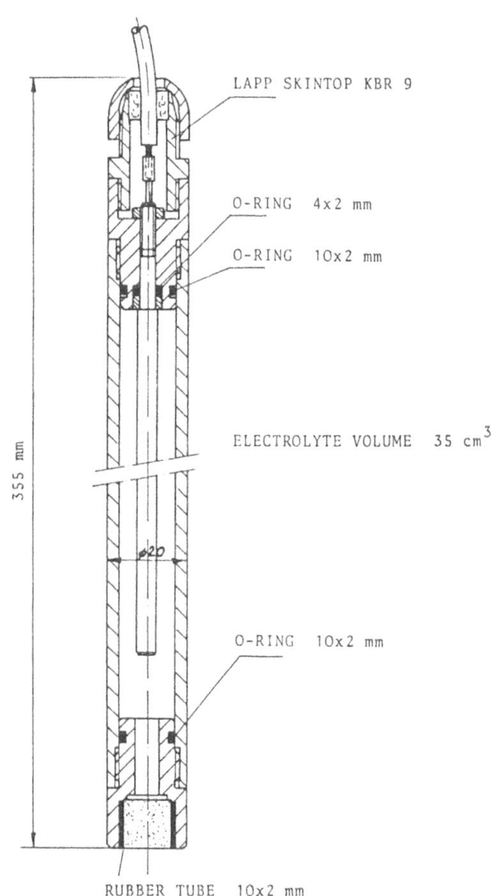

Figure 1: Construction drawing of the developed copper-copper sulfate electrodes

Lecture Notes in Earth Sciences, Vol. 27
G.-P. Merkler et al. (Eds.)
Detection of Subsurface Flow Phenomena

1. The electrode is made of transparent plexiglass and allows thus easy control of the electrolyte content.

2. The diameter of 20 mm allows an unproblematic placing of the electrodes into the ground with sufficient mechanical stability.

3. A copper rod is immersed into the electrolyte, positioned through the top end.

4. The soldered joint is protected against humidity by a cable screwing.

5. The contact between the copper sulfate solution and the earth consists of porous alumina ceramic. This diaphragm has a cylindric shape and is pushed into the lower top end of the electrode by a rubber tube. The elasticity of the rubber is sufficient to keep the ceramic in position in all cases of severe tests.

The following experience was obtained by several measurements:

1. By the use of O-ring gaskets leakage of the electrode was prevented effectively. Leakage was approximately 1% of the electrolyte content within 2 weeks.

2. Measurements of electrode resistances in various media gave values of about a few kΩ.

3. Additional tests showed that the period of adaptation of the electrodes to the environment was within 10 s. After this period the potential stayed constant within 1 mV in more than 3 h.

4. An electrode-specific noise of 0.5 μV at 1 Hz was noticeable.

SELF-POTENTIAL MEASUREMENTS TO DETERMINE PREFERRED WATERFLOW IN FRACTURED ROCKS

H. Hötzl and G.-P. Merkler[1]

Abstract

The determination of the permeability behaviour of fractured rocks is of essential importance for the judgement of necessary grouting measures. Geoelectrical self-potential measurements in connection with water pressure tests and grouting experiments in rock masses were carried out to determine preferred flow paths of water or cement in relation to the joint system.

The self-potential measurements proved to be closely related to the amount of water or cement absorption in the pressure-gauge borehole. The differences between the self-potential measuring values, calculated for various pressure/water or cement absorption proportions in the pressure-gauge boreholes, indicate an anisotropic system of water or cement pathways, especially in the statistical evaluation of the directions of self potential isolines. Use of this method may help to define the main water paths in fractured rocks and to determine the anisotropy of rock permeability.

1 Introduction

In subsoil exploration the determination of the permeability behaviour of the rocks concerned is an essential task for hydrogeology and engineering geology, particularly with respect to water supply or planned engineering constructions. Rock permeability is of decisive significance in cases where hydraulic constructions, for example dams, are projected, i.e. when circulation systems must be detected, reduced or prevented as far as possible by means of grouting measures. In many

1 Applied Geology, University of Karlsruhe, FRG

Lecture Notes in Earth Sciences, Vol. 27
G.-P. Merkler et al. (Eds.)
Detection of Subsurface Flow Phenomena

cases it is even necessary to stabilize the subsoil by means of injections, thus increasing the mechanical strength of the rock mass.

Within the scope of a research program, a partial program concerning the problem of improvement of disaggregated granites in the transition range between solid and loose rock, hydrogeological and engineering geological in situ investigations were carried out in order to characterize the permeability properties with respect to future sealing measures (Metzler et al. 1985).

The investigated area is situated in an Upper Carboniferous granite complex of the middle Black Forest in southwest Germany. Three test areas were selected for the experiments. The first test site is a relatively widely jointed and unweathered granite, in the second a closely intersected granite is exposed. The third test field represents the highest degree of loosening of the granite up to granite grus.

2 Hydrogeologic Field Investigations

2.1 Statistics of Joint Patterns

In fractured rocks water paths and permeability are strongly bound to the joint pattern. In hydrogeologic studies it is important to know the hydraulic efficiency of single joints and joint groups, defined by approximately the same strike directions. Therefore, the first step in determining the anisotropic permeabilitiy behaviour is to start with a geologic statistical measurement and to evaluate the joint pattern.

In order to determine the characteristics of rock loosening and permeability properties, a detailed survey of the existing structural constituents was made for the three different test areas. Here, not only strike and dip directions, joint spacing of the individual joint sets and the degree of separation were registered, but particular attention was given to the description of the joint planes, types and distribution of joint covers and to joint fillings (Fig. 1). In some cases the asperity of the joints could also be determined. The dimensions of the joint widths measured in the superficial zone could not be registered representatively because of rock loosening caused by blasting operations or relaxation procedures. However, it was possible to indicate, by means of an optical probe (Eastman) and a special measuring arrangement, more precise values of the joint widths. In addition, optical sounding confirmed the information on joint properties, which had been obtained above ground.

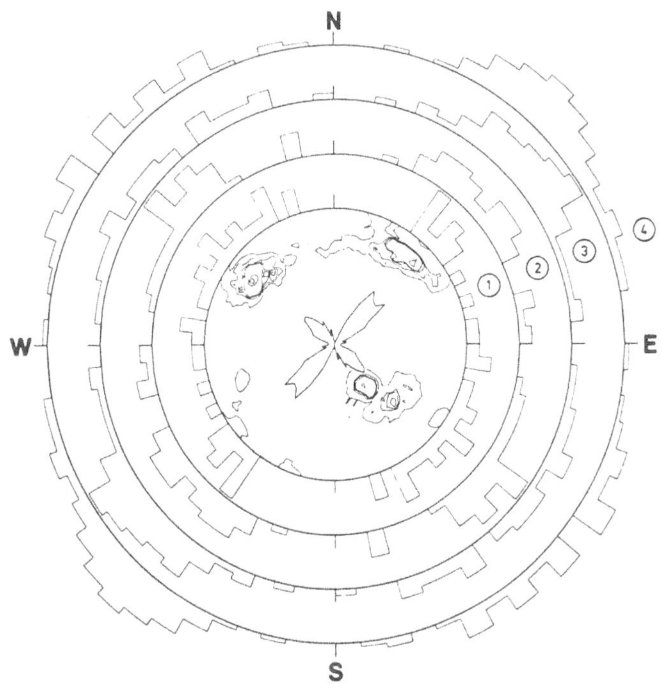

Fig. 1 Joint diagram and joint pattern in Schmidt's net (inner circle) together with percentage of oxidative discolouration on joint planes related to strike directions of granite (1=1-25%; 2=26-50%; 3=51-75%; 4=76-100%; outer circles)

2.2 Joint Tracing

The statistical evaluation of the joint pattern does not always give reliable information on the hydraulic anisotropy of fractured rocks. Additional information can be gained by tracing experiments which help to define the main discharge direction. But this is mainly dependent on the hydraulic gradient, so that the real joint directions of the water, still remain unknown. A new tracing technique was used.

In order to mark those joints, which are actually in use, joint tracing experiments (Hötzl et al. 1982) were carried out by applying highly adsorptive dyes (rhodamine, methylene blue and astra diamond-green). The dyes, dissolved in water, were injected at a constant pressure head into boreholes where they infiltrated the rocks (Fig. 2). On adjacent rock walls or after exposure of tracer-affected rock zones (for example exposed within the scope of the construction activities) the marked joints could be mapped and analyzed directionally due to the typical colouring of the joint planes. On the basis of the actually marked joints within the individual joint sets it was then possible to detect and define the hydraulically active rock masses.

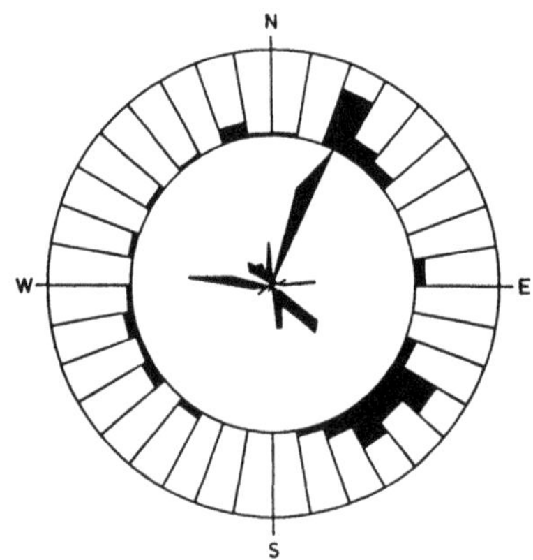

Fig. 2 Distribution of the coloured joints after tracing experiments with dyes (Rothengatter 1982)

The disadvantage of this method is that it requires good outcrop condition or subsequent excavation to define the joints which are marked by the dye tracer.

2.3 Water Pressure Tests

In connection with injection measures to be carried out within the range of dam construction, in international practice the results of permeability tests, according to Lugeon (1933), are the main reference. Water is pumped with relatively high pressure into defined segments of the borehole. During our packer tests the latest methodological knowledge was used. The water pressure test was performed with an electronic test device. The injection pressure was measured directly in the test section with a piezoresistive pressure gauge and was then transmitted to a recorder. Within the pressure pipe system a second pressure gauge was installed and the measuring results were registered synchronously in order to determine later the possible frictional losses. On the third track of the recorder the corresponding flow quantities were registered synchronously with a constant paper advance of 1 cm·min^{-1}. The packer tests were performed in steps of 0.8, 1 or 2 m, from bottom to top. Pneumatically extendable hose packers of 1 m length were used. Direct packer circulation could not be observed in the single packer tests which were mainly performed.

The water pressure tests were carried out in view of the stationary behaviour of joint flow and long-term effects (Blinde et al. 1981; Hötzl et al. 1982) with increasing and decreasing pressure steps using a shockless working Mohno-pump.

In order to detect the modifications of the water table within the test areas during packer tests, circular gauge drillings were sunk around the individual grouting holes (diameter 76 mm). Thus, the relevant propagation area and the possible active permeabilitiy anisotropy of the water could be detected by means of gauge level modifications (Metzler et al. 1985).

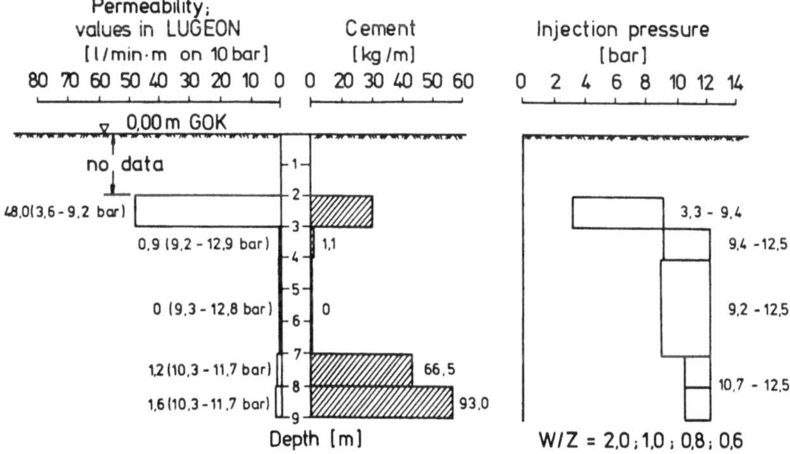

Fig. 3 Water level changes and pressure-quantity relation of water absorption in a packer test, showing an erosional process of joint filling material

Disadvantages of this method are the high costs of the boreholes and the test equipment as well as the fact that only a vertical inhomogeneity of permeability can be determined.

2.4 Grouting Experiments

The experimental grouting measures following the packer tests mainly served a comparative purpose with regard to the results obtained in hydrogeological and engineering geological investigations. Moreover, the intention was to follow up questions of the spread of different cement mixtures, the degree of joint filling and the geometry of the grouted fissures, on the basis of subsequent mapping of exposed areas of grouted rocks. The grouting experiments were carried out analogous to water pressure tests, i.e. with a single packer in steps from bottom to top by means of a variable adjustable Mohno-pump. The grouting pressure was measured electronically within the test section. In addition, pressure was measured at the top of the packer. The pressure data obtained were registered and processed by the recorder of the packer test device.

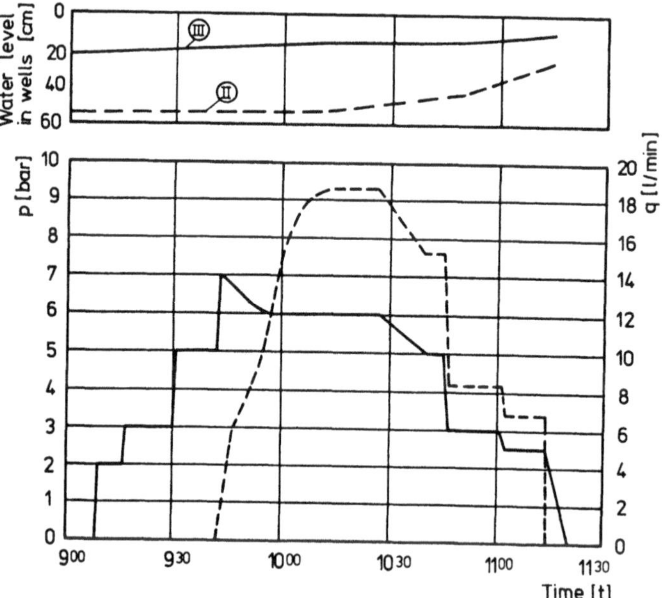

Fig. 4 Depth-dependent water and cement absorption

3 Geophysical Studies

3.1 General Remarks

Due to the incompleteness of hydrogeologic measurements in determining the anisotropy of the permeability or due to the high costs of the measurements, i.e. water pressure test, we were looking for other methods, which could help to define the preferential water paths in fractured rocks.

The self potential method, one of the oldest geophysical and geoelectrical prospection techniques, has been increasingly applied when borehole measurements and foundation testing had to be carried out.

The electrokinetic potential fields due to the passage of aqueous solutions (electrolytes) through a porous or jointed medium can be registered at the surface using a highly sensitive voltmeter by means of special electrodes (Merkler et al. 1970). Temporal modifications of these potential fields are mainly provoked by modifications of the electrolyte content, the flow velocity and the flow volume.

Within the scope of water pressure tests only the absorption capacity and the depth-dependent permeability of rocks are determined on the basis of the Lugeon values obtained (Blinde et al. 1981). The azimuth-dependent determination of water passages is difficult.

In order to establish a detailed analysis of the flow processes and the preferential, principal water-flow directions in the subsoil, the water pressure tests (Blinde et al. 1983, Merkler et al. 1985) were accompanied by self-potential measurements.

3.2 Results

Results of water load observations and SP-measurements have been elaborated by statistical analysis of the observed changes during the water pressure test by referring to the differences in the direction of the isolines.

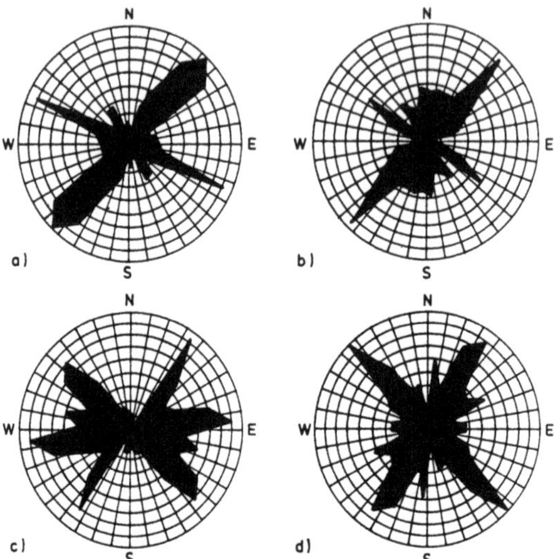

Fig. 5 Statistical analysis of the measured self-potential; isovalue of the differences (a, b), and of the water level differences measured in boreholes (c, d)

Figure 5 illustrates the results obtained in the self-potential measurements. These values present the time-dependent modifications of the pressure p (bar) and the quantity of water q ($l \cdot min^{-1}$) during a water pressure test. The value differences observed in self-potential measurements for different experimental grouting phases have been presented in isovalue maps (Blinde et al. 1981). With an angle-statistical evaluation of these maps (Neumann 1954) the detection of the preferential water-flow directions was possible. A direct angle-statistical comparison of the results with the water level observations in boreholes (Fig. 5a-d) shows that not only the self-potential method but also the water level observations in the dwells of two different experimental time phases indicate identical tendencies. The results agree with the geologically detected principal fissure directions (Figs. 1 and 5).

In the water pressure tests of the example given an absorption of up to 9 $l \cdot min^{-1}$ was found. From the values of the self-potential and their differences (Fig. 5a, b) it can be concluded that the NNE (40°) fissure direction has to be considered as the preferential water-flow direction and that the NNW (130°) direction has to be regarded as less relevant. Both directions are also outstanding features in connection with the level value differences found in the boreholes (Fig. 5c, d). However, with respect to these value differences it is not possible to designate one of those two directions as the preferential water-flow direction, since due to reduced time resolution, short time-dependent modifications cannot be registered with sufficient precision in water level measurements. Model experiments, which are being carried out at present, should provide data by means of direct comparisons between self-potential measurements and flow processes (flow volume, flow velocity, conductivity of the electrolyte etc.), thus helping to directly quantify this measuring method.

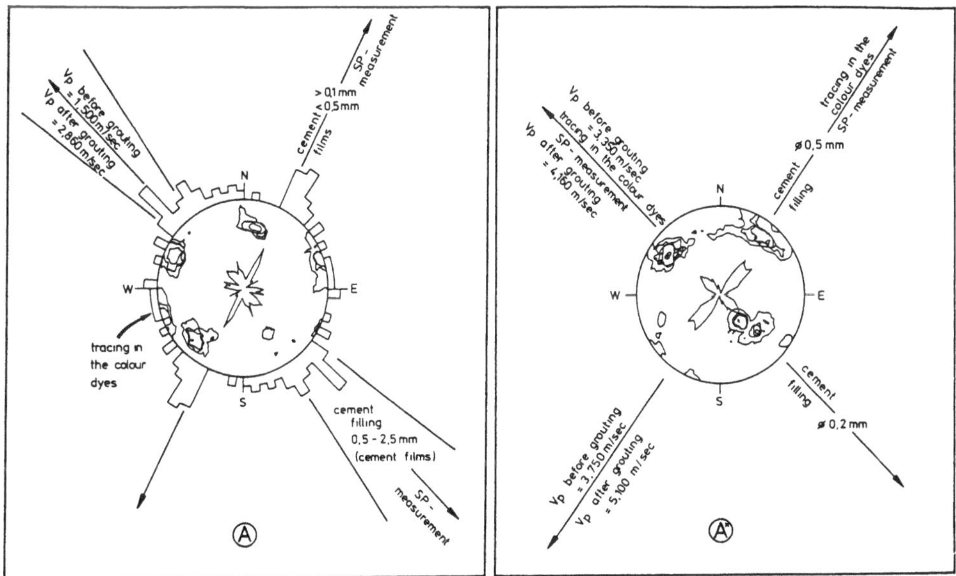

Fig. 6 Directional dependence of permeability and grouting properties (representations outside of the circles) where two different granite types are represented. Within the circles the joint maxima and the symmetric joint diagrams are shown

4 Conclusions

The self-potential measurements performed confirm that this method (accompanied by supplementary pressure tests) is an appropriate technique for establishing a differential analysis of the joint-dependent rock permeability. They are in a good agreement with the hydrogeological investigations (statistics of joint pattern, tracer experiments and grouting of dyed cement). A summarizing graph is presented in Fig. 6. Statistical evaluation of water level differences in boreholes and self-potential value differences over an identical time interval of a water pressure test lead to the same results. It should be noted that the analysis of preferential water-flow directions can be carried out more effectively on the basis of self-potential measurements.

References

BLINDE A, HÖTZL H, MERKLER G (1981) Ingenieurgeophysikalische Untersuchungen der Auflockerungsanisotropie von Gesteinskörpern. Ber 3 Nat Tag Ing-Geol, 201-207, Ansbach

BLINDE A, HÖTZL H, METZLER F (1983) Hydrogeological and engineering geological in situ investigations for the assessment of grouting purposes of disintegrated granites. Proc Int Symp Soil-Rock Invest, Paris

HÖTZL H, METZLER F, ROTHENGATTER P (1982) Die Kluftmarkierung - eine Anwendung der Markierungstechnik zur Ermittlung von Durchströmungseigenschaften klüftiger Gesteine. Beiträge z. Geologie der Schweiz, Hydrogeologie 28 II, Bern

LUGEON M (1933) Barrages et Géologie. Librairie de l'Université Lausanne

MERKLER GP, BLINDE A, ARMBRUSTER H, DÖSCHER HD (1985) Field investigations for the assessment of the permeability and identification of leakages in dams and dam foundation. Proc XV Congr High Dams, Q 58, R 7, Lausanne

MERKLER GP, MOLDOVEANU T (1970) Einige Beispiele zur Anwendbarkeit geophysikalischer Messungen bei Baugrunduntersuchungen. Proc Second Congr Int Soc Rock Mech, vol 1, Beograd

MERKLER GP, VLADUT T, METAXA V (1970) Geophysikalische Messungen angewandt zur Ermittlung einiger Materialkennwerte des Gebirges. Ergebnisse und Problematik dieser Messungen. Proc Second Congr Int Soc Rock Mech, vol 1, Beograd

METZLER F, BLINDE A, HÖTZL H. (1985) Durchlässigkeits- und Injektionsverhalten aufgelockerter Granite. Ingenieurgeologische Probleme im Grenzbereich zwischen Locker- und Festgesteinen. K-H Heitfeld (Editor) Springer-Verlag, Berlin, Heidelberg

NEUMANN W (1954) Praktische Untersuchungen zur Isanomalen-Richtungsstatistik. Freiberger Forschungshefte C 13 (Geophysik), Berlin

ROTHENGATTER P (1982) Ingenieurgeologische und hydrogeologische Untersuchungen an Graniten in Übergangsbereichen zwischen Fest- und Lockergestein zur Beurteilung der Wasserdurchlässigkeit und Injizierbarkeit. Diplomarbeit, Angewandte Geologie, Universität Karlsruhe

LABORATORY STUDIES ON THE CHARACTERISTICS OF ELECTRODES USED FOR STREAMING-POTENTIAL MEASUREMENTS

A, Kassel[1], S. Faber[2] and G.-P. Merkler[1]

1 Introduction

Results of laboratory experiments are presented which were carried out in order to investigate electrode systems placed in a homogeneous sand body, concerning their reaction to changing temperature at the surface and inside the sand body and to different water levels within the medium. In addition, electrical potential data recorded while water streamed through the sand model is discussed.

Streaming-potential data are of the order of several millivolts to several tens of millivolts. The effective signals are often contaminated by noise signals of the same order of magnitude or occasionally even much higher. This prevents the reproductibility of the data and makes an interpretation of self-potential measurements in terms of streaming potentials very difficult.

In order to estimate the order of magnitude of a variety of noise sources, several electrode systems were tested concerning their:

1. Stability versus time: electrodes were placed in an electrolyte of the same chemical composition and ionic concentration as that included in the electrodes themselves.

2. Electrical potential as a function of temperature: electrodes were placed in dry sand as well as in water-saturated sand.

3. Electrical potential as a function of temperature and water level in the sand-filled box. This implies a possible contribution of io-

1 University of Karlsruhe, Applied Geology, Kaiserstr. 12, 7500 Karlsruhe, FRG

2 University of Karlsruhe, Geophysical Institute, Kaiserstr. 12, 7500 Karlsruhe, FRG

Lecture Notes in Earth Sciences, Vol. 27
G.-P. Merkler et al. (Eds.)
Detection of Subsurface Flow Phenomena

nic transport to the electrical potential due to evaporation as a function of temperature.

2 Stability Versus Time

Figures 1 and 2 demonstrate two examples of stability, respectively instability versus time for two different electrode systems: the first one being calomel electrodes (Fig. 1), the second one copper-copper sulfate (Fig. 2). Both electrode systems, each consisting of five

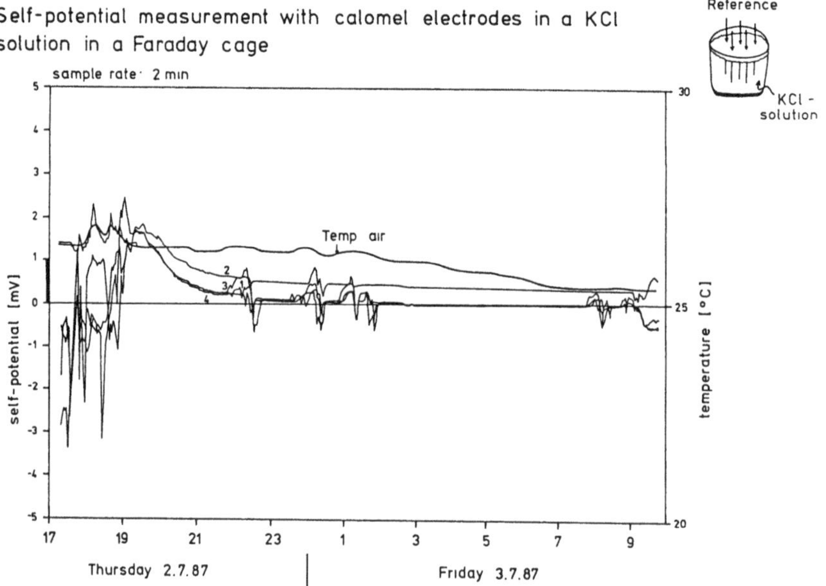

Fig. 1: Self-potentials (mV) versus time of four calomel electrodes placed in a KCl solution and recorded with respect to a reference electrode placed in the same solution itself. The experimental device was installed in a Faraday cage. The surrounding air temperature was recorded as well.

electrodes, were placed in an electrolyte identical to that included in the electrodes themselves. The electrical potential of 4 of the electrodes was recorded with respect to one reference electrode. The potential differences were automatically recorded by an HP scanner controlled by an HP86 desk computer. The sampling rate was 2 min, the duration of each experiment about 1 day and 1 night.

For the calomel electrodes we noted a slight stabilization after a partly irregular drift of 4 h, disturbed further on by irregular noise pulses with an amplitude of 1.5 mV (Fig. 1). Placement of the whole system in a Faraday cage did not change the irregular noise contributions. We assume that they were due to uncontrollable electromagnetic influences from outside.

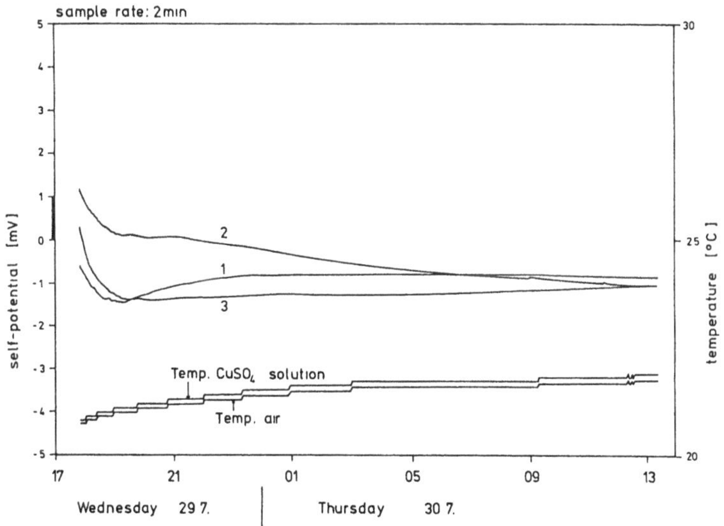

Fig. 2: Self potentials versus time of three copper-copper sulfate electrodes placed in a $CuSO_4$-solution and recorded with respect to a reference electrode placed in the same solution. Temperature curves of the surrounding air and of the $CuSO_4$-solution are plotted as well.

Contrary to the calomel electrodes, the copper/copper-sulfate electrodes show a far better stability versus time after a primary drift of several hours after placement (Fig. 2). They reveal a constant drift with temperature which will be more closely investigated below.

Due to missing irregular noise signals and due to the fact that $CuSO_4$ electrodes have been used in our field experiments for years, we prefer to use these electrodes for our basic laboratory experiments.

3 Experimental Arrangement

The model consists of a homogeneous sand body installed in a small channel (Fig. 3). Different experiments (constant water level, hydraulic gradients with water flow through the sand) could be carried out without any changes in construction:

Fig. 3: Experimental device used for the experiments described throughout figures 4 to 14.

1. The model can be heated through six infrared lights, 150 W each, with adjustable intensity.

2. The water level can be varied continuously and determined accurately.

3. The sand model can be streamed by water with variable hydraulic gradients.

The positions of the electrodes and temperature-sensors can be recognized in Fig. 4 showing a view onto the surface of the model.

The dimensions of the sand body are 0.93 x 0.70 x 0.35 m. The model is equipped with one reference electrode, six symmetrically located electrodes as well as five temperature lances, each equipped with four platinum-resistivity temperature sensors, one at 2 cm above the sur-

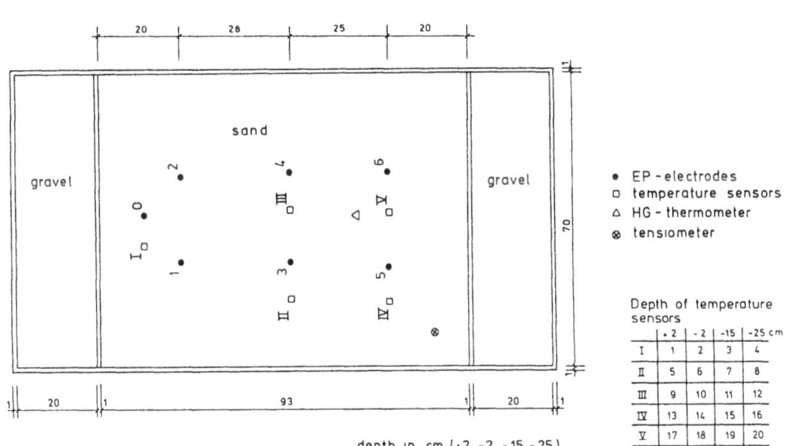

Fig. 4: Schematic view onto the surface of the model, showing the locations of electrodes, tensiometer and temperature lances.

face, and one at 2, 15 and 25 cm respectively below the sand surface. Sensors and electrodes were connected to HP-Data Acquisition System and recorded continuously with sampling rates of 5-10 min.

4 Data and Results

4.1 Dry Sand

Figure 5 shows the temperature distribution at the centre of the model at different depths at which the sensors of the centre temperature lance were located. Three heating cycles took place; during each cycle infrared heating was switched on for 2 h and remained switched off for 3 h. Reduction of amplitudes as well as phases shifts with increasing depth show up clearly.

At a depth of 2 cm beneath the surface temperature changes of 10°C occured during each cycle, while at a depth of 15 cm the temperature variation was about 1°C during each cycle and only 2°C from the beginning to the end of the experiment. As the SP electrodes have a length of 21 cm, the lower end remained stable with temperature.

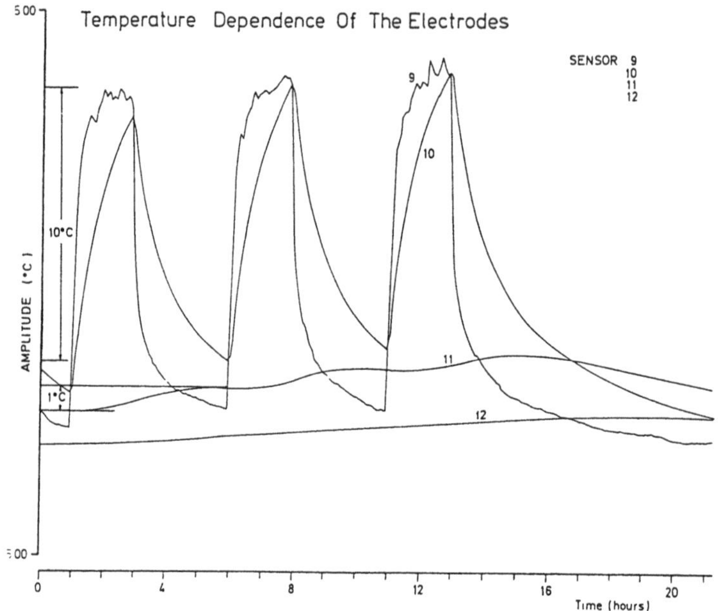

Fig. 5: Temperature distribution at four different depths in the centre of the model during three heating cycles.

The corresponding potential differences of the electrodes revealed an amplitude/temperature dependece symmetrical to the centre of the model (Fig. 6). Electrodes 3 and 4 located at the central part of the model (Fig. 4) reacted to the temperature increase with an amplitude increase of 5 mV, while for the remaining four electrodes amplitude increases were nearly 1.7 mV. The difference in amplitudes was due to the highest temperature at the centre of the model. In addition, we recognized a superposition of high-frequency noise on the electrode signals, probably caused by electromagnetic influence. Since the resistance of the whole circuit was very high in dry sand, it was extremely sensitive to noise.

For the following experiment, self-potentials were recorded while the sand was water-saturated. Two heating cycles with the same duration as those presented previously took place. Again the electrodes at the centre of the model, where the temperature variations were highest (7°C at a depth of 2 cm) had the largest amplitude variations (Fig. 7). High-frequency noise vanished probably due to the decreased resistivity of the whole system.

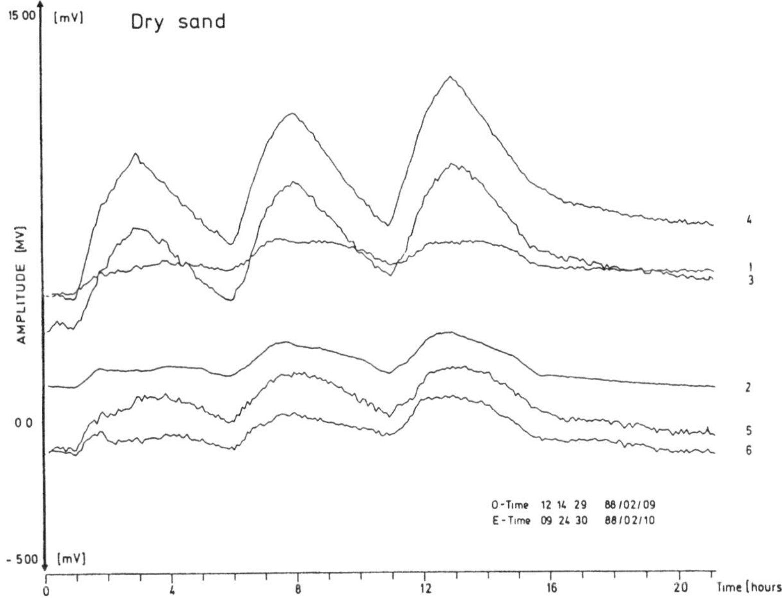

Fig. 6: Self-potentials (mV) of electrodes 1 to 6 (see Fig. 4) versus time during the heating cycles. The sand body was dry.

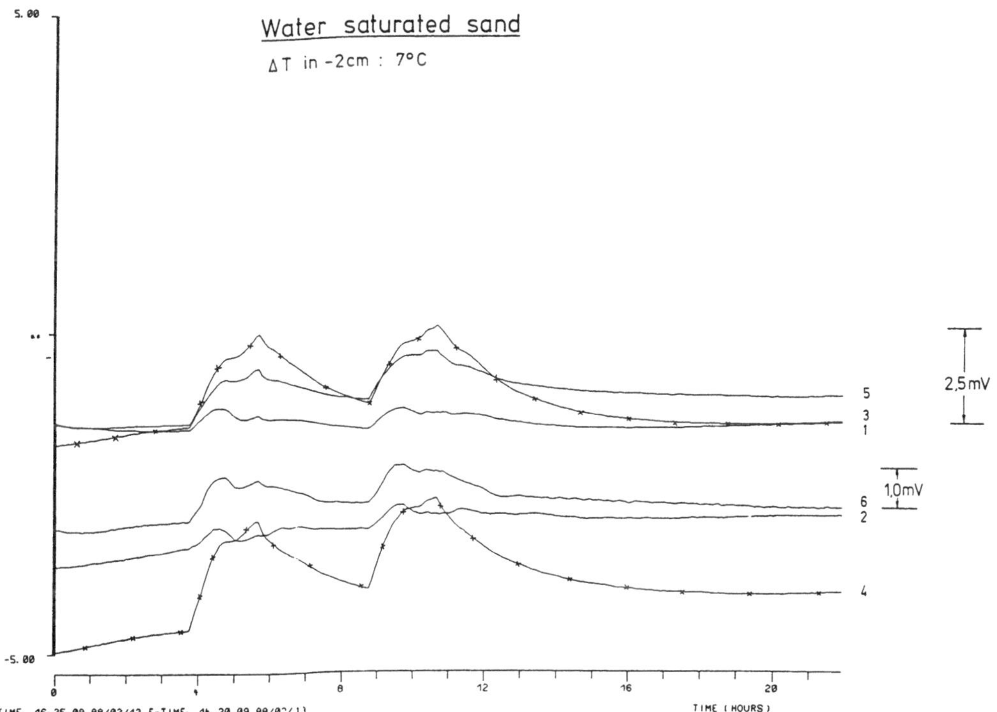

Fig. 7: Self-potentials (mV) of electrodes 1 to 6 versus time during two heating cycles. The sand body was water saturated.

The foregoing experiment was repeated while the reference electrode was isolated against heat radiation at the surface. All of the electrodes showed higher potential differences (Fig. 8). The shape of the signal resembles the shape of the temperature curve at a depth of 2 cm.

This means that the temperature dependence of the self-potentials was due to the electrode-specific reactions like separation of electric charges within the copper rod as a function of temperature or temperature-dependent potentials at the boundary between the metal and the electrolyte. In addition, the diffusion potentials which arise

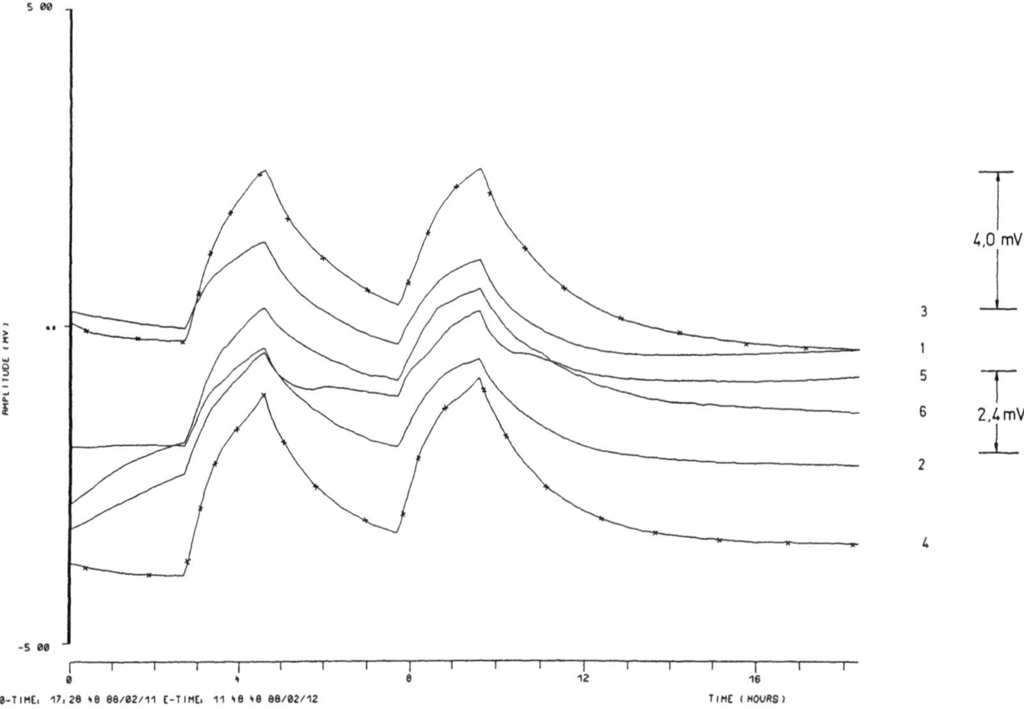

Fig. 8: Same experiment as described in connection with figure 7, except that the reference electrode was isolated against heat radiation from the infrared lights above the model.

between the $CuSO_4$ solution and the soil electrolyte depend on the T-gradient between the two media.

For the accurate determination of temperature coefficients it is necessary to keep the reference electrode under stable and constant temperature which is not an easy task, especially since electric contact with the surrounding medium, which changes temperature through conductivity, must exist.

5 Temperature Influence at Different Water Levels

This section will deal with self-potential measurements in a humid sand body including different water levels above the bottom of the model. The problem to solve is whether vertical ionic transport produced by evaporation has a measurable influence on electric potentials. We expected that these experiments would provide a better understanding of the temperature dependence of self-potential recordings carried out continuously over several months at an outdoor model dam on the BAW territory of the Federal Waterway Engineering and Research Institute (BAW). These recordings revealed periodic daily variations of the electric potentials with an amplitude of 20 mV.

Figure 9 shows the electric potential as a function of time during two heating cycles for the electrodes placed in humid sand with a 10-cm water level at the bottom of the model. The temperature dependence of the electric potential is analogue to that for water-saturated sand.

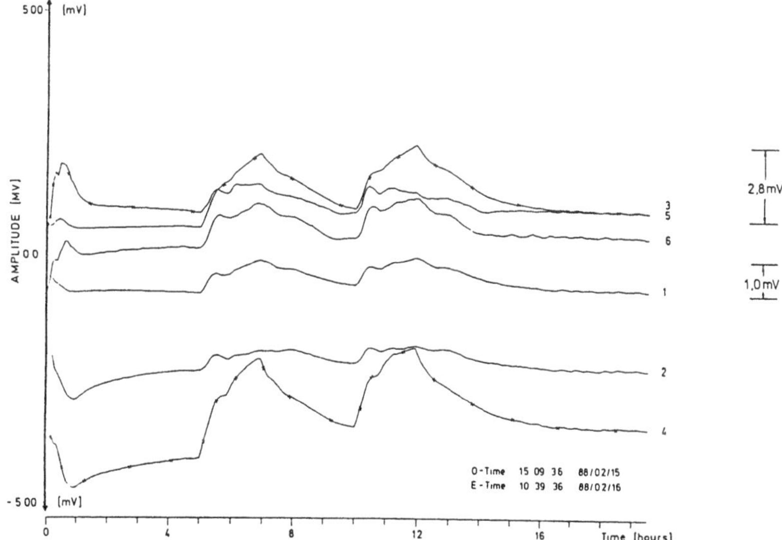

Fig. 9: Self-potentials (mV) of electrodes 1 to 6 during 2 heating cycles. The sand body was water-saturated up to 10 cm from the bottom and humid above.

These two experiments did not reveal an influence of the evaporation mechanism of the temperature dependency of the recorded self-potential field. If the mechanism of evaporation gives rise to a change in electric field its contribution is far smaller than the above discussed direct temperature dependency of the electrodes.

In summary, some plots are given (Fig. 10) which clearly show the relationship between temperature and the amplitude of the potential field recorded by the electrodes. For these plots temperatures at the electrode locations were derived by linear interpolation of the neighbouring temperature sensors at a depth of 2 cm. One axis shows the temperature value, the other one the corresponding electrode amplitude during the first heating cycle of each experiment.

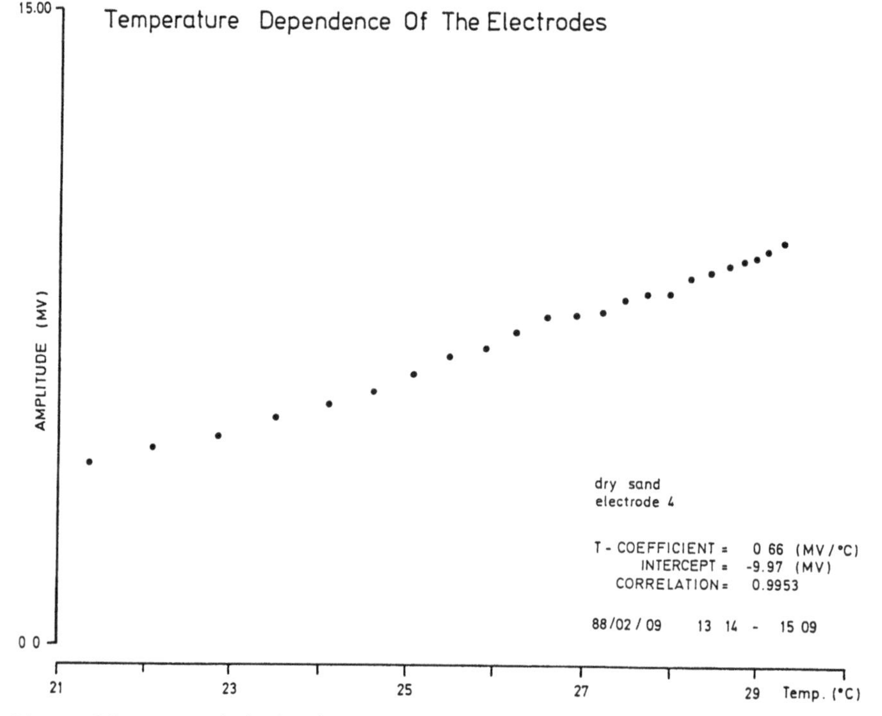

Fig. 10: Self-potential (mV) versus temperature (°C) in dry sand (see also Fig. 6).

The diagrams reveal to a linear relationship between temperature and electric potential.

The "temperature coefficient" is 0.66 mV/°C for dry sand, which corresponds approximately to temperature coefficients found in the literature for $Cu/CuSO_4$ electrodes.

For the experiment carried out in water-saturated sand the coefficient is lower (0.28 mV/°C), while for the case of the reference electrode isolated against heat radiation it amounts to 0.46 mV/°C (Fig. 11, 12).

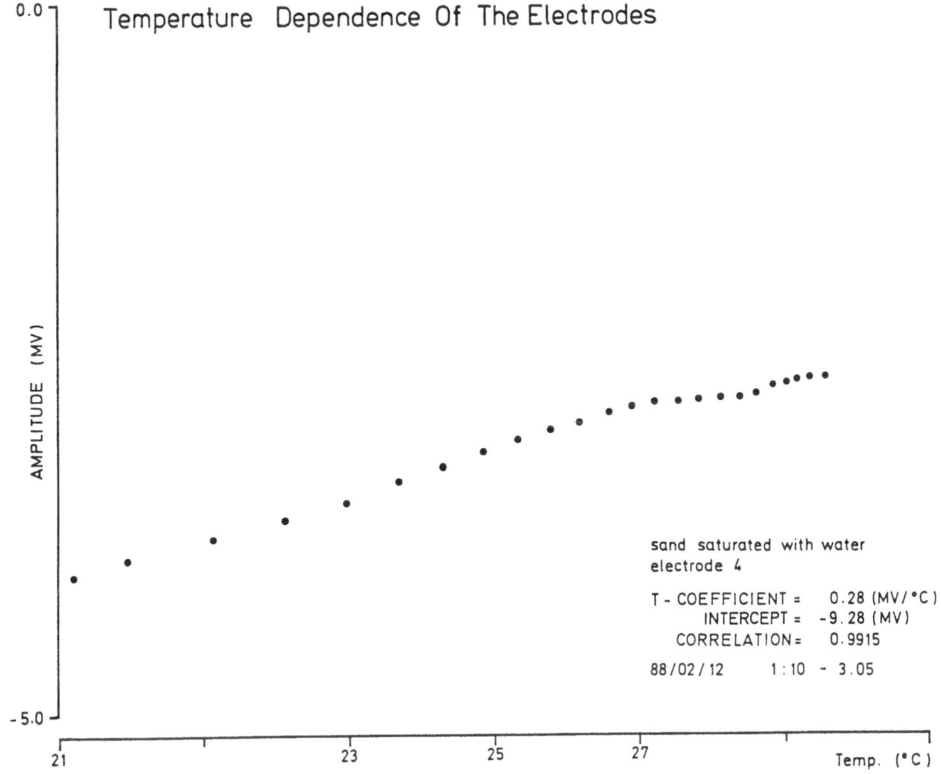

Fig. 11: Self-potential (mV) versus temperature (°C) in water-saturated sand (see also Fig. 7).

A similar temperature coefficient results from only partly saturated sand, with different water levels at the bottom of the model. Figure 13 shows one example for a water level of 10 cm.

6 Streaming-Potential Experiments

An additional short-term experiment in which water flowed through the sand body was carried out (Fig. 14).

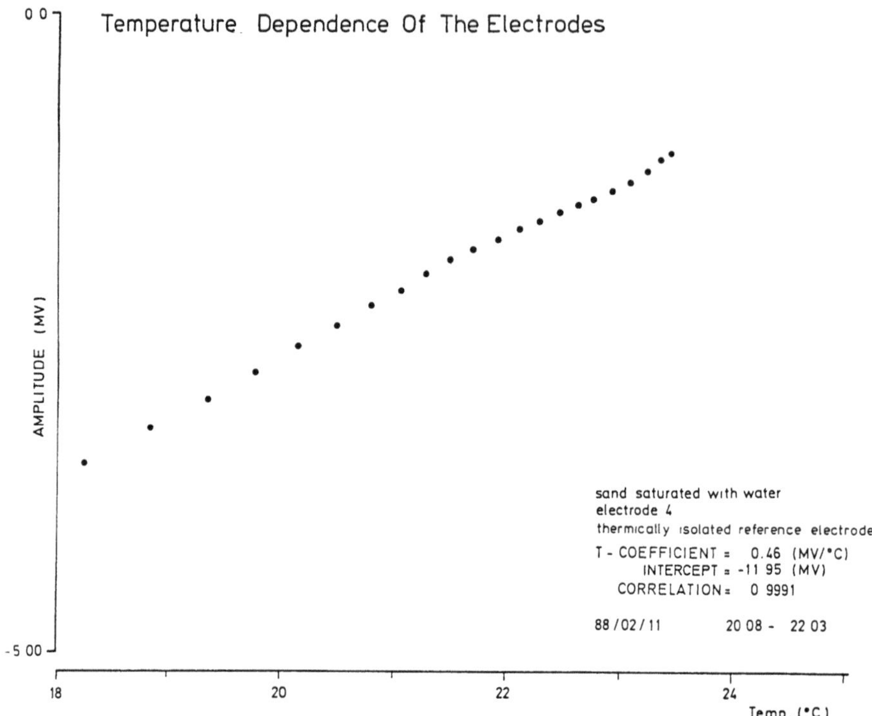

Fig. 12: Self-potential (mV) versus temperature (°C) in water-saturated sand with reference electrode isolated against heat radiation from above (see also Fig. 8).

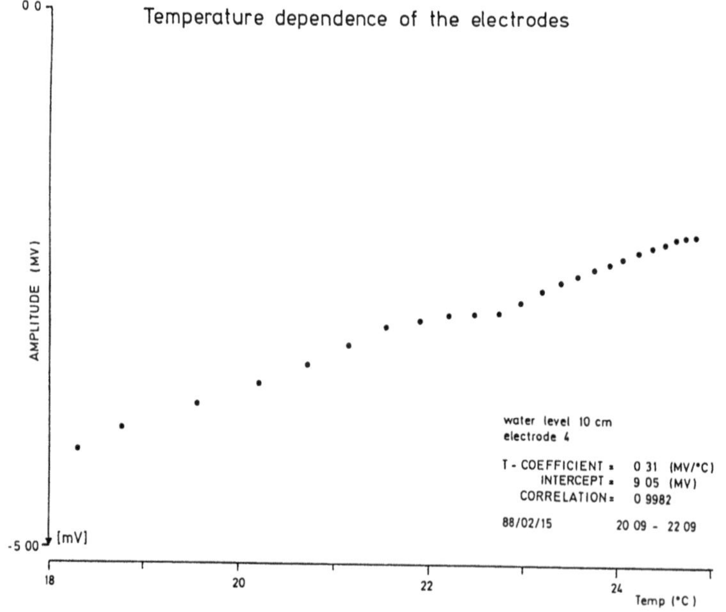

Fig. 13: Self-potential (mV) versus temperature (°C) for the experiment described in connection to figure 9.

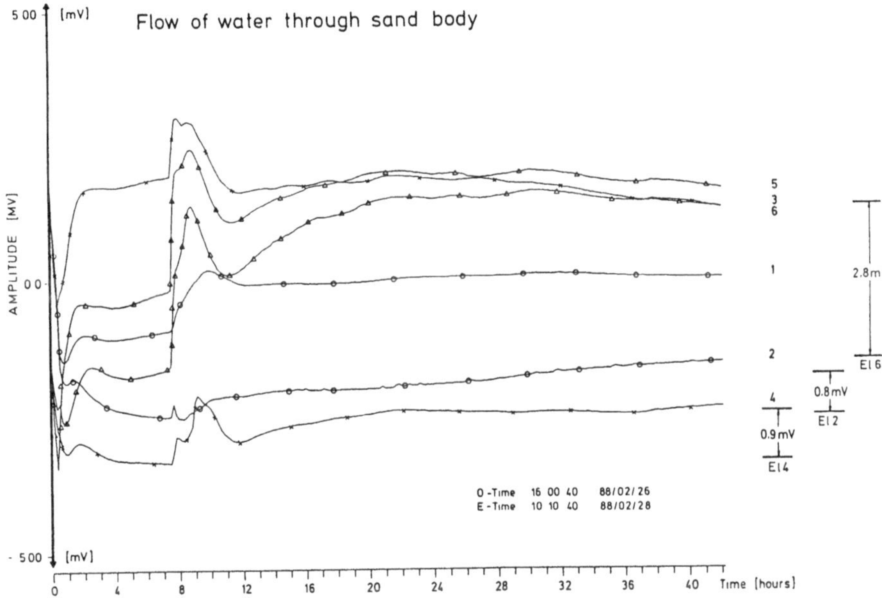

Fig. 14: Self-potentials (mV) versus time during a water-flowing experiment. Electrodes show different values of their self-potential depending on their distance from the reference electrode.

After a 7-h stabilization time of the electrodes, the water level at the inlet of the model rose to 16 cm and sank to 10 cm after 16 h. Five electrodes showed an increased electrical potential during the flow of water after a stabilization period of several hours. Electrodes 5 and 6 which were placed at the largest distance from the reference electrode showed the largest potential increase which correlated to the expected potential difference in a homogeneous electric field parallel to the sand body.

7 Conclusions

Though electrode systems, in this case copper-copper sulfate electrodes, are sensitive to hydraulic gradients, their temperature dependence is pronounced and represents a severe problem with regards to the noise contribution to streaming-potential measurements. In addition, the temperature coefficient is dependent on the state of humidity of the surrounding medium. An influence of evaporation could not be deduced from our experiments.

Since changes in moisture content of the investigated area cannot be avoided during field experiments, an effective but time-consuming way of dealing with these problems are long-term recordings of self-potentials at fixed positions followed by data processing including numerical filtering and correlation techniques.

References

Armbruster, H.; Merkler, G.P. (1983): Measurement of subsoil flow phenomena by thermic and geoelectric methods.- Bull. of Int. Assoc. of Eng. Geology, No. 26-27, Paris.

Armbruster, H.; Merkler, G.P.; Wagner-Ambs, M. (1986): Sachstandbericht II - Analytische Zusammenhänge zwischen geohydrologischen Vorgängen an unterirdischen Strömungsfeldern und daraus resultierenden thermischen und geoelektrischen Feldern an der Oberfläche.- Technical Report, Applied Geology, University of Karlsruhe.

Armbruster, H.; Degen, F.-P.; Faber, S.; Mazur, W.; Merkler, G.P. (1987): Bericht zum Forschungsvorhaben Nr. 30170/86 - Durchsickerung von Dämmen und Deichen bei Dichtungsleckagen und Methoden der Erkennung von Sickervorgängen.- Technical Report, Institute of Soil and Rock Mechanics, University of Karlsruhe.

Merkler, G.P.; Blinde, A.; Armbruster, H.; Döscher, H.D. (1985): Field investigations for the assessment of permeabilitzy and identification of leakage in dams and dam foundation.- 15th Congr. of Large Dams, Q. 58, R. 7, Lausanne.

Militzer, H.; Weber, F. (1985): Angewandte Geophysik Bd. 2: 174-187, Springer-Verlag Wien-New York.

Long-Term Self-Potential Data Acquisition and Processing

Burkhard Wurmstich and Sonja Faber[1]

Abstract

An example of data acquisition and processing due to measurements of self-potential effects of a leaky dam model (scale 1:1) is presented. Every alteration of the leak causes a change in the self-potential anomalies, which is superposed by the effects of temperature and noise. Measurement of the self-potential anomalies was done at 40 different locations on the surface of the dam with the aid of unpolarizable electrodes. In addition to the self-potential measurements, measurements of temperatures, water levels, and the quantity of the water leakage were also carried out. All of the above-mentioned data were recorded by a data acquisition system.

Time series records were frequently disrupted by failures of the measuring instruments and the data acquisition system. Such failures caused peaks and gaps in the record, which must be corrected. Other problems were the relativly low signal-to-noise ratio and the influence of drift effects, caused by oscillations of the temperature. The data processing consisted of interpolating gaps and smoothing peaks, Fourier analysis, digital filtering and noise reduction.

[1]Universität Karlsruhe, Institut für Geophysik, Hertzstr. 16, 7500 Karlsruhe 21, FRG

Lecture Notes in Earth Sciences, Vol. 27
G.-P. Merkler et al. (Eds.)
Detection of Subsurface Flow Phenomena

1. Introduction

The object of interest was the investigation of electrical self--potential anomalies connected with water movements, measured at a full-sized model of a leaky dam and recorded over long periods. First we will describe the conception of our experiments, explain some details of signal processing and finally discuss some results and unresolved problems of our experiments.

2. Description of Fundamental Principials, the Experiments and the Data Acquisition

For principal considerations on the structure of the dam we refer to the publication of Brauns et al. (1989). The structure of the dam is shown in Fig. 1. For our experiments it was important that the quantity of water leakage and the hydraulic pressure head could be changed by control. We assumed that we would obtain an approximately homogeneous streaming field with an angle of inclination to the outer surface of 3 to 6 %.

We carried out self-potential measurements to analyze the natural electrical fields connected with water movements, caused in particular by electrokinetic potentials. The streaming potential is given by the Helmholtz formula (1879) which is valid for the ideal capillary tube in Eq. (1):

$$U = c_f \, \Delta p, \tag{1}$$

where U is the electrical potential and
Δp is the pressure difference of a through streamed capillary tube,
c_f is the electrofiltration-coefficent given by Eq. (1a):

$$c_f = (\zeta \, \varepsilon \, \varrho)/(4 \, \pi \, \mu), \tag{1a}$$

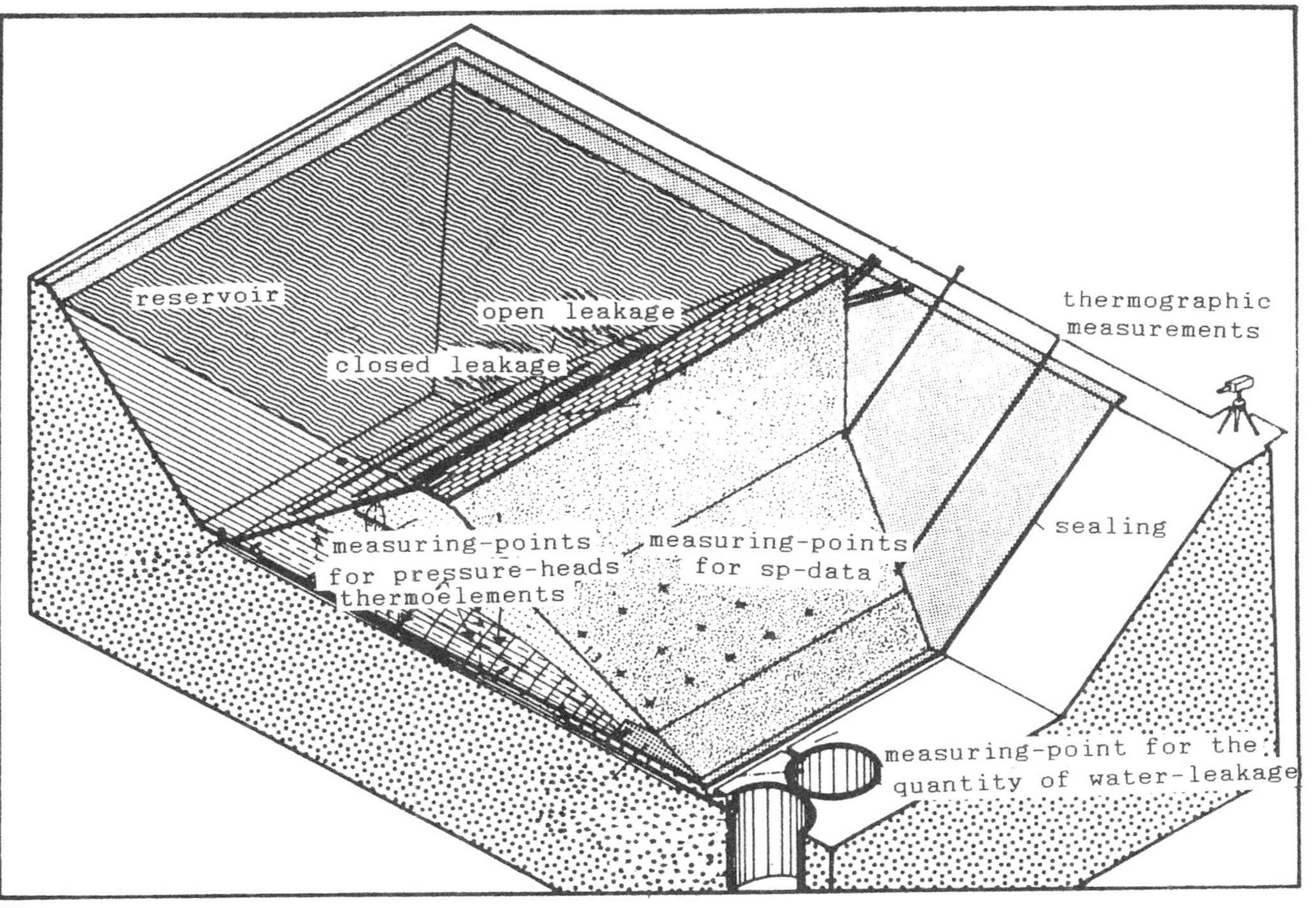

figure 1: schematic cross-section through the model of the dam

where ζ is the zeta-potential;
ε is the fluid dielectric constant;
ϱ is the specific electrical resistance;
μ is the fluid viscosity.

The measured electrical potentials were not due to the streaming potentials since the potential U depends in a complex way on further parameters such as grain particle size, grain shape, porosity. Moreover, the shape of the pore volume and effects of disturbance and noise in the data should be mentioned (Militzer and Weber, (1985)).

The measurement of the self-potentials was carried out at 40 points on the surface of the dam by the aid of unpolarizable Cu--$CuSO_4$ electrodes. The arrangement of the electrodes is shown in Fig. 1: There are five rows of electrodes with eight electrodes in each row (for example: row 1 contains the electrodes 1-8, row 4 contains the electrodes 25-32). We obtain the electrical potentials between a reference electrode, which is placed in the middle of row 1, and one of the electrodes of the system. A schematic cross-section of the electrodes is shown in Fig. 2.

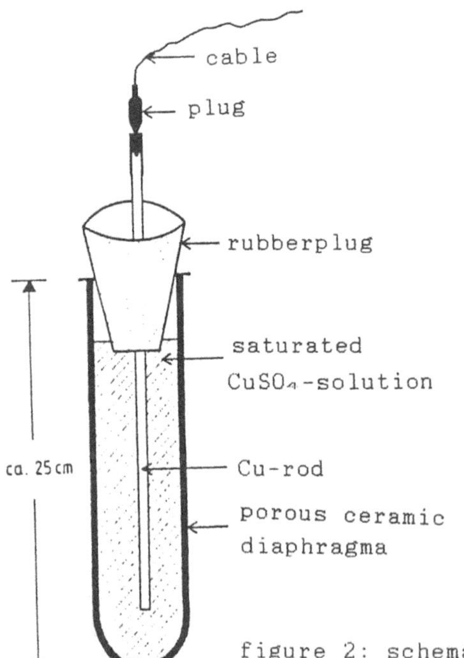

figure 2: schematic cross-section of an electrode

The data acquisition system for the measurement of the self-potential data consisted of a Hewlett-Packard 3421 A scanner with an internal voltmeter for automatic data acquisition and a Hewlett-Packard 86 B Personal Computer with periphery equipment to control the data acquisition. The measurement of the temperatures was done by a high-ohmic voltmeter (Siemens B1042), a Siemens B8102 scanner and the same PC. We always took three values of the self-potentials at one time. The median of the three values was recorded. The accuracy of the measurement was 1 mV. The system was tuned to a maximal sampling rate of 7.5 min, but in the data records time intervals of different length are found. The hydrostatic pressure head and the quantity of water leakage were recorded by another data acquisition system.

3. Data Processing

In Fig. 3 we present an example of the original data of self-potential registrations. We selected three hydrostatic experiments, which were carried out between 25 March 1987 and 15 May 1987. Fig. 4 shows the associated quantity of water leakage and the hydrostatic pressure heads of the moving water. In Fig. 4 the duration of the three experiments is also indicated. The streaming potential data were affected irregularly by negative anomalies, as can be recognized especially in the data of electrodes 12 and 14 (see Fig. 3). These anomalies cannot be correlated between the electrodes. In general, the anomalies were negative and no valid explanation can be given. We suppose, that these anomalies occur because of bad electrical contacts at the plugs of the electrodes. For further data processing these anomalies were replaced by gaps.

The available data for processing were not equidistantly sampled, because of different sampling intervals, failures and restarts of the data aquisition system, and gaps due to the above-mentioned negative anamolies. Since it is nessecary to complete the signal for the application of Fourier methods, a smoothing spline-interpolation was applied to acquire equidistant data for the signal processing. Spline functions are a class of piecewise polynominal functions. They are found to have highly desirable characteristics as interpolating functions and it is possible to determine the efficiency of the smoothing.The spline functions were used for resampling the data to equidistant 2-h intervalls and for interpolating gaps. The smoothing factor brings about an anti-aliasing filter and smoothes the peaks without affecting the phase of the signal (Flach et al., (1975)).

For further processing the approximate homogeneity of the streaming field was assumed. Using this assumption, we can stack the signals of all electrodes of one row. The next step in processing is Fourier analysis:

The Fourier transformation is given by Eq. (2):

$$H(f) = \int_{-\infty}^{\infty} h(t)\, e^{2\pi ft}\, dt\ ;$$

$$h(t) = \int_{-\infty}^{\infty} H(f)\, e^{-2\pi ft}\, dt. \qquad (2)$$

Digital frequency filtering is useful when the signal and the noise consist of different frequency-contents. A Butterworth-type filter was chosen as a good compromise between a sharp cutoff and no occurrence of 'damped ringing' in the filter-response function caused by sharp edges in the filter function.

The absolute value of the filter-response function is given by Eq. (3):

$$|f(\omega)| = 1 \,/\, [1+(\omega/\omega_1)^{2N}], \qquad (3)$$

where ω_1 is the cut-off frequency;
ω is the frequency;
N is the order of the filter.

The order of the filter determines the sharpness of the cutoff of the filter function.

The discrete crosscorrelation of two sampled functions g_k and h_k is defined by Eq. (4):

$$corr(g,h)_j = \Sigma\ g_{j+k}\ h_k \qquad (4)$$

It corresponds to the discrete Fourier transform pair in Eq. (5):

$$corr(g,h)_j = G_k\ H_k^* \qquad (5)$$

where G_k and H_k are the discrete Fourier transforms of g_j and h_j; the asterisk denotes the complex conjugation. The cross--correlation measurement is generally of importance when the functions g and h are represented by different, but generally similiar data sets. We assume, that the self-potential data are closely related to the pressure-head data in addition to noise.

4. Data Analysis

These processing steps will be elucidated with regards to the processing results.

Figure 5 shows the spline-interpolated data of the self--potentials and Fig. 6 an example of temperature data, which were measured at different depths of the dam subsurface.

Comparing these figures, there is a distinct influence of temperature in the self-potential data, which occurs for several reasons. We suppose, for example, direct influence of the temperature on the electrodes (0.36 mV / 1° C (Petiau and Dupis (1980)), or even higher), and the influence of descending and ascending capillary water, caused by temperature-dependent evaporation. It is difficult, however, to develop a physical model to quantify the different influences. Refering to the publication of Kassel et al. (1989), one can say that the temperature causes the highest amplitudes in the data.

Comparing the Fourier spectra of the self-potential data (see Fig. 7) and the temperature data (see Fig. 8) both the temperature wave and its harmonics with a period of 24 and 12 h, respectively can be distinctly recognized. Since the relation of the Fourier amplitudes, calculated from the two dominant temperature waves in the self-potential data, is not a constant, it is assumed that the temperature influences the electrodes individually. Since there is no physical model available of all temperature effects, the different influences of the temperature cannot yet be distinguished.

Because the signal of the pressure heads has a different frequency content than the influence of the temperature, a numerical low-pass-filter was applied to both the stacked and nonstacked data of the self-potentials in order to eliminate the temperature variable. Using a low-pass filter of 1/30 cycles per hour, all periodic functions of the temperature will be diminished.

Figures 9, 10, and 11 show a comparison between the interpolated and the filtered self-potential data with the signal shape of the pressure head for electrodes 10 and 16 and the stacked signal of row 2 (electrodes 9, 10, 11, 13, 15, 16). Experiments 1 and 2 reveal a correlation between the filtered self-potential data and the long period signal shape of the pressure head in duration as well as in the amplitude variation.

Figures 12, 13, and 14 show a comparison between the a few cross--correlation functions, calculated from the unfiltered self--potential data and the pressure heads and the filtered data and the pressure heads respectively. In these figures "P8" denotes the pressure-head signal from point 8 and "EL13" electrode 13. The cross-correlation function between the pressure head of point 8 and stacked self-potential data is shown in figures 15, 16, and 17 for each of the three experiments. "SUM2" denotes the stacked self-potential data of row 2 and "FIL2" the stacked, filtered self-potential data of the same row. The cross-correlation function is a good indication for the dependence and possible delay between the two correlated functions. It has large values at some values of t, if the function of self-potential data is very similiar to the function of pressure heads, but it lags in time by t. For our measurements, no lags lags of time were expected. The two functions should be correlated best at t=0. In general, there is a good correlation for experiment 1, some electrodes also show a good correlation in experiment 2, which can be improved by eliminating the influence of temperature with low-pass filtering in a few cases (see, for example, electrodes 10 and 11 in Fig. 12). Experiment 3 does not reveal any correlation, which can be explained by the low amplitudes of the pressure-head.

Two more aspects not previously mentioned are the influence of precipitation, which does not seem to be very well correlated with the self-potential measurements, and the influence of potentials of diffusion. These potentials occur at the boundary between the electrodes, especially if they are electrodes of the first kind, as Cu-$CuSO_4$ electrodes and soil. This may define the accuracy of self-potential-measurements to 10 mV (Fischer et al., (1988)).

5. Summary

The streaming-potential data are affected irregularly by negative anomalies which could not be correlated between different electrodes. In principle these anomalies are negative up to - 150 mV.

There is a distinct influence of temperature concealed in the electrical potential data, however, the influence of the temperature cannot be estimated quantitatively.

In some cases there is a good correlation between the pressure head and the streaming potentials, which can be improved by low--pass filtering (eliminating influences of temperature).

For efficient use of the self-potential method in order to detect leaky dams, and to improve the accuracy and reproductibility of long-term measurements, it is nessesary to improve the long-term stability of our electrodes and to develop physical models for various kinds of noise and disturbing influences, since they cannot be avoided as well as in short-time registrations.

Acknowledgements

The investigation of streaming potentials by methods of time series analysis is a part of the thesis of B. Wurmstich in geophysics and we would like to thank Prof. Hubral, Prof. Wilhelm (Institute of Geophysics, Karlsruhe) and Dr. Merkler (Institute of Applied Geologie, Karlsruhe) for their support and advice.

References

Brauns J, Armbruster H, Blinde A, Degen F-P, Mazur W, Merkler G-P (1989): Effects of leaks in dams and trials to detect leakages by geophysical means. Proceedings from detection of subsurface flow phenomena, Springer, Berlin Heidelberg New York Tokyo

Fischer W, Hildebrandt H, Prinz W, Schwenk W (1988): Problems related to the measurement of IR-drop free potentials in the presence of compensating currents, point 4: Elektrodenfehler(21-22). Werkst und Korros 39: 18 - 22

Flach D, Jentzsch G, Rosenbach O (1975): Interpolation and smoothing of tidal records by spline functions. Dt Geodät Komm, Reihe B

Kassel A, Faber S, Merkler G-P (1989): Laboratory studies of electrodes used for streaming potential measurements. Proceedings from detection of subsurface flow phenomena, Springer, Berlin Heidelberg New York Tokyo

Militzer H, Weber F (1985): Angewandte Geophysik, Vol 2. Springer, Berlin Heidelberg New York Tokyo; Akademie, Berlin

Petiau G, Dupis A (1980): Noise, temperature coefficient and long time stability of electrodes for telluric observations. Geophys Prosp 28: 792-804

Press W H (1987): Numerical Recipies. Univ Press Cambridge

Späth H (1973): Algorithmen für elementare Ausgleichsmodelle. Oldenbourg, München

Stearns S D (1975) : Digital signal analysis. Hayden Rochelle Park, N J, USA

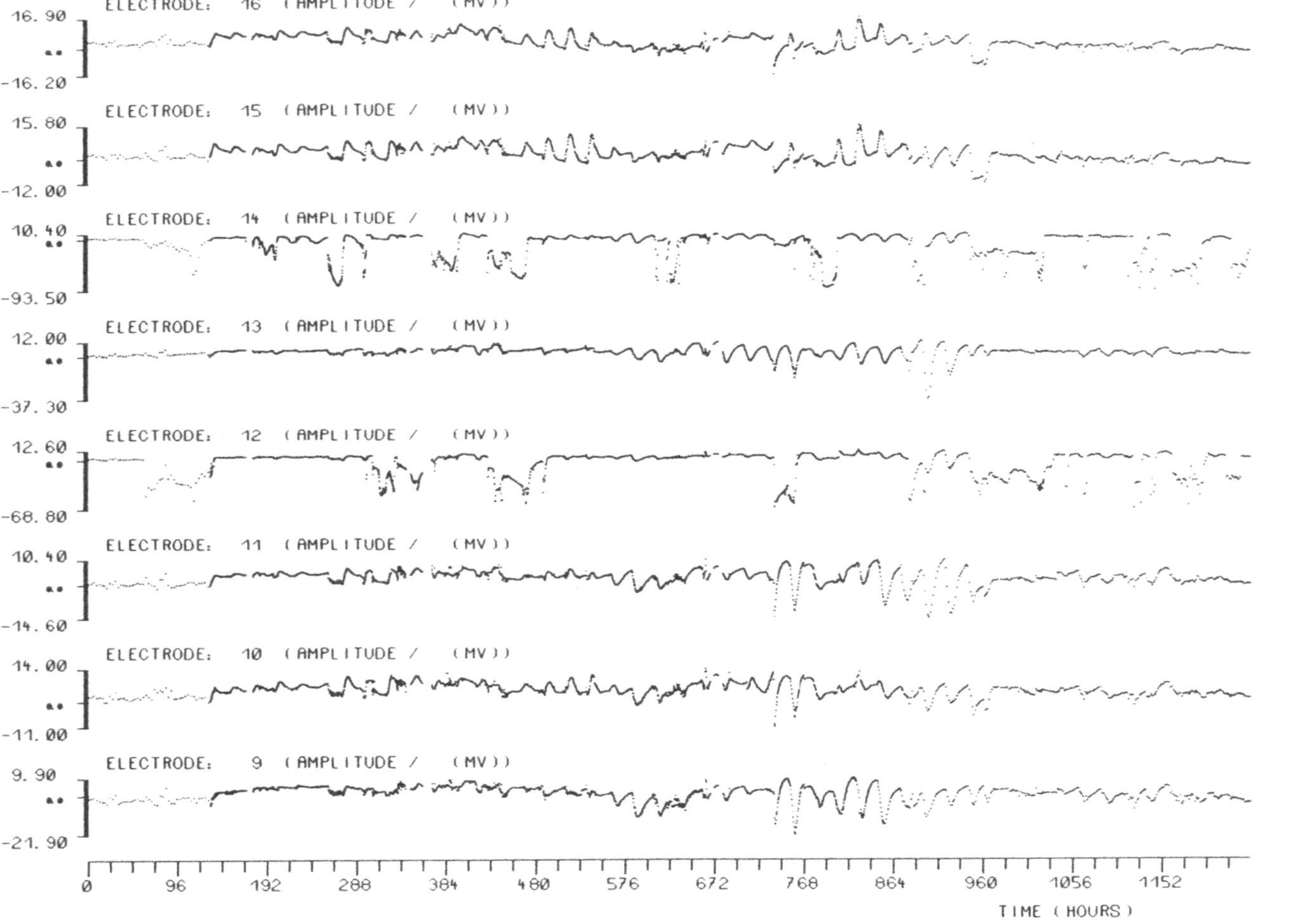

figure 3: non-processed measured self-potential data

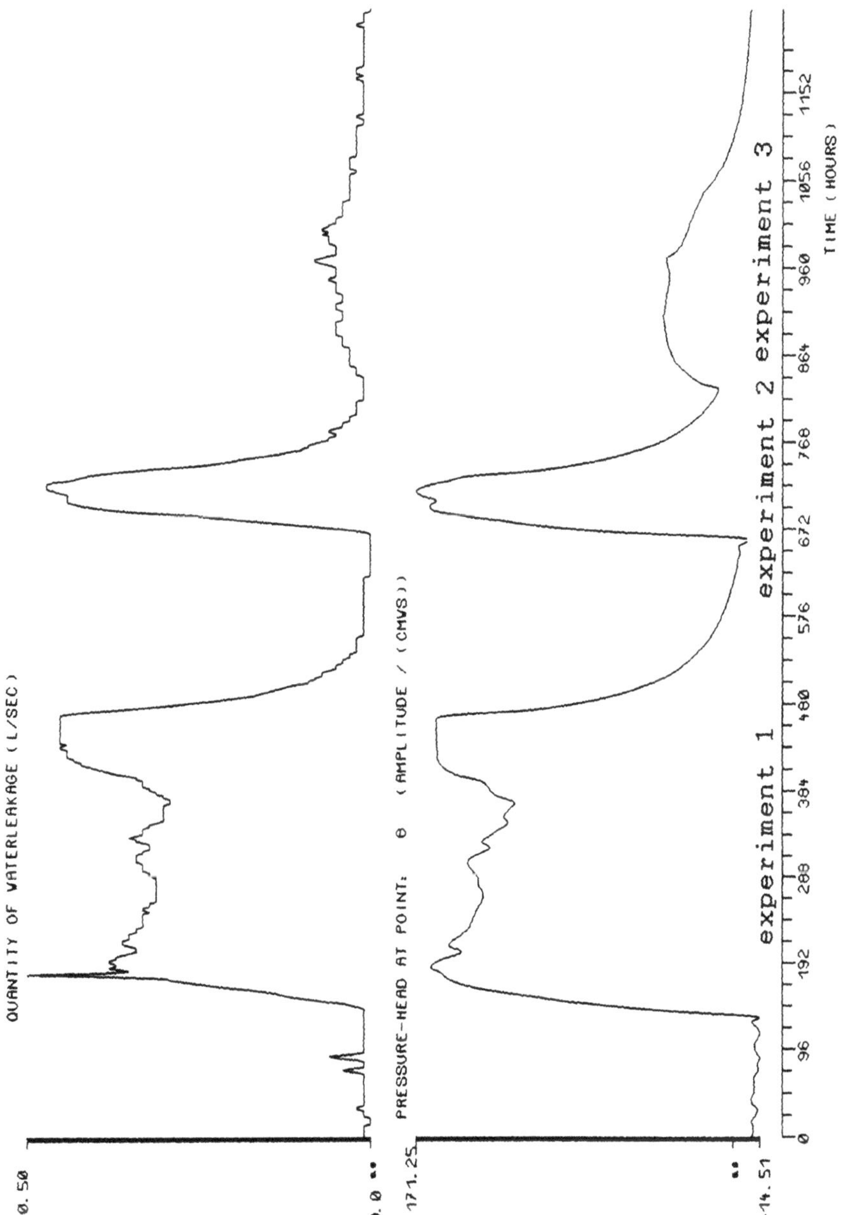

figure 4: quantity of waterleakage and pressure-head data.

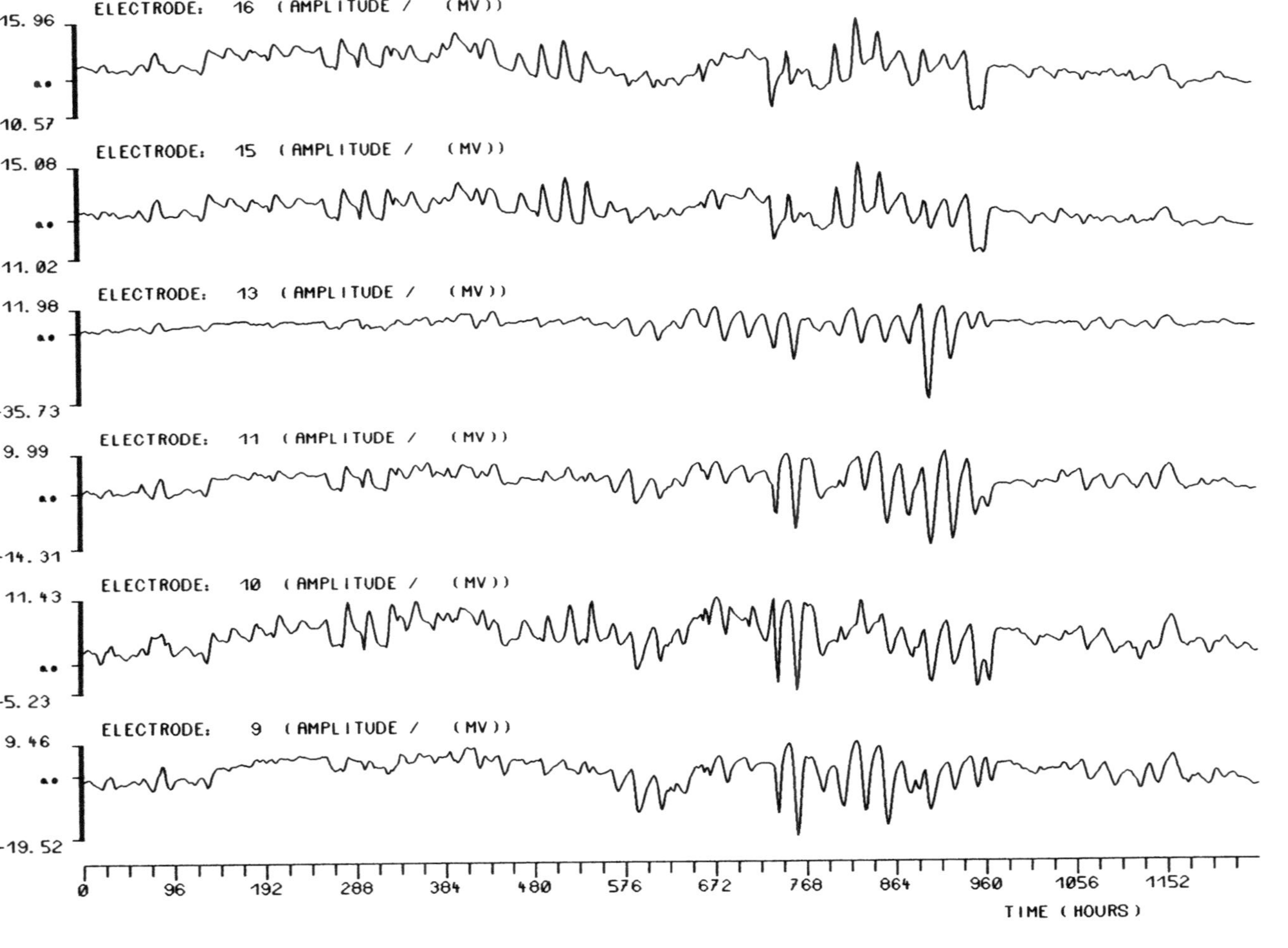

figure 5: spline-interpolated self-potential data

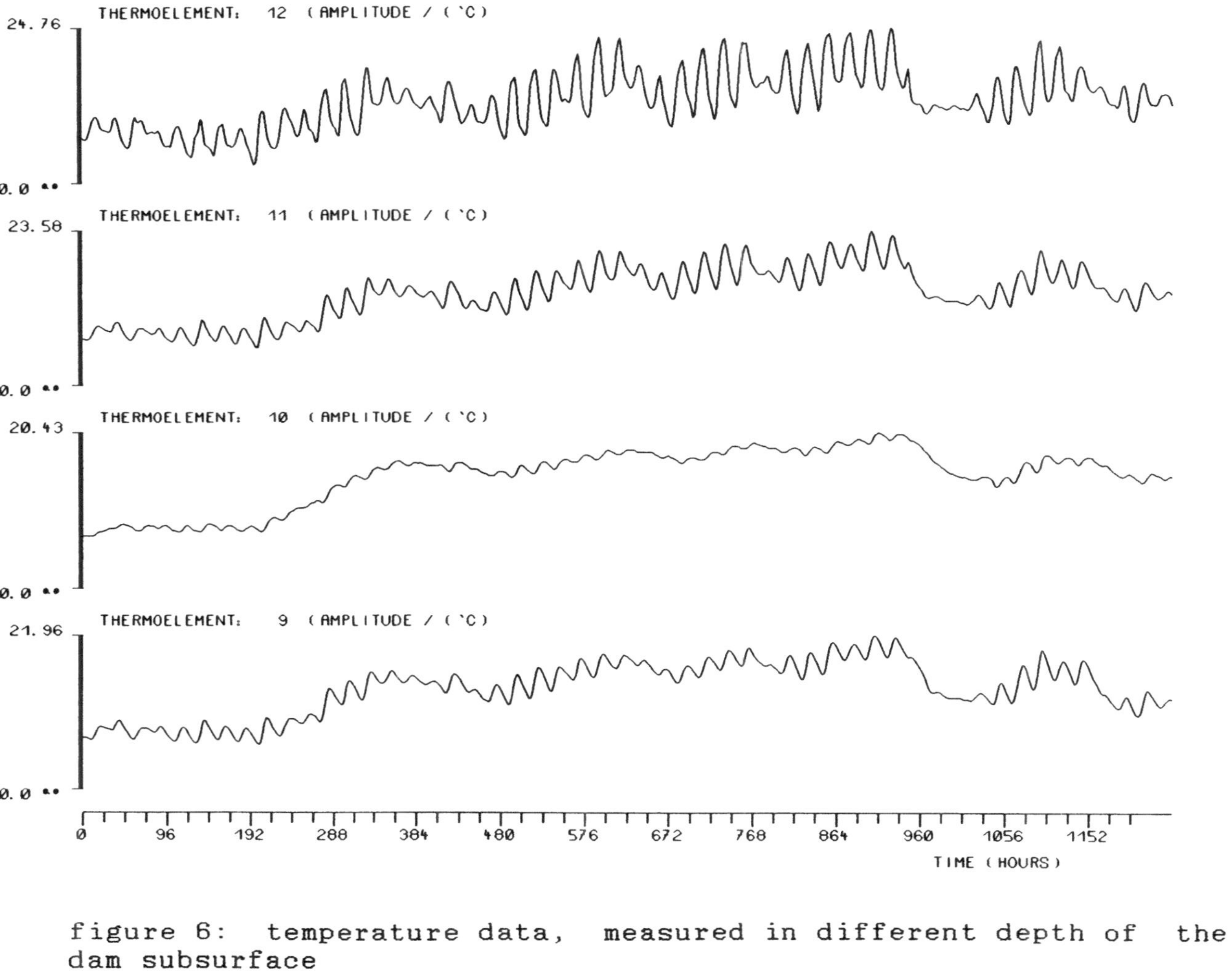

figure 6: temperature data, measured in different depth of the dam subsurface

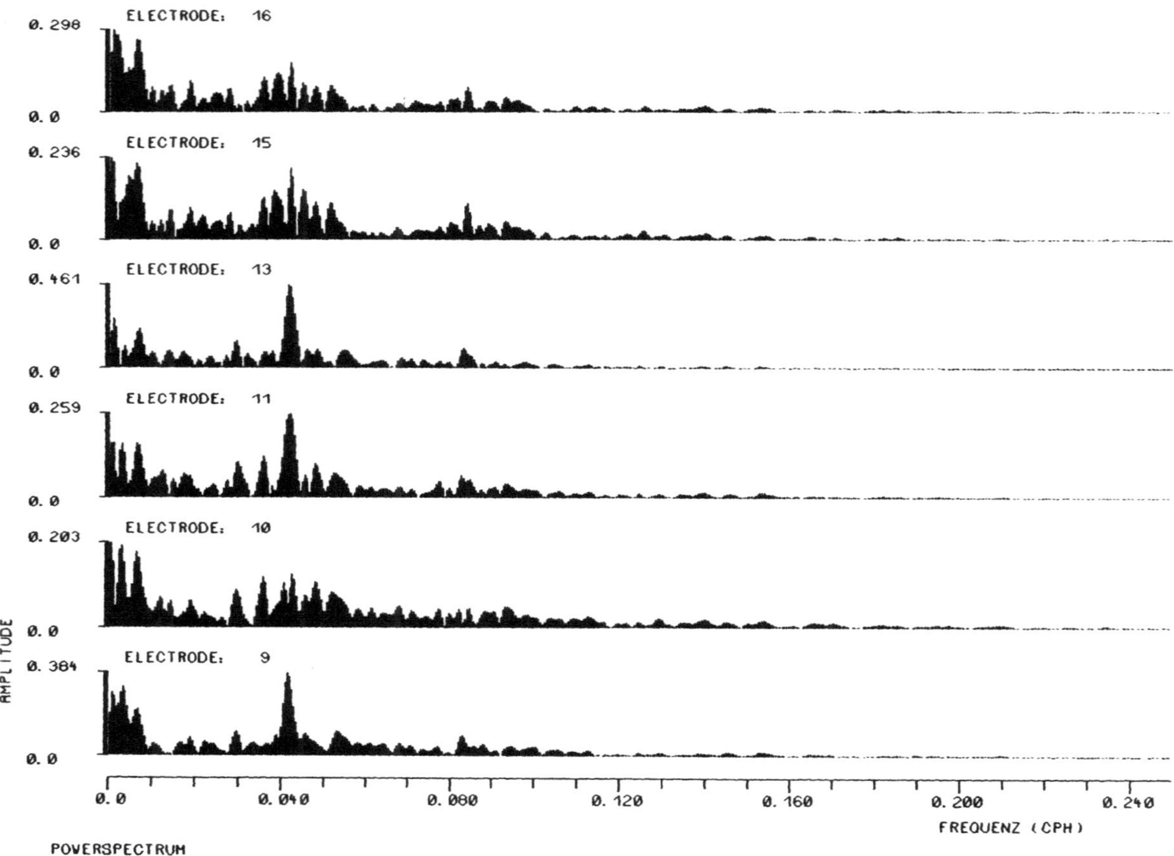

figure 7: fourier-spectra of self-potential data

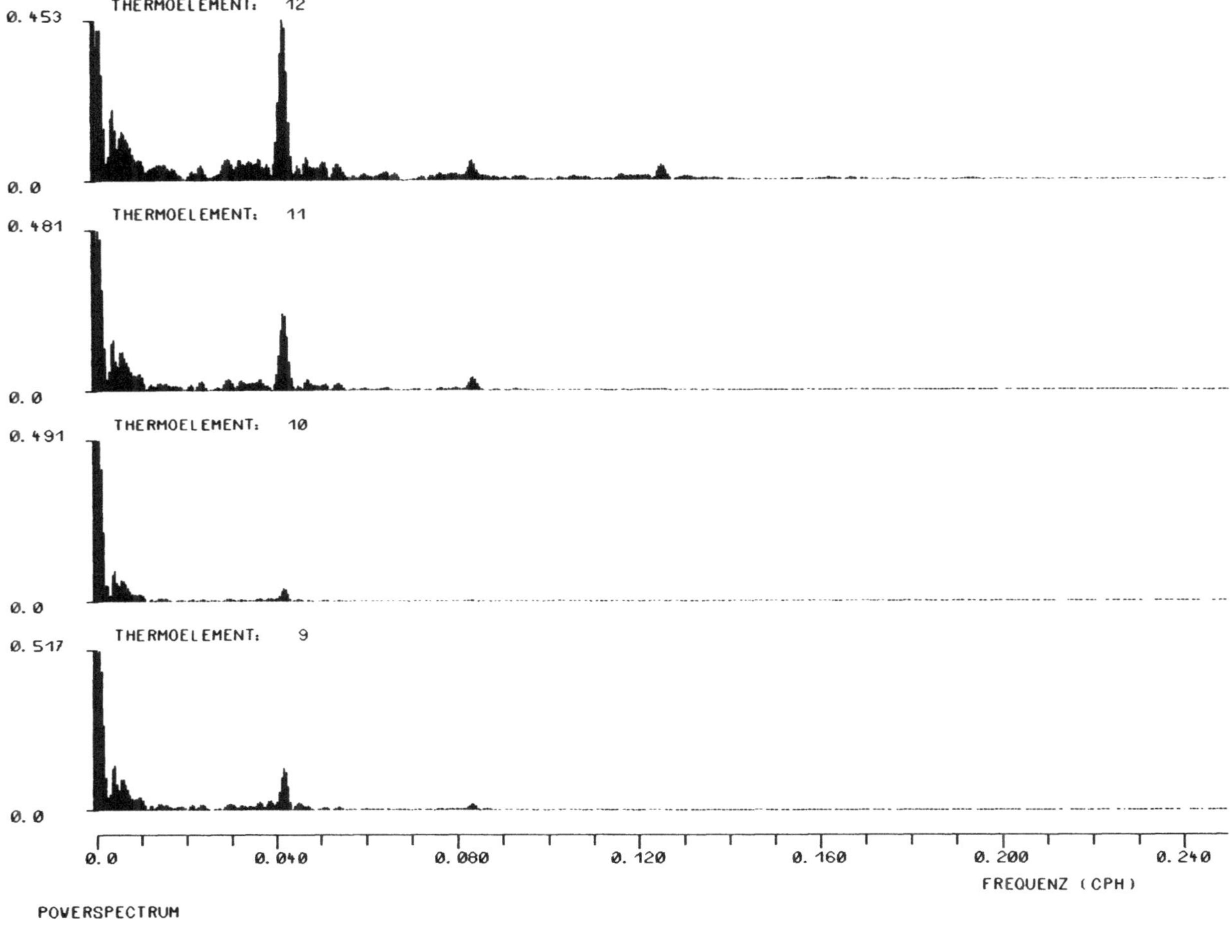

figure 8: fourier-spectra of temperature data

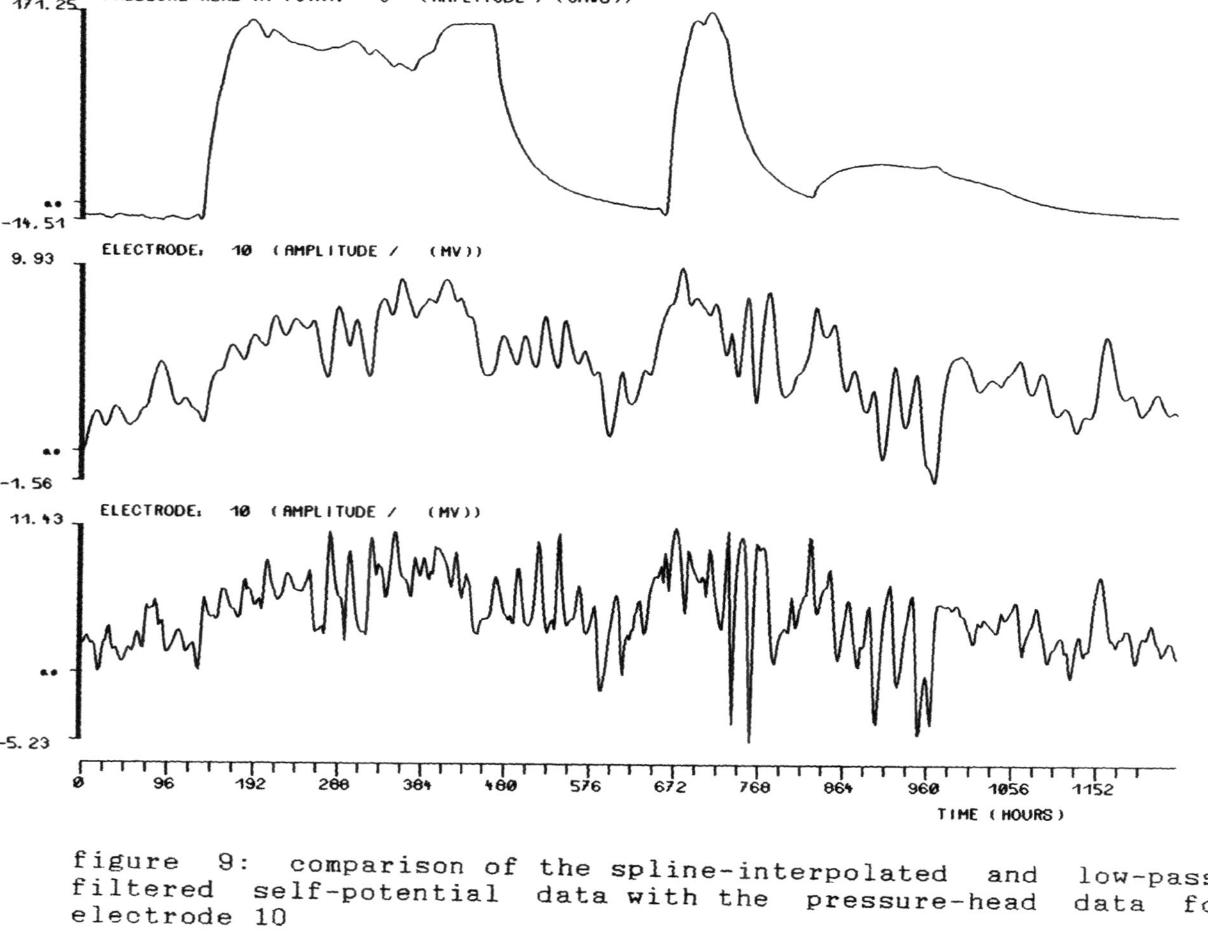

figure 9: comparison of the spline-interpolated and low-pass-filtered self-potential data with the pressure-head data for electrode 10

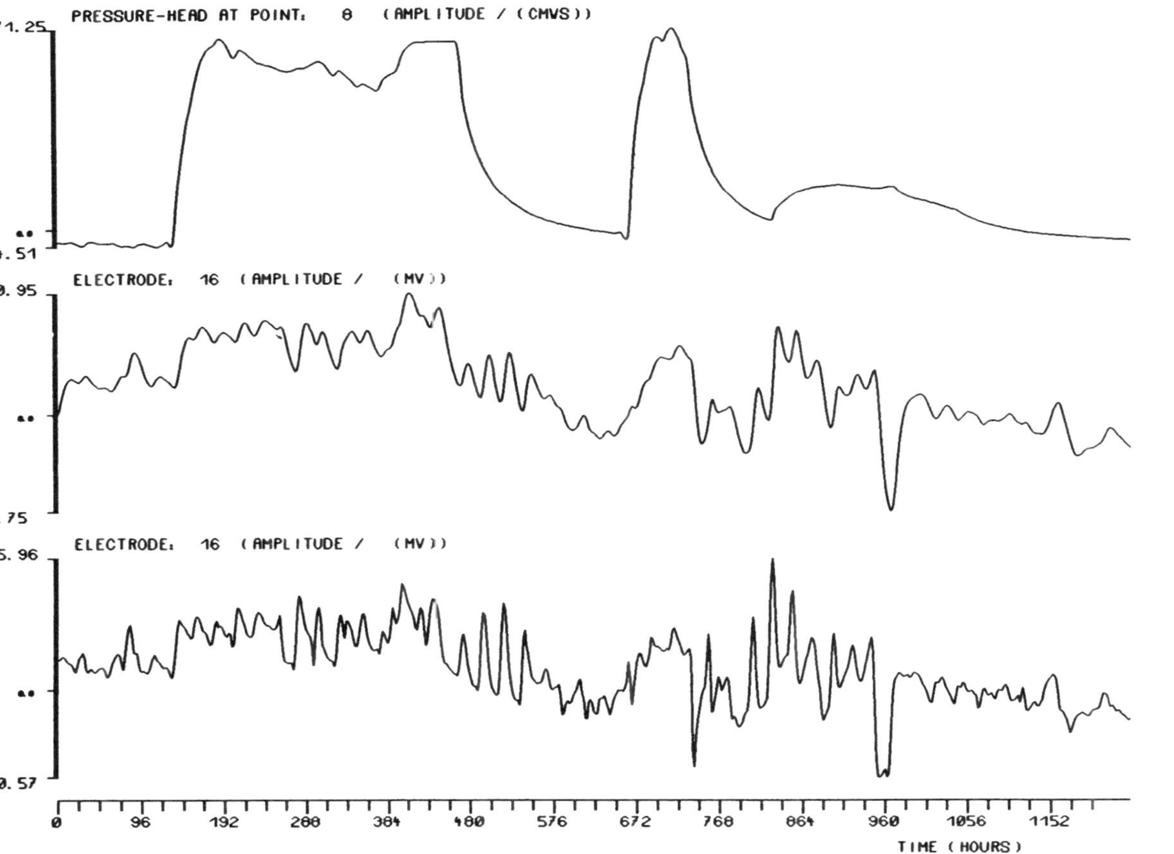

figure 10: comparison of the spline-interpolated and low-pass-filtered self-potential data with the pressure-head data for electrode 16

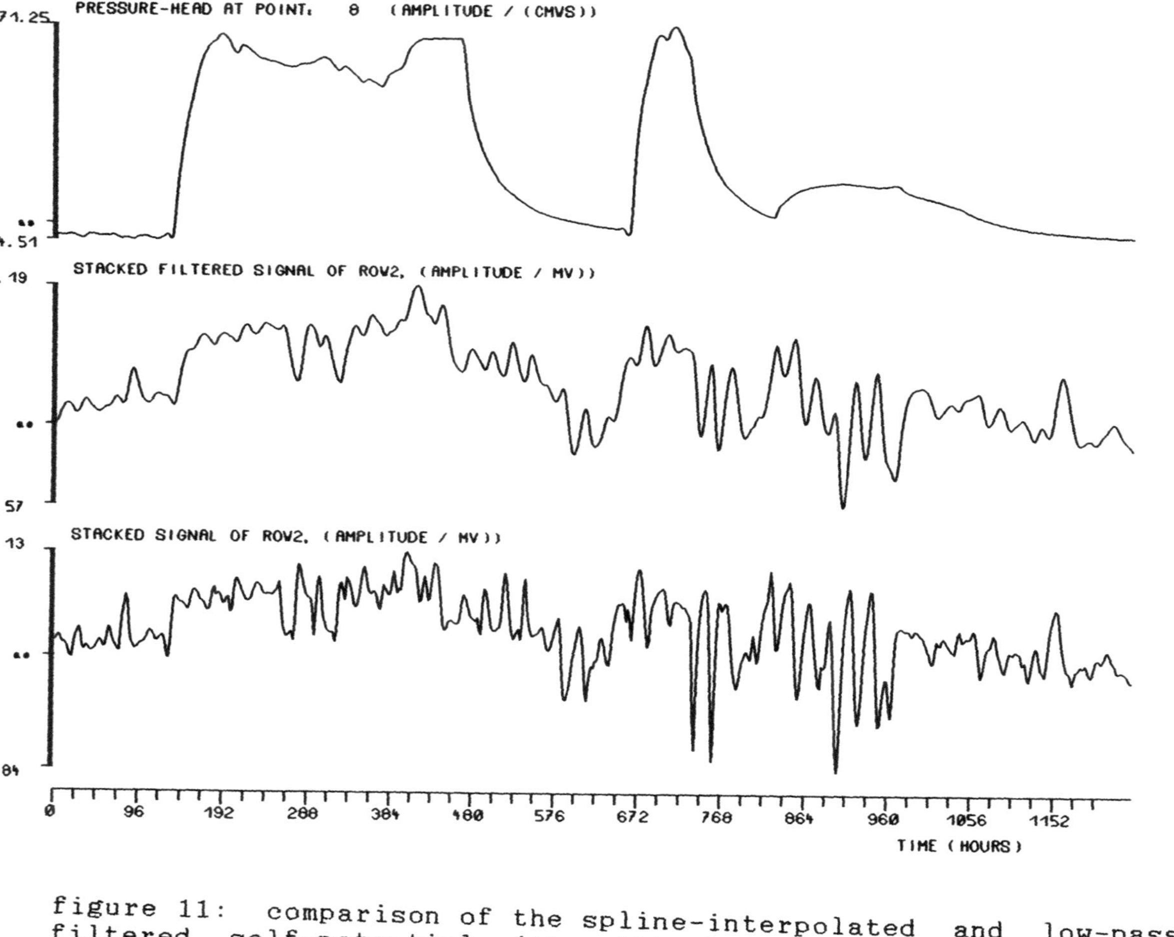

figure 11: comparison of the spline-interpolated and low-pass-filtered self-potential data with the pressure-head data for the stacked data of row 2

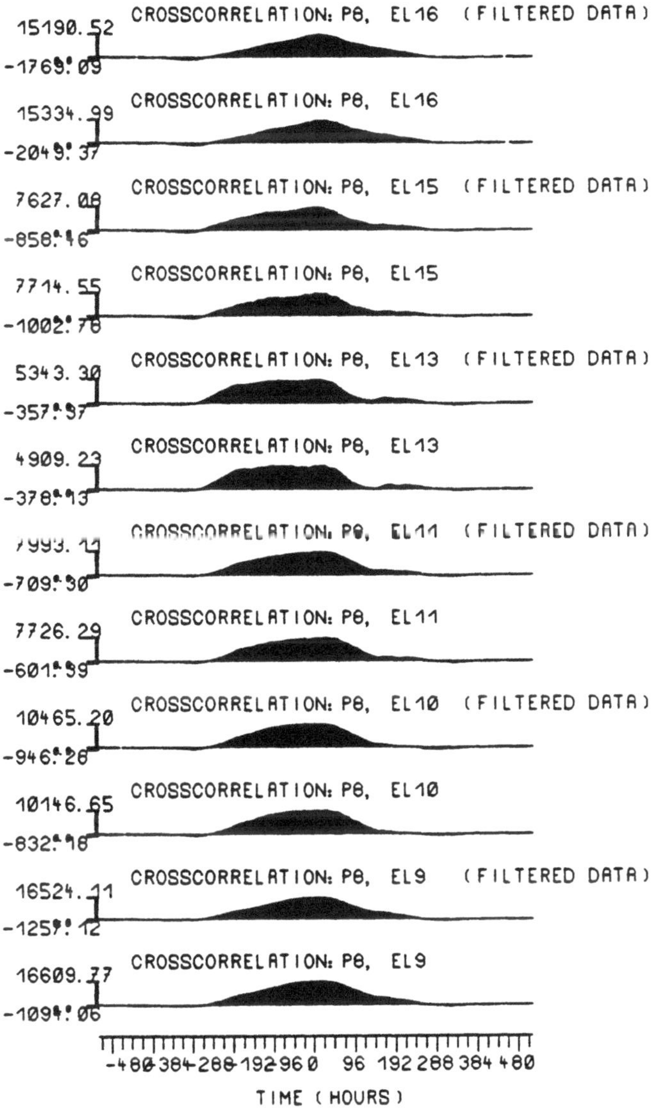

15190.52 CROSSCORRELATION: P8, EL16 (FILTERED DATA)
-1769.09

15334.99 CROSSCORRELATION: P8, EL16
-2049.37

7627.08 CROSSCORRELATION: P8, EL15 (FILTERED DATA)
-858.46

7714.55 CROSSCORRELATION: P8, EL15
-1002.78

5343.30 CROSSCORRELATION: P8, EL13 (FILTERED DATA)
-357.37

4909.23 CROSSCORRELATION: P8, EL13
-378.13

7993.14 CROSSCORRELATION: P8, EL11 (FILTERED DATA)
-709.30

7726.29 CROSSCORRELATION: P8, EL11
-601.39

10465.20 CROSSCORRELATION: P8, EL10 (FILTERED DATA)
-946.26

10146.65 CROSSCORRELATION: P8, EL10
-832.16

16524.11 CROSSCORRELATION: P8, EL9 (FILTERED DATA)
-1257.12

16609.77 CROSSCORRELATION: P8, EL9
-1094.06

-480 384 288 192 96 0 96 192 288 384 480

TIME (HOURS)

EXPERIMENT 1

figure 12: comparison of the crosscorrelation between the pressure-head and the unfiltered and filtered self-potential data respectively for experiment 1

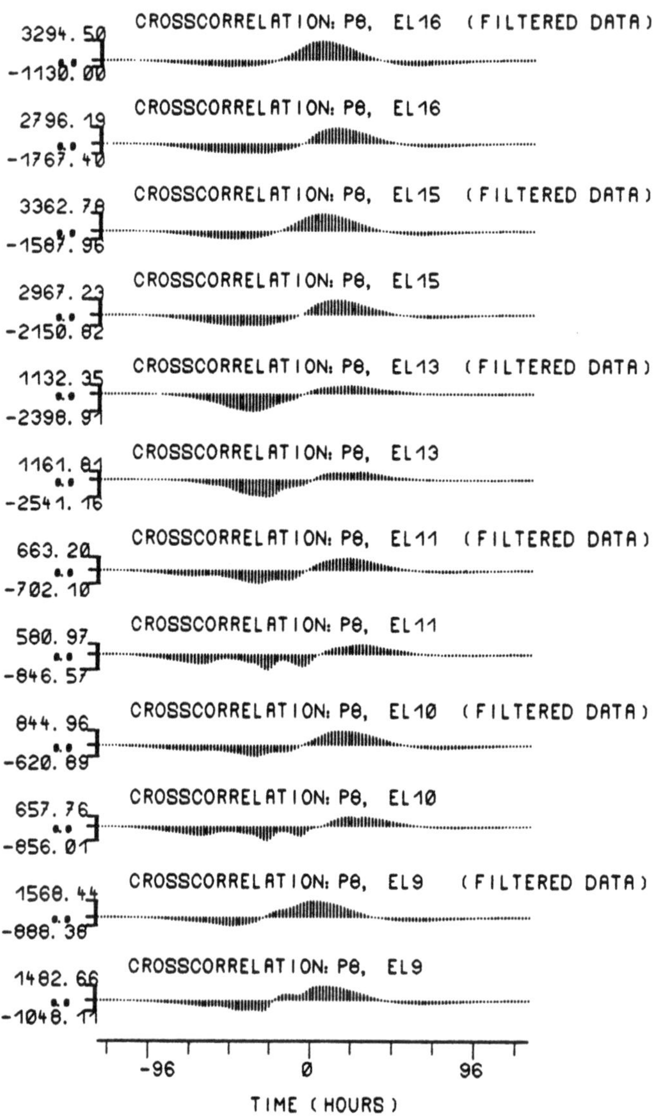

figure 13: comparison of the crosscorrelation between the pressure-head and the unfiltered and filtered self-potential data respectively for experiment 2

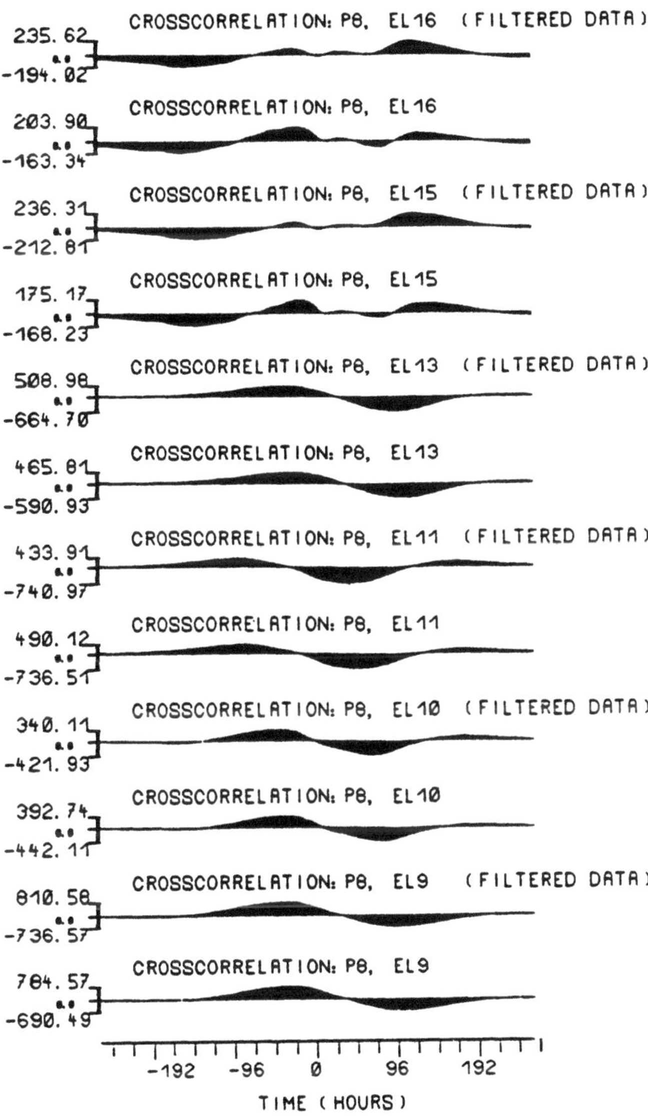

figure 14: comparison of the crosscorrelation between the pressure-head and the unfiltered and filtered self-potential data respectively for experiment 3

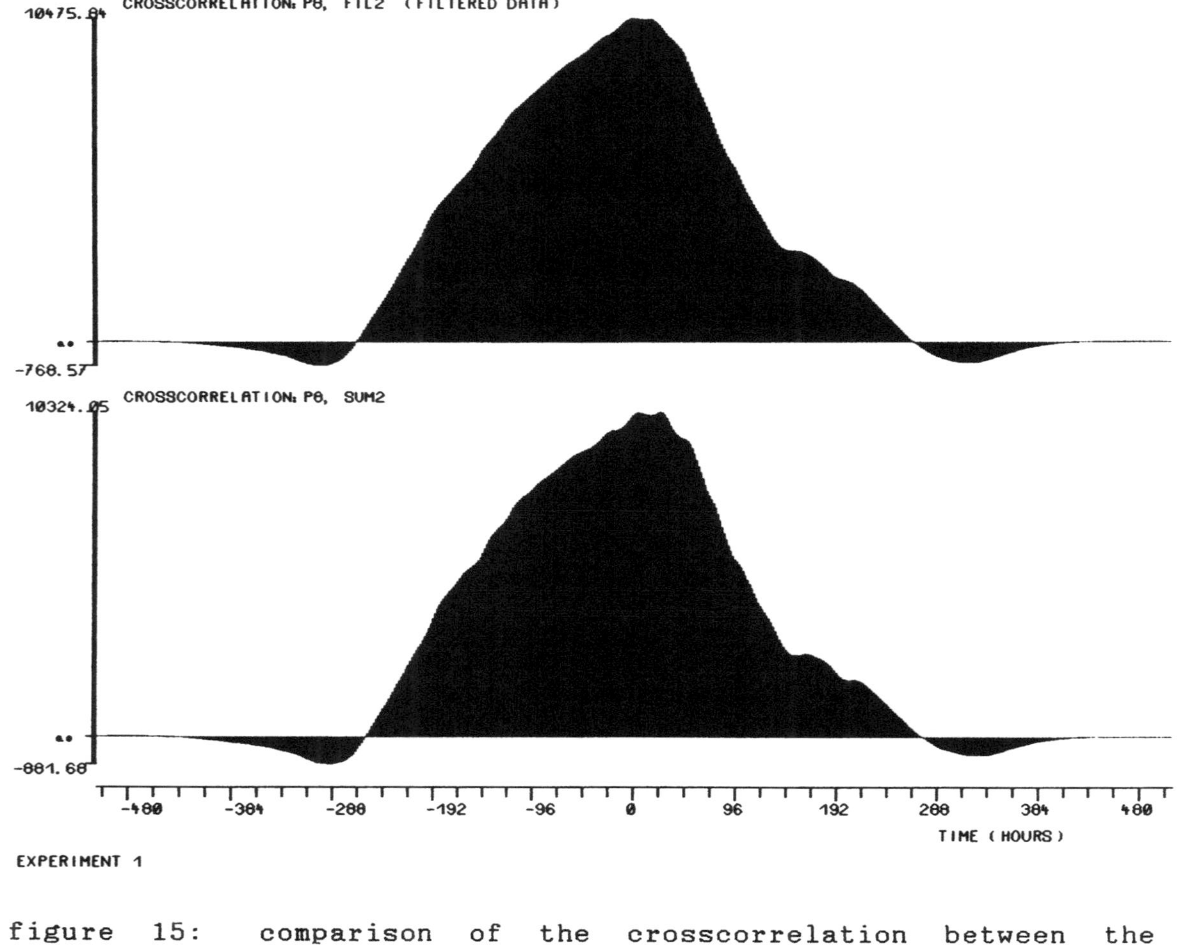

figure 15: comparison of the crosscorrelation between the pressure-head and the stacked unfiltered and filtered self-potential data of row 2 respectively for experiment 1

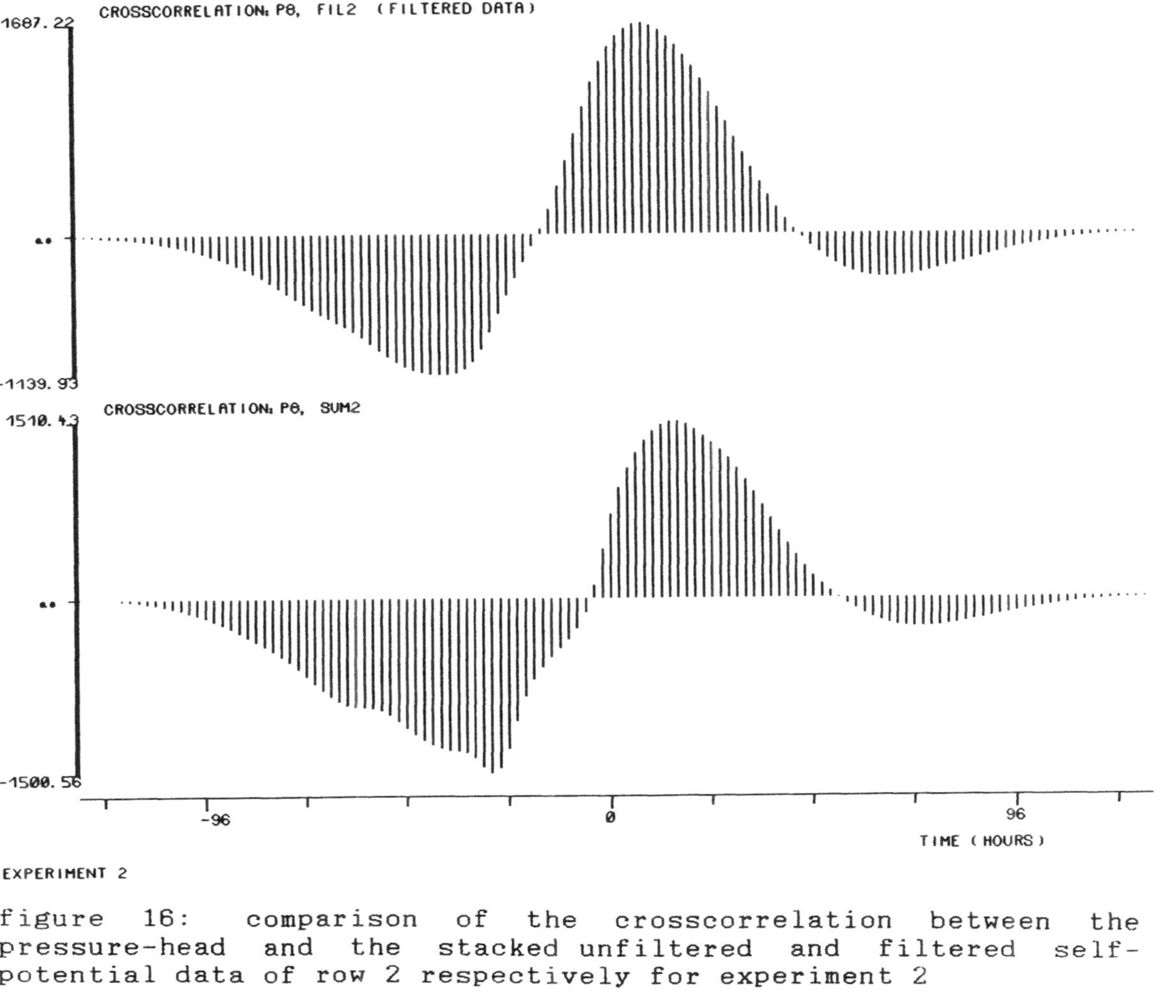

figure 16: comparison of the crosscorrelation between the pressure-head and the stacked unfiltered and filtered self-potential data of row 2 respectively for experiment 2

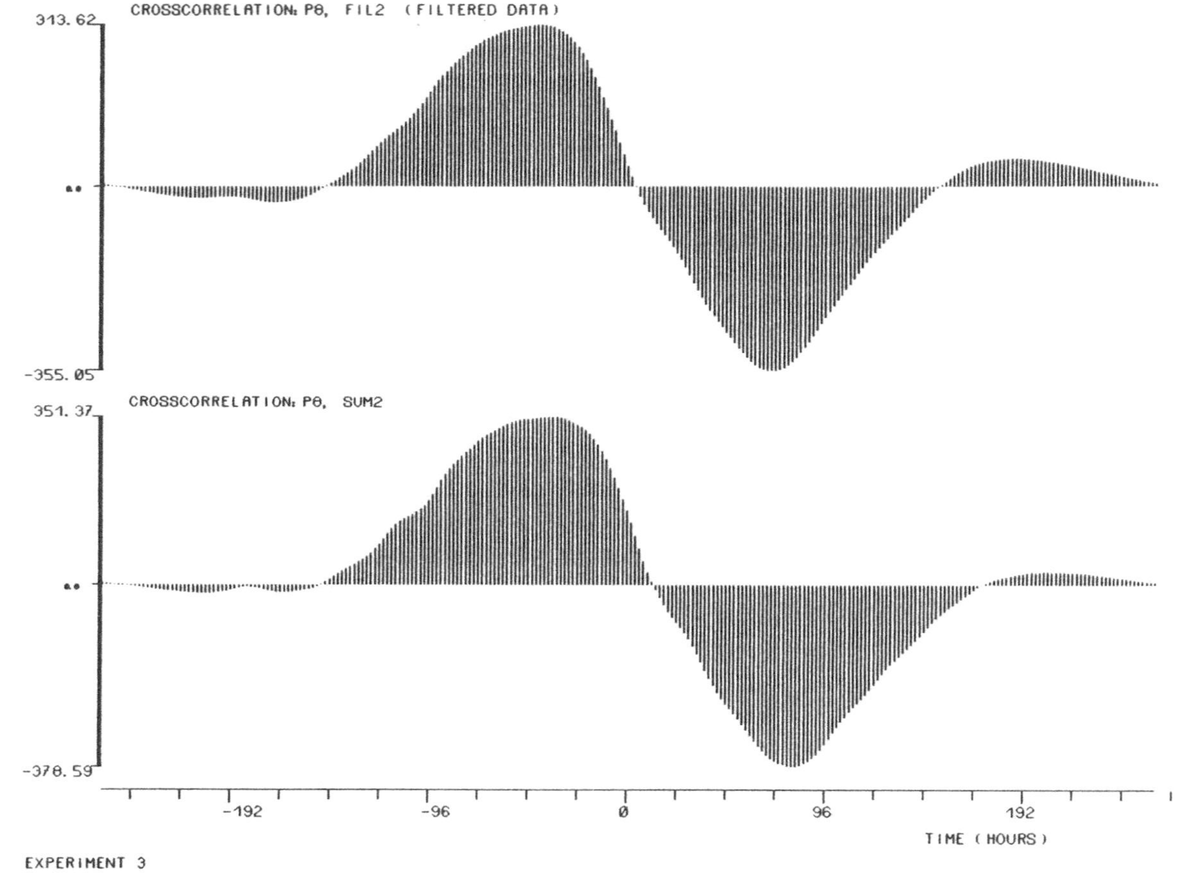

figure 17: comparison of the crosscorrelation between the pressure-head and the stacked unfiltered and filtered self-potential data of row 2 respectively for experiment 3

MODEL EXPERIMENTS ON A SMALL TEST CHANNEL. EMPIRICAL CORRELATIONS BETWEEN FLOW POTENTIALS AND THE HYDRAULIC FIELD

G.-P. Merkler and H. Hötzl[1]

Abstract

In connection with the geoelectrical investigations to locate leakages in dykes and dams, tests were carried out on a model made of sand in a small channel to measure the electrical streaming potential.

By using a PC-computer scanner-system it was possible to record simultaneously the values from 20 platinum electrodes and therefore also the electric field depending on the hydraulic streaming processes.

A direct correlation between the measured electric streaming potentials and the water flow processes could not be detected. With regard to the measuring results, it was shown that even under different measuring conditions the streaming potentials and the measured quantity of water seepage corresponded well.

Taking Darcy's law and Helmholtz equation into consideration, empirical methods of calculation were derived which made it possible to determine the quantity of water flow via the electrical streaming potentials measured.

During the experiments carried out on the model, the actual water flow quantities were consistent with the water flow quantities calculated from the electric field.

The empirically determined correlations between the electrical streaming potentials and the quantities of water flow, as well as the experiments performed will be briefly described.

---

[1] Lehrstuhl für Angewandte Geologie, Universität Karlsruhe, Kaiserstr. 12, D-7500 Karlsruhe

Lecture Notes in Earth Sciences, Vol. 27
G.-P. Merkler et al. (Eds.)
Detection of Subsurface Flow Phenomena

Introduction

Recent measurements and experiences have shown that self-potential anomalies in dykes and dams usually are caused by seepage phenomena (Bogolovsky and Ogilvy 1970, Armbruster and Merkler 1982 and 1983, Sill 1983, Merkler 1981, Merkler et al 1985).

It is not in all details clarified which physico-chemical phenomena cause electrical fields which usually occur under natural conditions. For this reason, no direct correlations between the degree of anomalies and the flow quantity or flow velocities in the underground could have been determined yet.

In order to simulate streaming potentials under natural conditions a dam model of quartz sand was built in an experimental channel. The water level and the degree of seepage were controlled and kept up constantly during the several intervalls of the experiments.

The electrical potentials were measured by using platinum electrodes which were fixed within the quartz sand. A simultaneously recording of the self-potential measurements was possible by using a PC-computer-controlled scanner-system.

This paper introduces briefly the experimental plants, the laboratory experiments and their results. It also reports about the deduced correlations between the quantity of water seepage and the measured self-potential values.

General Principles

Self-potential anomalies associated with subsurface flow phenomena are usually small in extent ($\leq$ 10 mV) and are often influenced by disturbing potentials, e.g. electrochemical, redox or diffusion potentials. It is difficult to quantify streaming potentials under this conditions in the field.

Fig. 1 Experimental plants
1: plexiglass channel; 2: PC-computer-controlled HP-scanner; 3: PC-computer; 4: printer; 5: floppy drive; 6: water conductivity measuring gauge

During seepage experiments with laboratory models it was tried to avoid those influences. The aim was to find correlations between measured self-potential respectively streaming potentials, and water seepage, water flow quantity and flow velocities.

The plexiglass experimental channel was 1,40 x 0,75 x 0,55 m in extent. The dam model (Fig. 1) was built with quartz sand (Ø= 0,2 - 1,0 mm).

To keep the hydraulic values constant, it was tried to avoid water level changes in the area of the reservoir exceeding ± 0,5 cm .

The measurements in the sand model in the channel were carried out with 18 to 20 platinum electrodes. The electrodes were put into the sand 2 cm below the surface and were fixed by punched strips.

Fig. 2 shows the top view of the channel with the distribution of electrodes.

In Fig. 3 the distribution of electrodes is shown in a vertical and a horizontal section.

Fig. 2 Sand model and distribution of electrodes

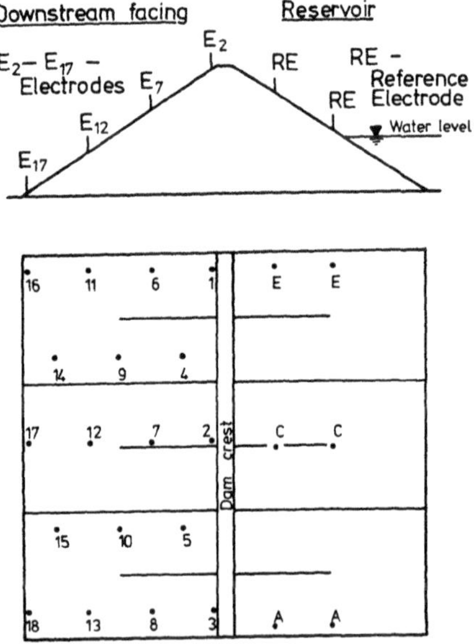

Fig. 3 Schematic view of distribution of electrodes.
A, C, E: different positions of the reference electrode

Implementation of Experiments

The upstream of the model was very carefully filled with water until the maximum water level of 23 cm to ensure stationary streaming conditions and to avoid disturbance values caused by vertical (capillary) water movement. Before the measuring took place, the water level was kept constant for about 1 hour. For the measuring under lower water conditions, the water level was reduced slowly and kept constantly for again 1 hour before the measurements were started again.

The period of one measurement of the whole net of electrodes was limited to 20 msec per electrode using the scanner system. By measuring in a time interval of maximum 400 msec (20 electrodes) a simultaneous registration of the electrical field at the downstream face of the dam model is given. Including the saving and the listing of the results one measurement cycle lasted between 3 and 6 minutes.

Analyzing the results it was taken into consideration that the electrodes which were near to the channel margin could cause disturbed values resulting from different permeability in the contact zone between sand and plexiglass.

Due to this, the examples described later on, refer only to the middle row of electrodes which is fixed in the longitudinal axis (electrodes 2, 7, 12, 17).

Laboratory Studies

The results of the experiments are shown in Fig. 4a and b.

To get an impression of the course of seepage lines in the model the seepage line for a homogeneous dam on an impermeable underground was calculated according to Pavlowsky-Dachler (Davidenkoff 1964).

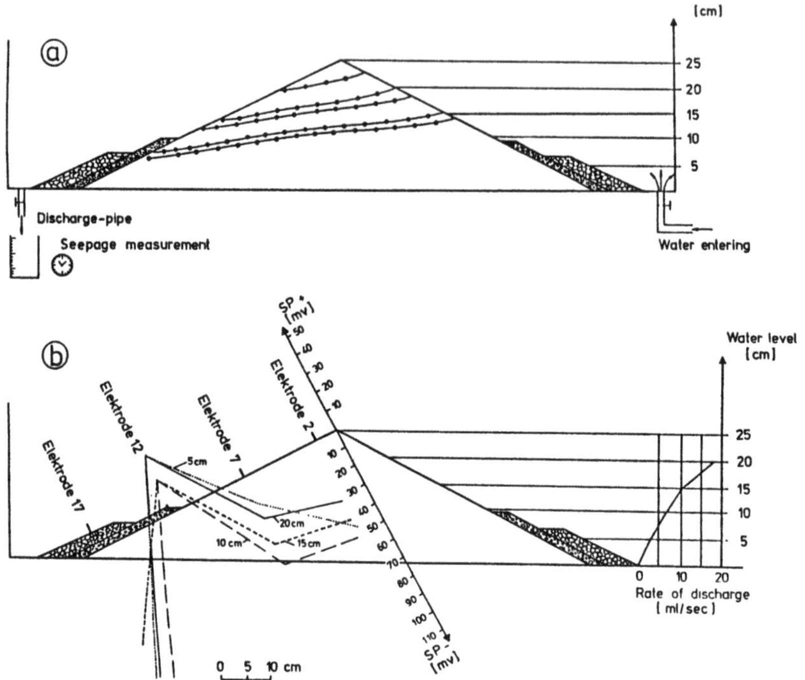

Fig. 4a Computed theoretical seepage line according to Pavlowsky-Dachler

4b Diagram of self-potential values depending on water levels and water discharge

The calculation of the theoretical seepage line was effected using Eq. (1):

$$j^2 = (H-h_I)^2 - 2\ h_{III}/m_b \cdot x = h^2 - 2\ h_{III}/m_b \cdot x \qquad (1)$$

Fig. 5 helps to recognize the quantities used in Eq. (1).

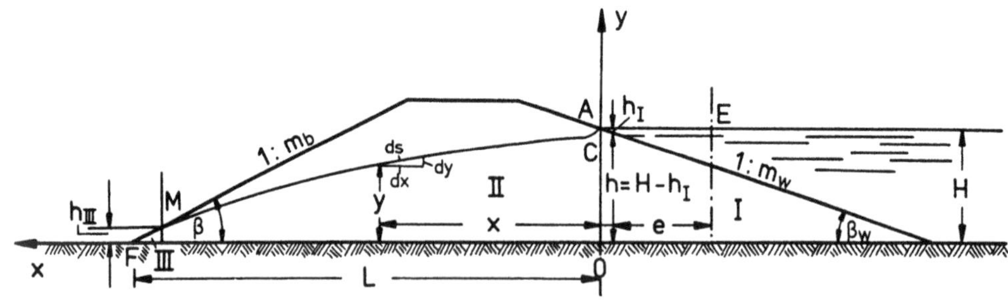

Fig. 5 Values used for the calculation of the seepage line after Pavlowsky-Dachler (Davidenkoff 1964)

From a scaled drawing of the model (Fig. 5) the following values can be derived: L and H and also m_w and m_b from the correlations m_w = cot β_w and m_b = cot β_b. For the symmetrical model is $m_b = m_w$ = cot β.

From an auxiliary diagram (Fig. 6) one gets the additional values h_{III} and h_I, which are necessary for the calculation. Other values which the diagram contains, such as L/m_bH and $m_b[1,12 + (1,93/m_w)]$, can be computed.

With the two computed values for the corresponding seepage line the values for h_{III}/H and h_I/H can be taken from the auxiliary diagram in Fig. 6; $h = H - h_I$ can be computed, too. Therefore, all necessary values to compute Eq. (1) for the seepage line are known, and for each X-value the corresponding Y-value is available. The seepage lines computed in Fig. 4a correspond to water levels of 13, 15, 17, 20 and 23 cm.

It is obvious (Fig. 4a), that one can assume the seepage line of a water level at the upstream face exceeding around 15 cm tends to issue above the toe drain.

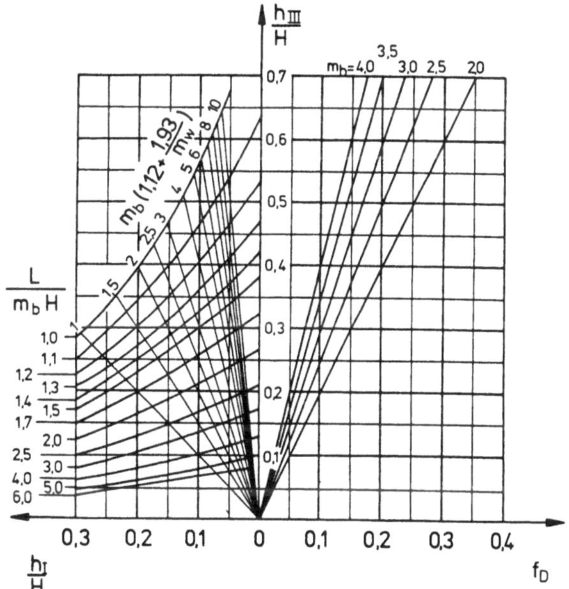

Fig. 6 Auxiliary diagram for computing of seepage lines (Davidenkoff 1964)

Below a water level of 13 cm the computing of the seepage line for the predescribed model and conditions (homogeneous dam on impermeable underground) is not possible.

The changing of the self-potential values for different water levels (5, 10, 15 and 20 cm) is shown in Fig. 4b. It is obvious that there is a hint of parallel course of self-potential values and a steep gradient/changing from negative to positive potentials, where the computed seepage lines (Fig. 4a) intersect the downstream face of the dam surface. This corresponds to the theoretical basis (Helmholtz-Smolukovski) of the origin of streaming potentials and ion transport or concentration of positive ions, carriers in the issued area of seepage lines. It indicates a comprehensive dependency of the streaming potential measurements on the water level (> 10 - 23 cm). Therefore, there is a correlation to the water flow quantities and, indirectly, to the water flow velocity. The fact, that there are no results for the theoretical seepage line of water levels < 10 cm, has to be regarded in correlation to the unreproducible self-potential measurements received from this domaine. It is said that this is caused by vertical water movements in the unsaturated zone. The evaporation can be of great and main importance. High negative potentials at the toe drain are said to be caused by the drain effect of the toe drain of gravel.

In Fig. 7 the measured self-potential values are plotted as a function of the water level. The correlation is a straight line, that means, the higher the water level is the more positive the electrical potentials are. Repeating measurements could confirm this.

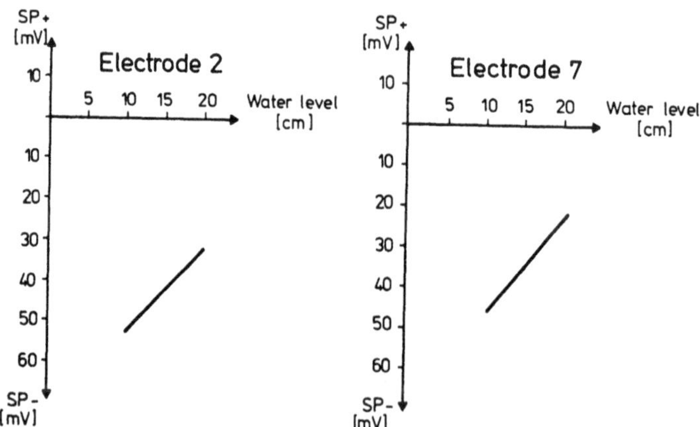

Fig. 7 Self-potential values plotted as a function of the water level at electrodes 2 and 7 (Bernhardt 1986)

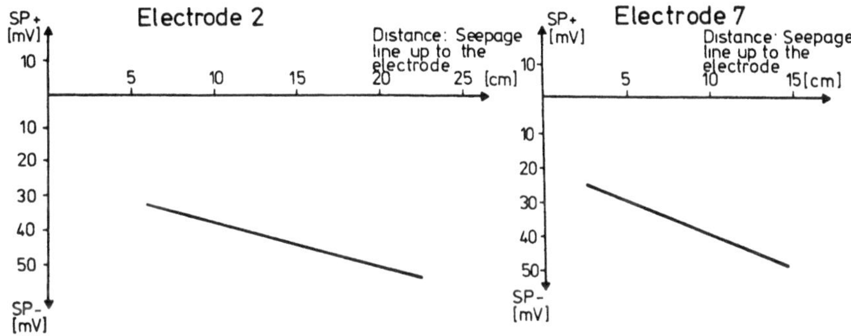

Fig. 8 Self-potential values for electrodes 2 and 7 as a function of the distance to the seepage line (Bernhardt 1986)

During experiments it was possible to note a dependency of the distance between the seepage line and the measuring electrodes on the self-potential values, but this correlation could not have been quantified.

As described before, for lower seepage lines the influence of soil water tension within the unsaturated zone, which can appear as disturbing potential, might be higher than the streaming potential itself.

From Fig. 8 the linear correlation between the distance of the seepage line to the electrode or to the surface of the model is evident. This correlation is valid for the range where the seepage lines are above 10 cm high. Therefore, the longer the distance to the seepage line the more negative the measured streaming potentials.

It was found out during measurements with varying positions of the reference electrode, that after a sufficient long time or with long measuring intervalls, respectively, the drift of all electrode values set in in the area of their start values.

Interpretation of Data

The measurements at the small experimental channel confirmed the dependency between the streaming potentials and the water level, respectively, the water flow velocity.

Those values which have confirmed the streaming potentials to be dependend on water level, water flow quantity and water flow velocity were taken as a reason to search for empirical correlations between the streaming potential and the hydraulic values of the experiments (Bernhardt 1986).

The Helmholtz equation (2) and Ohm's law for streaming potentials and Darcy's law for the water flow quantities were guiding the derivation of empirical calculation methods and the ascertaining of empirical correlations:

$$E = \frac{\xi \cdot \varepsilon \cdot P}{4\pi \cdot \eta \cdot \chi} \tag{2}$$

where E is the electrical streaming potential, ξ is the zeta potential, ε the fluid dielectric constant, P the pressure, η the dynamic fluid viscosity and χ the specific electrical conductivity of fluid, and

$$Q = F \cdot V_F \tag{3}$$

where Q is the quantity of discharge, F the surface and V_F the Darcian velocity.

Unfolding empirical correlations it was taken into consideration that they contain only values, which were directly measured during the experiments, e.g. the measured self-potential values, the measured conductivity of the electrolyte, the water level and the geometrical parameters of the model (height and surface of the dam).

Concerning Darcy's law the limits of validity, which are given by the Darcian velocity and the diameter of the capillars, have to be taken into consideration. Darcy's law only is valid for the laminar flow, not for high Darcian velocities (turbulent flow). It is also invalid for capillars $< 3\text{-}8.10^{-6}$ m. Within the limits of validity of Darcy's law the Darcian velocity has to be regarded as a linear function of pressure differences (Busch, Luckner 1974).

As the Darcian velocity is a linear function of the pressure difference, in the Helmholtz equation pressure was substituted by the quantity of Darcian velocity.

Under the described conditions the following derivations from Eqs. (2) and (3) proved to be suitable for the calculating of streaming potentials and water flow velocitites:

$$|E| = \frac{\rho \cdot V_F}{4\pi \cdot e \cdot \chi} K_B \qquad (4)$$

where $|E|$ is the absolute value of the streaming potential, ρ is the electrical resistivity, V_F is the Darcian velocity, χ is the specific electrical conductivity and K_B is the cross-coupling coefficient.

The cross-coupling coefficient, derived from experimental data, contains the geometry of the dam model which is defined for the surface where the water ran through (0,44 m^2), the ratio between the height of dam and the water level, the hydraulic potential (ΔH) and the porosity (%) of the model where the water ran through.

As described before the empirical correlations between the measured streaming potentials and the Darcian velocity was determined using Eq. (5):

$$E = \frac{|E| \cdot \chi \cdot \pi \cdot e}{\rho \cdot K_B} \qquad (5)$$

where V_F is the Darcian velocity, $|E|$ is the absolute value of the streaming potential, χ is the electrical conductivity of the fluid and ρ is the electrical resistivity.

To calculate the quantity of discharge with the help of the measured streaming potentials and the Darcian velocities (which can be calculated using this measurements) corresponding to Darcy's law it was necessary to solve the problem of the isolate self-potential measurements referring to the whole measured surface.

The electrodes at the downstream surface were distributed in a lattice. The surface was subdivided in such a manner, that each electrode represented its own raster field which is in extention dependent and proportional to the absolute value of the total potential field. Though the whole surface (downstream face of the model) was divided through the sum of absolute values of self-potential measured in each measuring cycle. This value made it possible to assign a certain surface to each electrode according to the measured self-potential value.

Height of the dam: 25.00 Date: 27.02.83
Water level: 20.00
Measured values: 17.10
[ML/SEC]

Measured Values	Calculated Values	Variation	Percentage
17.1	16.74	0.35	2.07
17.1	16.86	0.23	1.38
17.1	16.74	0.35	2.07
17.1	16.98	0.11	0.70
17.1	17.09	0.00	0.01
17.1	17.21	0.11	0.66
17.1	17.21	0.11	0.66
17.1	17.09	0.00	0.01
17.1	17.33	0.23	1.35
17.1	17.56	0.46	2.72

Height of the dam: 25.00 Date: 02.03.83
Water level: 15.00
Measured values: 9.80
[ML/SEC]

Measured Values	Calculated Values	Variation	Percentage
9.8	09.86	0.06	0.69
9.8	09.86	0.06	0.69
9.8	09.97	0.17	1.77
9.8	10.07	0.27	2.84
9.8	10.07	0.27	2.84
9.8	10.18	0.38	3.91
9.8	10.18	0.38	3.91
9.8	10.07	0.27	2.84
9.8	10.07	0.27	2.84
9.8	10.18	0.38	3.91

Height of the dam: 25.00 Date: 28.02.83
Water level: 10.00
Measured values: 5.30
[ML/SEC]

Measured Values	Calculated Values	Variation	Percentage
5.3	7.44	2.14	40.51
5.3	7.33	2.03	38.38
5.3	7.33	2.03	38.38
5.3	7.33	2.03	38.38
5.3	7.33	2.03	38.38
5.3	7.33	2.03	38.38
5.3	7.33	2.03	38.38
5.3	7.33	2.03	38.38
5.3	7.22	1.92	36.25
5.3	7.22	1.92	36.25

Tab. 1: Measured and calculated values compared by PC

To compare the quantities of discharge received by using the self-potential values, the quantity of seepage was regularly controlled manually.

Tab. 1 shows the calculated and the manually measured quantities of discharge.

There is a relatively satisfacting correspondence between the manual measured and - with the help of self-potential values - calculated quantities of discharge. The differences decrease if the quantity of discharge increases. This differences are partly said to be caused by inaccuracy in manual measurements of the quantity of discharge. The differences increase while the water level decreases, that indicates the dependency of the coherence of measuring values on the distance of the electrodes to the seepage line.

Repeated measurements at different times demonstrated that, if the water level is kept up constantly, the differences in measuring data are very small, even the differences in calculated quantities of discharge. They are caused by the variation of the water and air temperature influencing the streaming potentials. The influence of temperature has not yet been taken into consideration within the previous calculations.

Conclusions

In this paper the results of laboratory studies to determine empirical correlations between self-potential and quantity of water discharge are briefly presented.

The theoretical base of these correlations have to be proved, mainly by carrying through further model experiments.

Regarding the results, which have been received till now make us hope that further experiments in all probability could optimize this method, and that it would even be applied in in-situ-measurements in the future.

Literature

ARMBRUSTER H & MERKLER G P (1982) Möglichkeiten der Leckstellenortung an Erddämmen, Geotechnik, Vol. 1, 5, ISBN 0172-6945

ARMBRUSTER H & MERKLER G P (1983) Measurement of Subsoil Flow Phenomena by Thermic and Geoelectric Methods. Bull.Int.Assoc. of Eng.Geology, No. 26-27, Paris 1983

BERNHARDT R (1986) Hydrochemische und geophysikalische Untersuchungen zur Ortung von Leckagen und Sickerstellen an Dammabschnitten des Main-Donau-Kanals und an einem Durchströmungsmodell. Diplomarbeit, Lehrstuhl für Angewandte Geologie der Universität Karlsruhe, Karlsruhe 1986

BOGOSLOVSKY W A & OGILVY A A (1970) Application of Geophysical Methods for Studying the Technical Status of Earth Dams. Geophys. Prosp. Vol. 18, pp. 758-773

BUSCH K-F & LUCKNER L (1974) Geohydraulik für Studium und Praxis.- 442 p., 277 figs., 58 tabs.; Enke Verlag Stuttgart 1974

DAVIDENKOFF R (1964) Deiche und Erddämme.-115 p., 54 figs.; Werner Verlag Düsseldorf 1964

HELMHOLTZ H (1879) Studien über elektrische Grenzschichten.-Ann.Phys., (3), 7, pp. 337-382

KRAJEW A P (1957) Grundlagen der Geoelektrik.-358 p.; VEB Verlag Technik Berlin 1957

MERKLER G P (1981) Geoelektrische Messungen an Deichabschnitten im Bereich der Stauhaltung Iffezheim.-Unveröffentl. Bericht des Institutes f. Boden- und Felsmechanik, Abt. Erddammbau und Grundbau der Universität Karlsruhe

MERKLER G P, BLINDE A, ARMBRUSTER H, DÖSCHER H D (1985) Field Investigations for the Assessment of Permeability and Identification of Leakages in Dams and Dam Foundation. Proc. XV Congr. of Large Dams, Q .58 R7, pp. 125-141, Lausanne

MILITZER H (1953) Die elektrische Eigenpotentialmethode im Erzbergbau.-Bergbautechnik, 9: pp. 444-451

SCHUCH M & WANKE R (1967) Strömungsspannungen in einigen Torf- und Sandproben.-Z.f. Geophysik, 33, pp. 94-109

SCHUCH M & WANKE R (1968) Die zeitliche Variation der elektrischen Strömungsspannung in einem Fichtenstamm, verursacht durch die tägliche Änderung des Saftstromes.-Z.f. Oecologia Plantarium, 3, pp. 169-176

SCHUCH M & WANKE R (1969) Ein neues Verfahren zur Beobachtung der Wasserbewegung in Böden.-Deutsche Gewässerkdl. Mitteilg., Sonderheft 1969, pp. 37-40, 8 figs.

SILL W R (1983) Self Potential Modeling from Primary Flows.- Geophysics Vol. 48, No. 1, pp. 76-86

WILT M J & CORWIN R F (1988) Numerical modelling of Self-Potential Anomalies due to Leaky dams: Model and Field Examples. Proc.Detection of Subsurface Flow Phenomena by Self-Potential/Geoelectrical and Thermometrical Methods. Int.Symp.Karlsruhe, FRG, March 14-18th, 1988

ASPECTS CONCERNING THE RESULTS OF LABORATORY GEOELECTRIC MEASUREMENTS FOR THE STUDY OF HYDROGEOLOGICAL PHENOMENA

T.Moldoveanu[1] and P.Georgescu[2]

Abstract

This paper presents some technical and theoretical aspects of the study on simplified models carried out to establish possibilities for determining the fluid flow within the soil by means of surface electric measurements.

1 Introduction

The problem of achieving physical models at scale and their use in solving some technical and economical problems has recently become a current concern of researchers throughout the world.

The use of physical models in geophysics has gained increasing attention for a long time, but the achievements of scale models to aid in the geological interpretation of geophysical investigation data have not often been applied.

The geoelectric investigation, a geophysical method, in part self potential (SP) and vertical electric sounding (VES), has often enough made use of the physical model methods, especially in order to solve the problems raised by the methods of self-potential (SP)and vertical electric soundings (VES).

The present paper presents some technical and theoretical aspects of the study on reduced models carried out with the purpose of estab-

1 Institute of Hydroelectric Studies and Design, Bucharest, Romania

2 University of Bucharest, Romania

Lecture Notes in Earth Sciences, Vol. 27
G.-P. Merkler et al. (Eds.)
Detection of Subsurface Flow Phenomena

lishing possibilities for determining the fluid flow into the soil by means of surface electric measurements of SP.

2 Aspects Regarding Geoelectric Investigations in Modelling Tanks

The problem of geoelectric laboratory investigations arises from the necessity of knowing the distribution of the natural or created electric field or of other electric parameters under certain given conditions.

The knowledge of these distributions can facilitate the interpretation of the geoelectric investigation results under certain geological conditions.

The main way of performing the laboratory investigations is the following: a scale model is made and the geoelectric measurements are carried out using a work methodology similar to that used in the field.

The reduced models can be achieved in two ways:

1. Simplified models on the basis of a number of typical geological cases;

2. Building a model according to the geological structure assumed in the stage of interpretation of a geoelectric investigation.

An interpretation technique has practical interest if it can be rapidly applied and if the method is precise and not very expensive.

The model technique hardly meets the above conditions, therefore, it was used to a lesser extent.

The models are constructed as variable electric conductivity masses, installed in isolating tanks. The measuring electrodes are small rods of metal (Cu) or unpolarizable electrodes (Cu in oversaturated $CuSO_4$ solution) especially made for model measurements to thus observe the technical conditions of the point electrodes.

Prior to achieving a modelling tank, several technical-theoretical problems must be solved:

1. Size of the tank and quantity of necessary electrodes;

2. Marking of the points at which the measurements are made;

3. Influence of the isolating tank on the electric field.

The first and third problems are dependent on one another, since a sufficiently large portion of the tank, in which the electric field is to be determined, should remain undisturbed as far as possible.

The tests performed in a plastic tank of 140 x 120 x 60 cm showed that when the emission electrodes are closer than 25 cm to the tank walls, their influence on the central third of the emission device is greater than 10%, compared to the normal field distribution.

For solving the second problem, a system of points in a geometrical network is chosen in relation to a system of axes that are maintained unchanged for all the measurements.

It was considered hypothetically that the potential difference values due to fluid underground flow phenomena are superior to the potential differences occuring as a consequence of the small variations of temperature, moisture and pressure under laboratory conditions.

3 Self Potential Measurements on Reduced Dyke Models

3.1 Sand and Gravel Dykes

First, measurements were carried out on a model composed of a vertical sand wall (grain size of 0.15 - 1.0 mm) 1.6 m long, 0.25 m wide and 0.40 m high. The supporting sides of the wall are of pierced plastic material for providing free passage of the fluids. Under such circumstances the influence of the supporting walls on the potential distribution is quite large and almost impossible to estimate.

More determinations of the potential distribution at the upper part of the dyke were carried out for different hydrostatic levels. It was determined that at the same time with the level difference modifications between the two wall sides, significant changes of the self-potential distribution occur, but the maximum values of the potential differences remain almost unchanged (Fig.1).

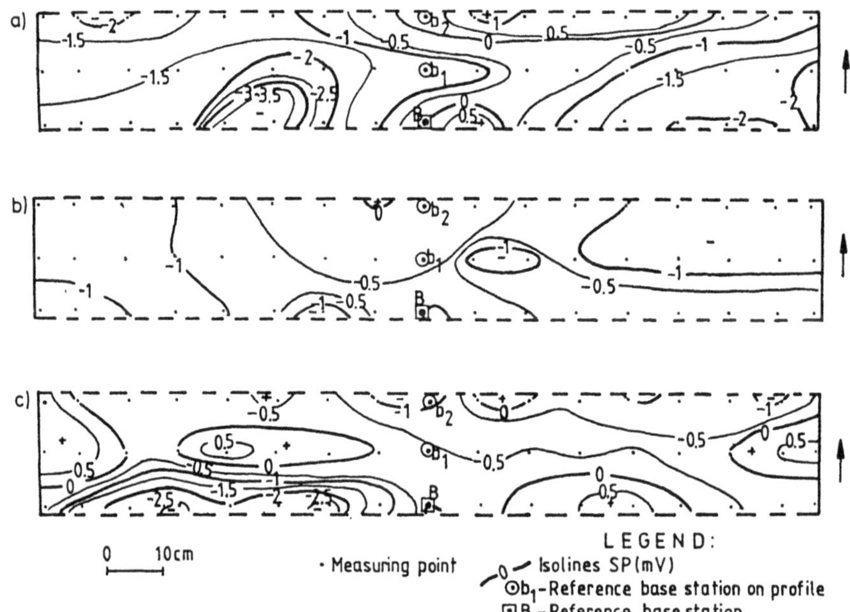

Fig. 1: SP maps obtained on a homogeneous sand model
a) dislevelling 0 cm
b) dislevelling 5 cm. Water flow direction from B to b_2.
c) dislevelling 9,4 cm. Water flow direction from B to b_2.

SP-value isoline deformation may be due to the porosity variation in the dyke assembly as a consequence of heterogeneous settlement. The measurements carried out on a gravel dyke also show SP-value deformations.

In order to estimate the environmental porosity influence on the potential distribution, a vertical sand-gravel interface was modelled in the very midst of the dyke. The measurements show a significant modification of the potential distribution.

It was found experimentally that in the previously presented cases the fluid flow velocity as a function of the environmental porosity is the parameter determining the essential modifications of the potential distribution.

It may be concluded that in the presented cases the predominant effect is generated by electrofiltration phenomena.

Since a uniform homogenization from the view point of the material settlement could not be achieved in the presented models, the correlation of the fluid flow velocity with the electrofiltration potential values is difficult. Subsequently, the necessity of obtaining and studying a dyke model on an advanced homogenization degree resulted.

3.2 Slag and Ash Dykes

The construction of slag and ash dykes appeared simultaneously with the construction of some large thermal power stations which burn inferior coal as a source of energy. The quantities of slag and ash resulting from burning have to be stored vertically, so as not to cover great amounts of ground surface. For the transport of scour and ashes, hydromechanical methods are used. Thus, settling ponds are necessary.

Due to the unsatisfactory results of rockfill dykes, this problem was intended to be solved by the construction of contour dykes and excessively increasing of slag and ash.

The purpose of laboratory investigations was to find an optimum dyke section consisting of slag and ash.

The main problem is represented by the velocity of the seepage flow which, from a certain value on, begins to wash up the fine fraction of ashes, thus endangering the local stability of the dyke and then of the entire dump of slag and ash.

The slag and ash model was carried out on a scale of 1:10, thus observing the nature of the material, the grain size distribution and the flow gradients known from the actually existing slag and ash dykes.

The foundation layer was achieved from slag and ash prepared by hydro-mechanical methods, permanently changing the point of slurry outlet. A stratification and a grain size distribution similar to the existing one were obtained.
As fill material for the dyke, slag and ash from the dump were used, and for the filtering zone, slag alone was used.

In order to establish the hydrostatic level, free-level piezometers were used, being installed in different sections of the model (Fig. 2, items 1, 2, 3, 4)

The main characteristics of the materials used in the model construction are presented in Table 1.

On the model achieved in this way, several tests were performed. The average hydrostatic levels and the average discharges are given in Fig. 2.

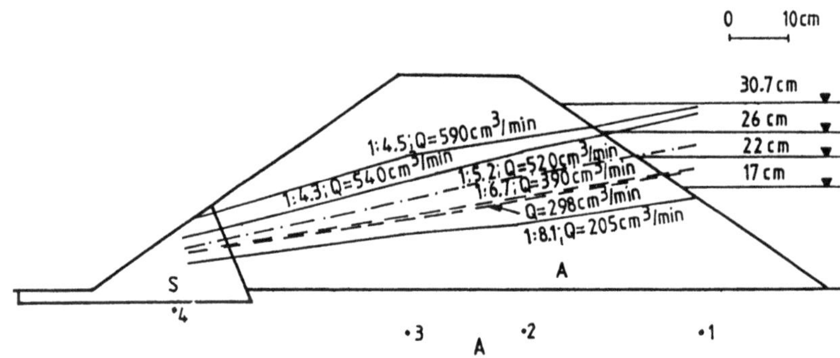

Fig. 2: Dyke hydraulic model made of ashes
A Ashes
B Slag
1 ...4 Piezometers

On this dyke model several series of measurements of self potential (SP) were made at different hydrostatic levels, taking into account that these different conditions could influence the spatial distribution of the natural electric field.

Table 1: Characteristics of the materials composing the model shown in Fig. 2.

Materials	Wet density (γ_W) (g/cm^3)	Porosity (n) (%)	Grain size 0-0.25mm (%)	0.25-1.0mm (%)	Permeability K_f (cm/s)	Ø*	c*
Hydromechanically laid ashes and slag	1.25	50	55	45	5×10^{-4}	30°	0
Compact ashes and slag (dyke)	1.25	50	55	45	2×10^{-4}	30-40°	0
Slag (filter)	1.33	50	-	100	2×10^{-3}	-	-
Ø* = Angle of internal friction; c = cohesion							

SP measurements were performed at the points situated in a square geometric network with 0,10 m sides. The measurement values are related to the "B" reference electrode, having 56 measuring points (Fig. 3).

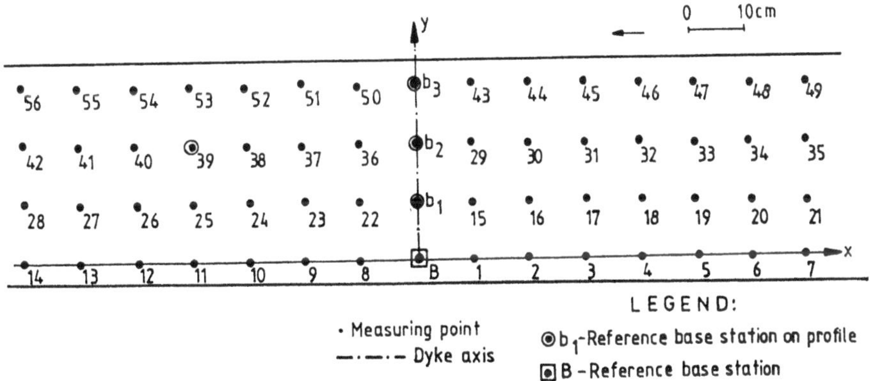

Fig. 3: Sketch with location of SP network measuring points.

Besides these measurements, other SP measurements were performed between two fixed points situated on the line of the steepest slope of the downstream abutment of the dyke ($b_2 \rightarrow 39$), in order to determine a variation in time of the potential difference measured with the variation of the hydrodynamic conditions.

The SP map (Fig. 4) obtained based on the measurements carried out before the occurence of the flow through the dyke body, presents two distinct ranges, a positive range located on the upstream field and a negative range located on the downstream slope. The zero line follows the dyke crest.

The SP-value variation range is -50 to 50 mV. Subsequently, there is a differentiation of the dyke moisture content. The measurements were carried out at the hydrostatic levels of: 22, 26 and 30,70 cm. All these measurements were performed twice, at a time interval of 24 h, the first measurement being carried out at about 30 min after stabilizing the piezometer (Figs. 5a, b; 6 a,b).

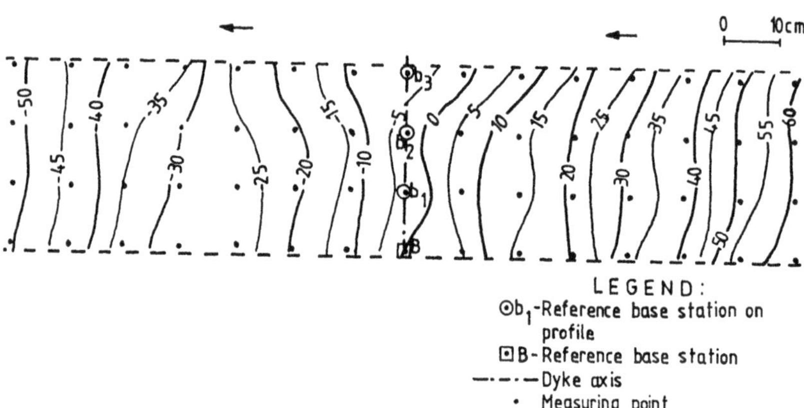

Fig. 4: Distribution map in the case of no water.

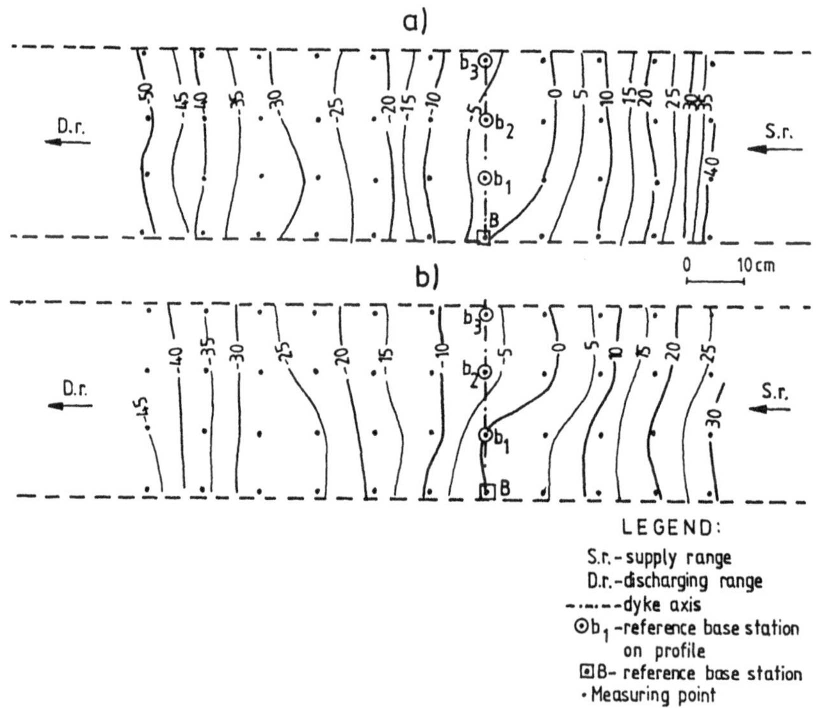

Fig. 5: SP distribution map at level +17 cm
a) at 30 min b) at 24 h

On all these SP maps, as well as on the map presented in Fig. 4, two ranges are evident: a positive range located the upstream slope and a negative one located on the downstream slope.

The analysis of the maps obtained on the basis of the SP measurements carried out immediately after stabilization (after about 30 min; Figs. 5, 6) shows that as the dynamic level increase, a more marked orientation of the field takes place.

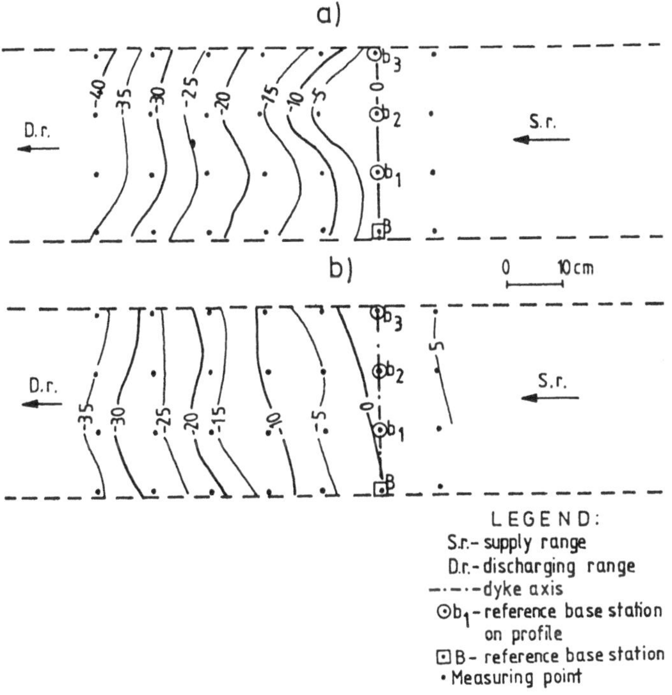

Fig. 6: SP distribution map at level +30,7 cm
a) at 30 min b) at 24 h

The SP map achieved at 24-h intervals (Figs. 5, 6) are characterized by the fact that the difference between the maximum positive and maximum negative values decreases by 5-10 mV. These maps are more similar to the maps of Fig. 4.

The results obtained on the SP maps indicate the existence of a certain connection between the flow velocity, the hydrostatic level height and the electric field distribution.

The second SP measurement group was intended to render evident the variation in time of the potential difference due to the modification of the hydrostatic level (namely of the dislevelling upstream of the dam) and due to the flow velocity.
In view of the short time necessary for the stabilization of the hydrodynamic regime, these measurements were performed every 2 min over a time interval of 20 min (Fig. 7). A jump of 5-8 mV was evident on the line of steepest inclination of the downstream slope between the points b2 und 39 (Fig. 3).

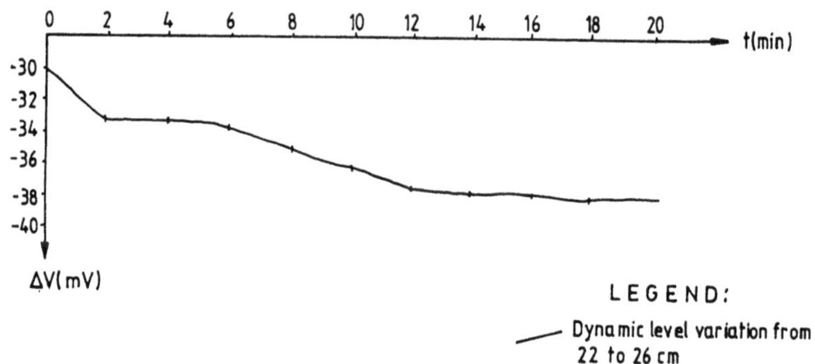

Fig. 7: SP time variation profiles.

From the analysis of the SP measurements performed on the model, the following conclusions can be drawn:

1. The natural electric field distribution is influenced by the flow through the dyke.

2. When the level of the seepage line is at greater depth, the SP values are lower.

3. By spatial displacement of the hydraulic field, an increase of the natural electric field occurs.

4 Conclusions

From the results obtained in the laboratory investigations, which were also confirmed by the field investigations, it is evident that the connection between the underground flow and the electric field distribution is measurable at the soil surface. The flow, to be determined, should not be at a great depth in relation to the surface on which the electric field distribution is measured and should have a sufficiently high velocity. In conclusion, the electric field distribution is influenced by both the flow and the environment through which it passes.

The measurements also show the existence of certain grain and settlement heterogeneities in the potential distribution. However, an SP anomaly cannot be related to the existence of fluid flow through the rock pores.

Significant SP anomalies were evident under flow-absent conditions and in the presence of moisture variation in the environment.

Thus, the SP measurements carried out on laboratory models may contribute to the investigations of hydrogeological phenomena and to more accurate interpretation of the field measurements.

References

Constantinescu, P.; Moldoveanu, T.; Stefănescu, D.; Vîjdea, V.; Visarion, V. (1979): Eng. Geophys, Ed. Tech., Bucharest.

Georgescu, P.; Ionescu, V.; Moldoveanu, T. (1974): Aspects of electric measurement use for hydrogeological phenomenon study.- 8th Symp. Geophysical investigations and earth physics, Bucharest, Rom.

Ionescu, V. (1974): Physical and mathematical modelling for hydrogeology problem solving.- Grad. Pap. Univ. Bucharest, Fac. Tech. Geol.

Merkler, G.P.; Blinde, A.; Armbruster, H.; Döscher, H.D. (1985): Field investigations for the assessment of permeability and identification of leakages in dams and dam foundations.- 15th Congr. Large dams, Lausanne.

MODELLING OF STREAMING POTENTIALS AND THERMOMETRICAL MEASUREMENTS AT A BIG LABORATORY CHANNEL

Merkler G P[2], Armbruster H[1], Hötzl H[2], Marschall P[3], Kassel A[2], Ungar E[2]

Abstract

The University of Karlsruhe, represented by the Department of Applied Geology, the Institute of Soil and Rock Mechanics and the Federal Waterways Engineering and Research Institute (BAW), Karlsruhe, have been cooperating in large-scaled research concerning the hydraulics and geoelectrics in thermic fields, with special interest on the detection of leakages of dykes and dams.

From 1984 to 1988 the Volkswagen Foundation supported a research project whit the intention to determine the relationship between the hydraulic field and the appertaining thermic and electric field.

The construction of a special experimental channel (6m x 2m x 1,5m) is described in this paper. Particular attention was paid to obtain optimal feasibility of measuring the relevant electric and thermic fields in connection with the change of the hydraulic field.

In a sand model 140 thermosensitive sensors (Pt 100), one water discharge measuring sensor, 80 pressure transducers (piezometers), 20 electrodes of platinum, 25 electrodes of copper sulfate and five tensiometers were installed. All the sensors and electrodes were connected with an automatic PC-controlled scanner system.

Several streaming tests were carried out in sand models built within the channel. They lead to ascertainment about the variation of the electric field (self potential) and the thermic field in relation to the hydraulic field.

1 Bundesanstalt für Wasserbau (BAW), D-7500 Karlsruhe

2 Angewandte Geologie, Universität Karlsruhe, D-7500 Karlsruhe

3 Institut für Wasserbau, Universität Stuttgart, D-7000 Stuttgart

Lecture Notes in Earth Sciences, Vol. 27
G.-P. Merkler et al. (Eds.)
Detection of Subsurface Flow Phenomena

The results as well as the technical problems occuring during the experiments are presented and discussed briefly in this paper.

Introduction

As it is generally known, flow phenomena of water, in the way they occur in nature or are caused by leakages in dams, are manifestated by disturbances of the natural electrical and thermal field. The obvious idea was to make assumptions on the state of dams by measuring geoelectrical fields. In case of success we would deliever an economical method which works quickly and without doing damage to the dam. The inhomogenity of dams caused by accretion, settlement and other uncontrollable processes, and the complexity of the physical and mathematical relations make it necessary for the quantification of the connections between hydraulic, electrical and thermic field to fall back to a homogeneous model. Therefore a few years ago in the Institute of Applied Geology (Merkler & Hötzl 1988) a channel was constructed of the dimensions 1.4m x 0.75m x 0.55m. A sand model was filled into this channel and used to carry out self potential measurements. Though the results seemed to be promissing, they were partly influenced by marginal effects due to the small dimensions of the model. Consequently, it was decided to built a bigger channel, a project sponsored by the Volkswagen Foundation.

How could we define our aims and, furthermore, in which way could they be achieved? Self potentials, hydraulic potentials and temperatures should be measured under controlled conditions. This was attained with the help of a laboratory model dam with a measuring device, which will be introduced later. The quantification of the correlations between the single fields is strived to verify those results at an open-air-model. This open-air dam exists at the Federal Waterways Engineering and Research Institute in Karlsruhe. The test channel is also located there, in a hall near the open-air model (Armbruster et al. 1988).

Experimental Plants

In Fig. 1 the channel is shown as it exists in the Federal Waterways Engineering and Research Institute. The dimensions of the channel should offer the possibility of modelling two- and three-dimensional processes without the difficulty of marginal effects. This modelling refers hydraulic processes and the thermal and electric field connected with them.

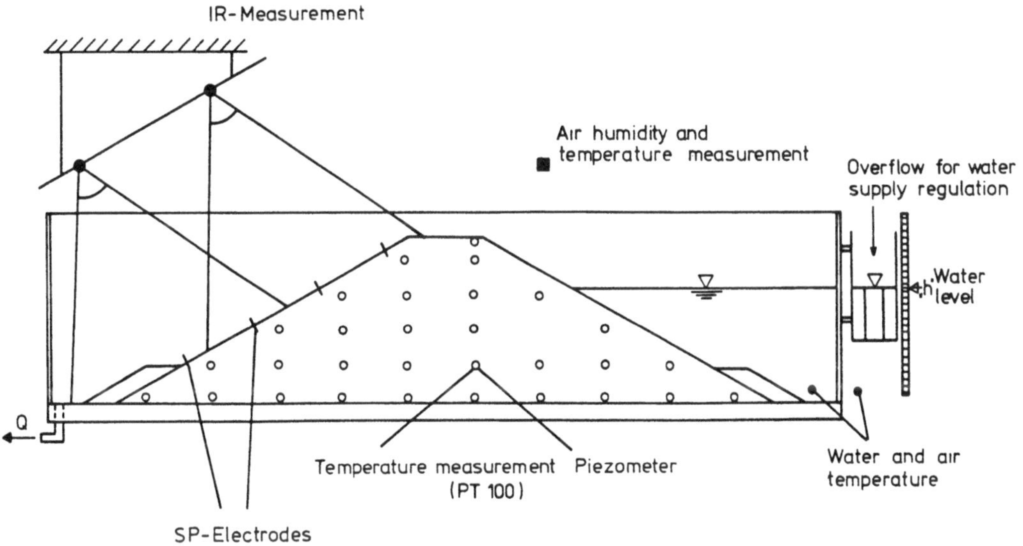

Fig. 1: Channel with the dam model and the measuring systems

The hydraulic field is controlled by piezometers or pressure transducers, respectively. With the help of an overflow for water supply regulation we are able to keep the heights and amounts of water in the channel relatively constant. A flow measuring device is located at the outlet.

Therefore it was possible to make statements on the amount of water flowing through the channel at any given time.

The hydraulic potentials in different profiles and heights are gathered along pipes directly at the channel. The hydraulic potential was registrated with the help of piezometer tubes and pressure transducers. This system enabled us to registrate automatically and simultaneously, as it was necessary.

Only deaerated water is led into the channel over a big deaeration equipement and a gravel filter to ensure that the permeability of the sand is not substantially affected.

To register the thermal field, thermoresistive sensors (Pt 100) had been installed. They were located next to the piezometers in the inner part of the sand body. These thermometers work with a precision of $\pm$ 0,5°C, as calibrating measurements have shown.

The surface temperature of the dam body was measured with the help of a infrared camera working in the range between 2 and 5 µm.

Kork plates of 10 cm thickness were installed at the bottom of the channel, between two plexiglass layers, for thermal isolation matters.

The electric field was measured with the help of 45 platinum electrodes. They were fixed in certain profiles at the dam bodies downstream side. The coupling was achieved by a wet clay layer (bentonite) of good conductibility. The potential difference was measured by platinum electrodes versus three reference electrodes in the dam body; they could be measured alternatively.

As the use of metal at the construction of the channel could interfere with the geoelectrical signals, no metal was used at the construction in the inner parts of the channel, in order not to disturb the electrical field.

This led to the following compromise:

A plexiglass channel in a steel girder construction was built, the plexiglass plates were weld together. The dimensions of the channel were 6 meters length, 1.99 meters broadth and 1.38 meters height. The volume of the empty channel was about 16.5 cubicmeters. The volume of the first sand body (sand model) amounted about 7 m^3. The volume of the water filled inside was about 2.5 m^3. In the dam model 80 piezometers and 124 resistance thermometers were applied. On the downstream face of the dam 45 self potential electrodes were installed. With this equipment we are able to model the hydraulic, thermic and electric fields by different cross- and vertical sections.

With the lack of metal within his inner parts and its dimensions our model channel is the biggest one of his kind.

Fig. 2 shows the front view of the channel with the dam model.

Fig. 2: The channel with the dam model

The technical devices for the measurement of the desired values. e.g. the water discharge quantities, as there are the pressure sensors, the temperature sensors and the self potential electrodes, are connected with a Hewlett Packard scanner with extender. The scanner is controlled by a Hewlett Packard 86A personal computer, hence the data can be stored on discs. Simultaneously these data are printed. The help of this measuring system guarantees an almost simultaneously registration of the three fields.

On the downstream face of the dam model, 45 platinum electrodes are installed (see Fig. 3).

Self potentials are measured versus a reference electrode located in the middle part of the dam crest.

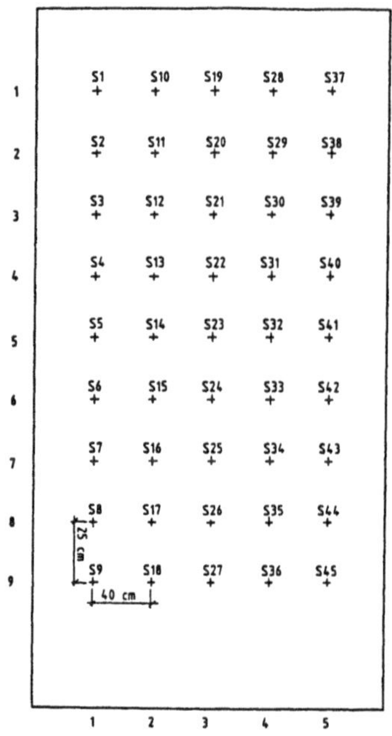

Fig. 3: Distribution of the self potential measuring points at the downstream face of the dam

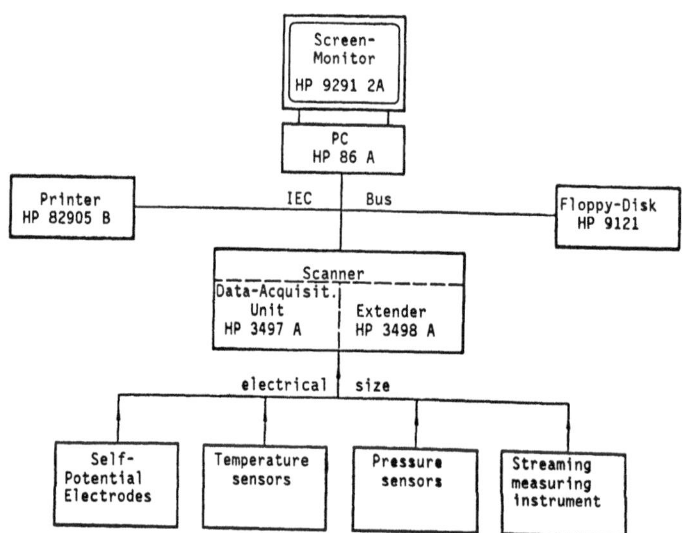

Fig. 4: Principle circuit diagram of the scanner system

In Fig. 4 the principle circuit diagram of the scanner system is figured. As described before, this system made it possible to record simultaneously all measuring values.

A second, plain model with a gravel drain was built after termination of the former dam model experiments. Fig. 5 is a schematic representation of this model.

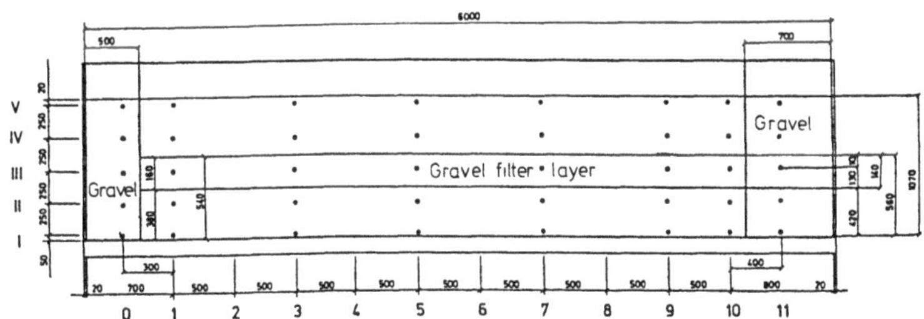

Fig. 5: Longitudinal section of the plain model of sand with a gravel drain

As illustrated in Fig. 5 the model consists of a sand body supported by a gravel body at the ranges of supply (right hand side) and outlet (left hand side) of water. A gravel drain of 16 cm height and ca. 40 cm broadth should guarantee a higher flow velocity. At the same time it was tried to achieve a low hydraulic pressure drop in the middle range of the cross-section, above the filter.

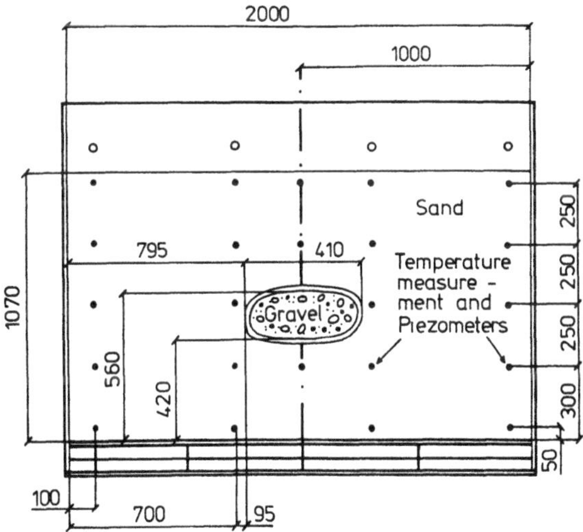

Fig. 6: Cross-section of the plain model

The installation of piezometers and temperature sensors within this plain model corresponded to that of the former dam model. The self potential electrodes were distributed allover the surface in a lattice.

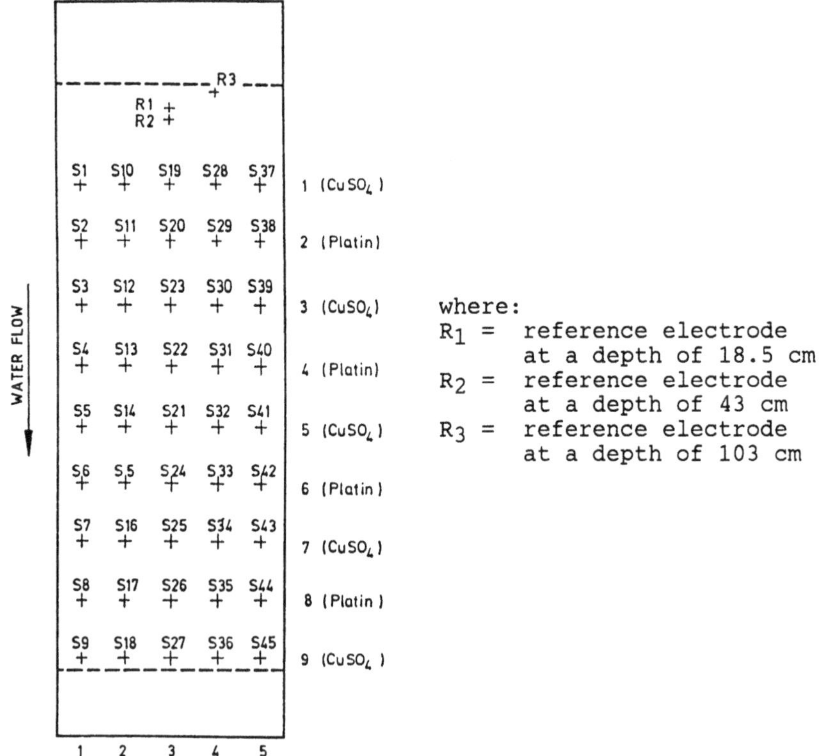

Fig. 7: Distribution of the self potential electrodes on the surface of the plain model

In this model 20 electrodes of platinum and 25 copper sulfate electrodes were used. Three reference electrodes ($CuSO_4$) were fixed in different depths (about 20, 45 and 103 cm) in the domaine of the water inflow.

The PC-controlled measuring system was able to registrate, save and print the almost 280 measuring values within about 3 minutes. Therefore, it was a simultaneous registration of the values again.

Fig. 8 illustrates the distribution of the thermoresistive sensors (Pt 100). More than half the thermometrical measuring points were also equipped with piezometers, mainly in the lower part of the channel.

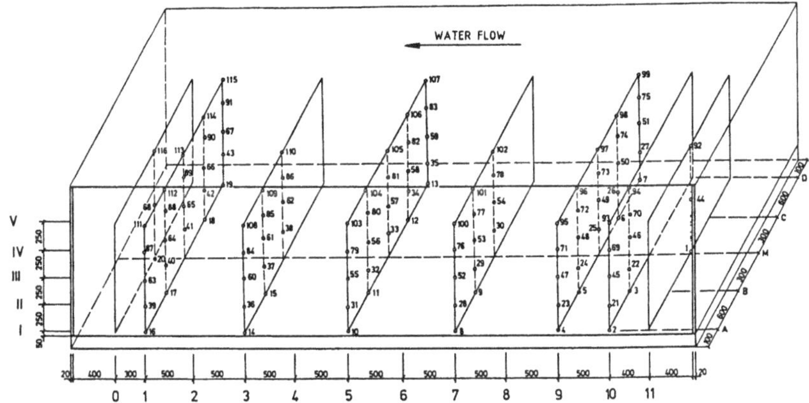

Fig. 8: Distribution of the thermoresistive sensors in the plain model

Results of the Model Studies

a) Dam-Model

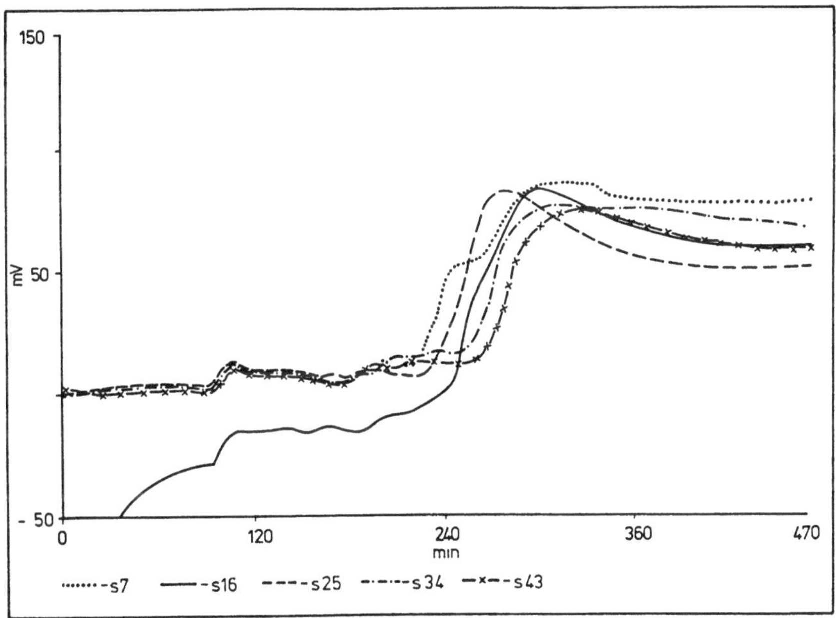

Fig. 9: Self potential values of the electrodes in row 7 of the dam model

In Figs. 9 and 10 the data of five electrodes each in row 7 and row 9 (see Fig. 3) are presented. The self potential values are plotted for a period of 470 minutes. They are offset corrected, that is to say, the initial values of the single electrodes were substracted. At the time "zero" the upstream face of the dam model was flood. A sudden increase of the self potential values after about 100 minutes accompagnied by a small drifting a short time later could be observed.

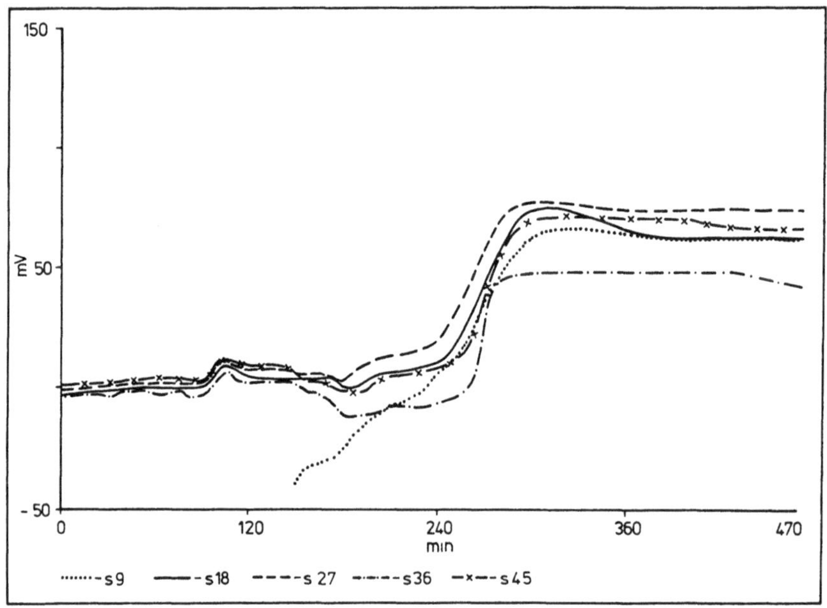

Fig. 10: Self potential values of the electrodes in row 9 of the dam model

In Fig. 11 the development of the values of hydraulic potentials measured at the bottom of the channel with the piezometers under the electrodes of rows 7 and 9 are presented.

At about 100 minutes after flooding the water reaches the piezometers arranged under the row 7 of self potential electrodes.

A comparison between Figs. 9, 10 and the Fig. 11 shows an obvious correlation between the increase of the water level and the change of the self potential values.

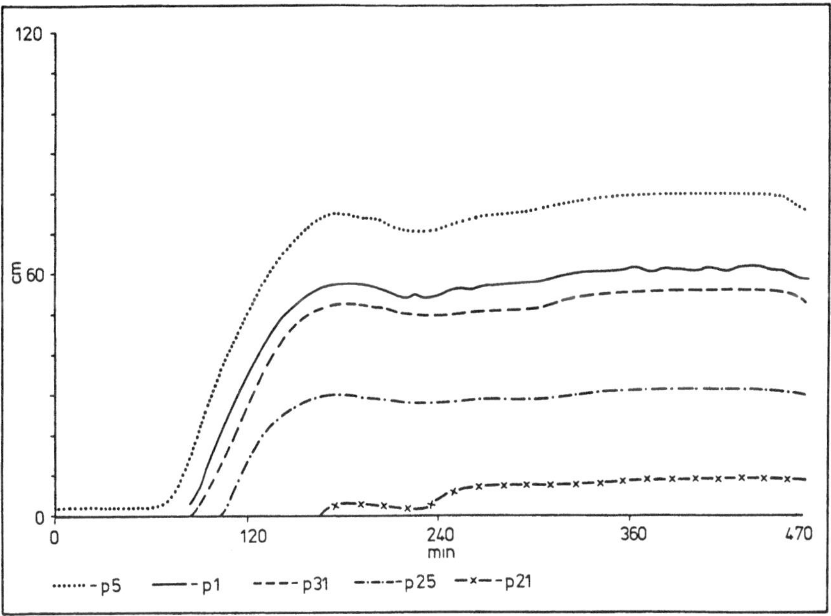

Fig. 11: Development of the piezometer values after flooding the upstream face of the dam model

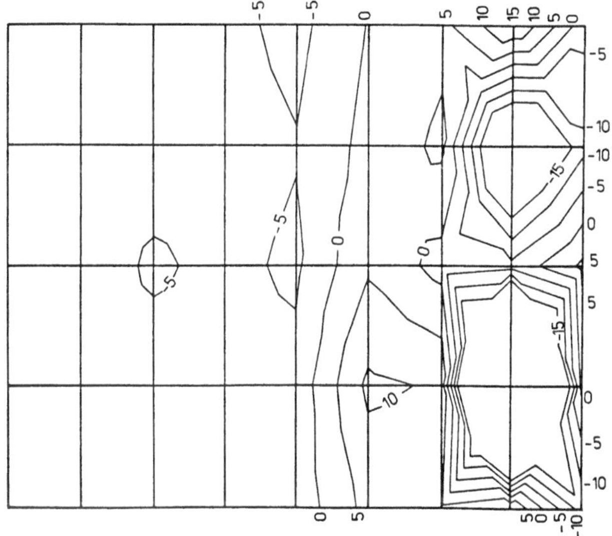

Fig. 12: Distribution of the self potential values at the downstream face of the dam model (water supply from the left hand side)

The distribution of the self potential field on the downstream face of the dam model at several times is shown in Figs. 12, 13 and 14.

The experiment was started at 11:48 a.m. Fig. 12 illustrates the distribution of self potential after 1 h 40'.

At the lower side of the dam model negative potentials are measured, whereas in the middle part of downstream face the values still remain undisturbed. The upper part of the upstream face of the dam is slightly negative (about -5 mV). These values depend on the time of water reaching the piezometers under the rows 7 and 9 (see Fig. 11).

Twentyone hours after starting the experiment and after keeping the water level up constantly at 90 cm, the lower part of the downstream face is more posititve (Fig. 13).

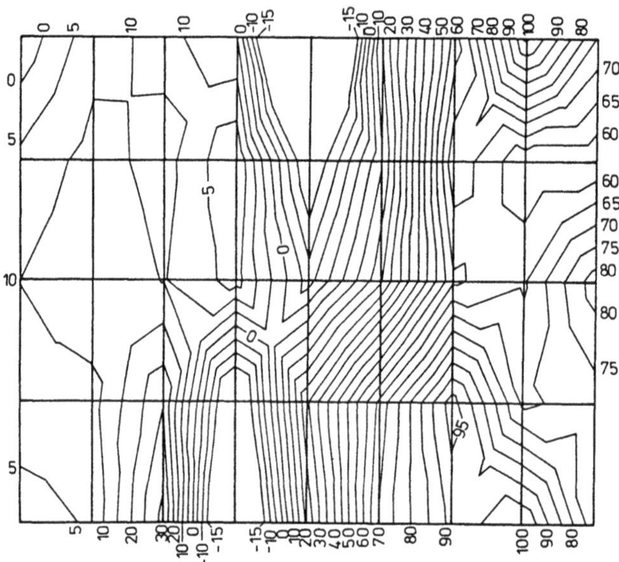

Fig. 13: Distribution of the self potential values at the downstream face of the dam model 21 hours after starting the experiment

The negative potential moves from the toe drain towards the middle range of the dam.

After 21 hours of the experiment and 14 hours of keeping up constantly the water level at 90 cm, the water level had been lowered.

Fig. 14 illustrates the extention of the self potential field at a water level of 52 cm. This water level had been kept up constantly for 2.5 hours.

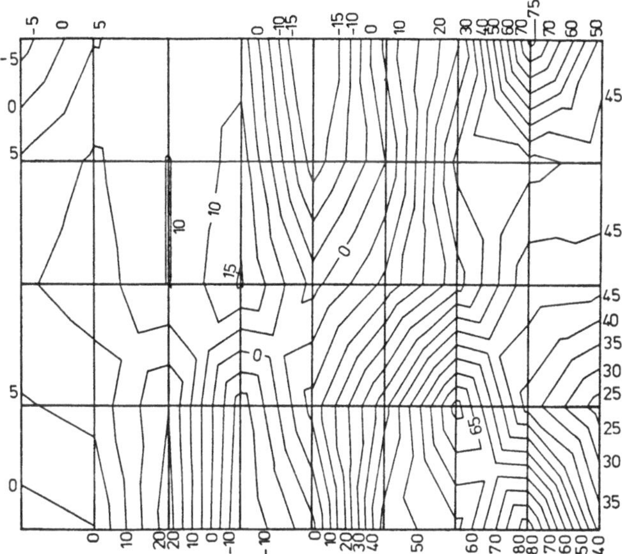

Fig. 14: Distribution of the self potential values on the downstream face of the dam model after starting to lower the water level

It is obvious that the positive potential fields in the range of the toe drain begin to decrease. Analogeously the negative fields in the middle part of the dam start to decrease. No substantial changes occur in the upper part of the dam model.

When the water is totally drained, the self potential values finally show a tendency of becoming uniform in the whole range of the upstream face of the model. The self potential values tend to become constant at about +10 to +25 mV.

The development of the thermic field in dependence upon hydraulic processes is illustrated in Figs. 15 and 16.

At the beginning of the experiment the average temperature of the sand was about 17°C. Differences here and there (up to 20°C) are caused by different room temperatures at different places.

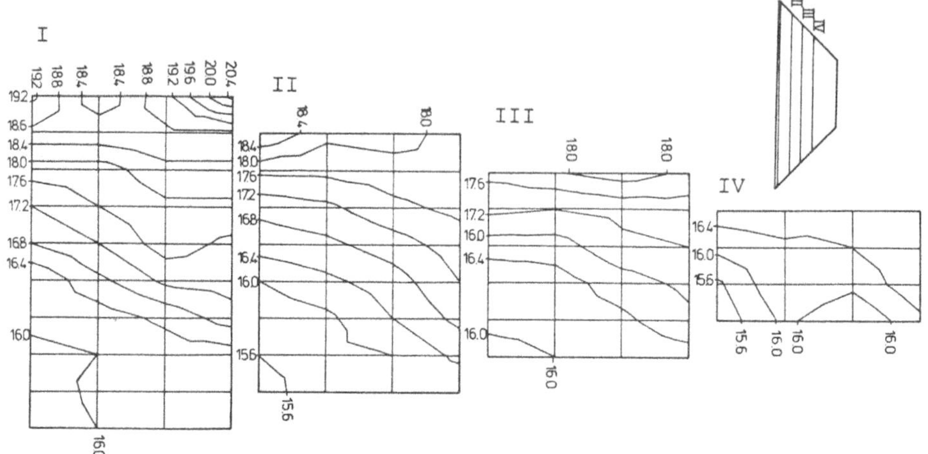

Fig. 15: Temperature field in the dam model before flooding, in different horizontal sections

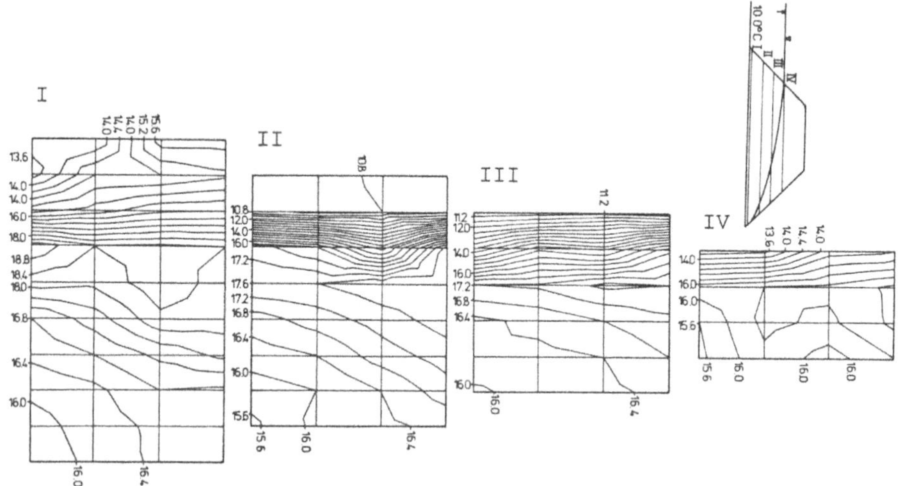

Fig. 16: Temperature field in the dam model, 1h 30' after the begin of flooding, in different horizontal sections

Fig. 16 shows that the cooling down of the dam occurs in all of the four measuring levels. The steep temperature gradient practically only applies to the range where the water ran through (the seepage line did not reach the toe drain totally). The parallel course and the temporal development can be regarded as proof of the homogeneity of the model.

Fig. 17 shows the hydraulic values of the described experiment.

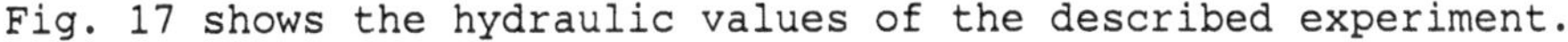

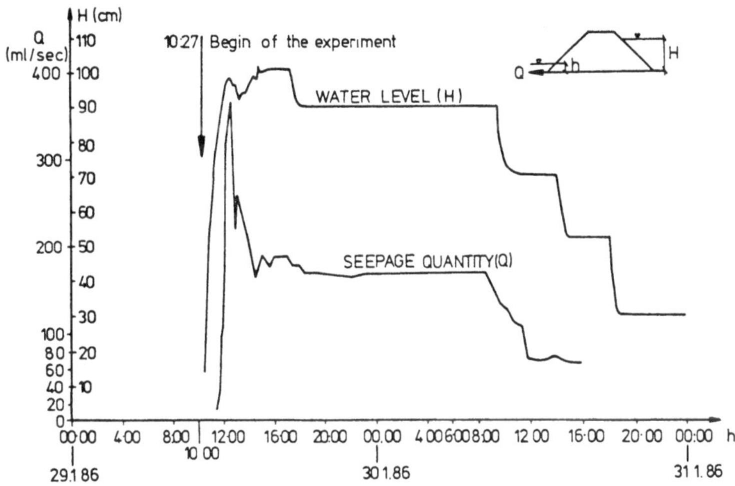

Fig. 17: Water level and quantity of seepage

As it is obvious from Fig. 17, the water supply was effected relatively quick. A water level of 90 cm was wanted and kept up constantly. Hydraulic measurements proved that stationary conditions were attained after ca. 1 hour. The distribution of the hydraulic potentials is relatively regular, giving a hint on a low pouring anisotropy within the sand.

b) Measurements on the Plain Model

With the help of these measurements it was intended to perceive different streaming processes, caused by the structure of the model, within the diagrams of self potential values, hydraulic and thermic potentials. The toe drain of gravel should play a decisive part.

For the following examples stationary situations and measuring cycles were analyzed, representing water levels at the supply of around 25, 27, 32 and 50 cm.

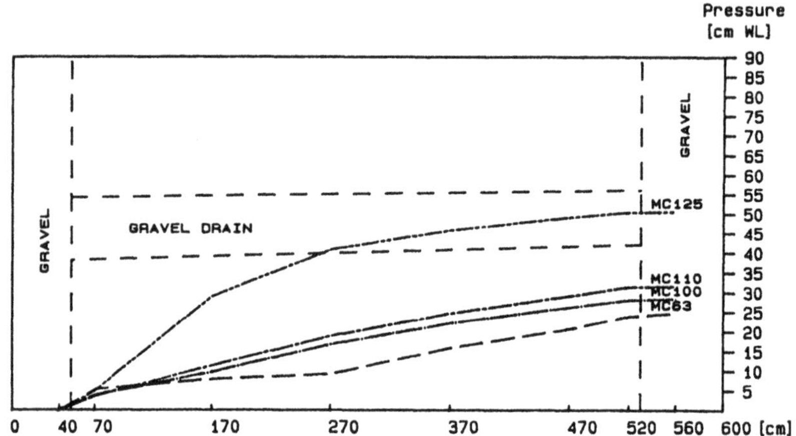

Fig. 18: Seepage lines for different water levles and measuring cycles, respectively

The influence of the gravel drain on the course of the seepage line is evident.

The course or the changes respectively of the self potential field is illustrated in the following figures.

Special attention should be drawn to the fact, that measuring value differences occur and that these values are offset corrected.

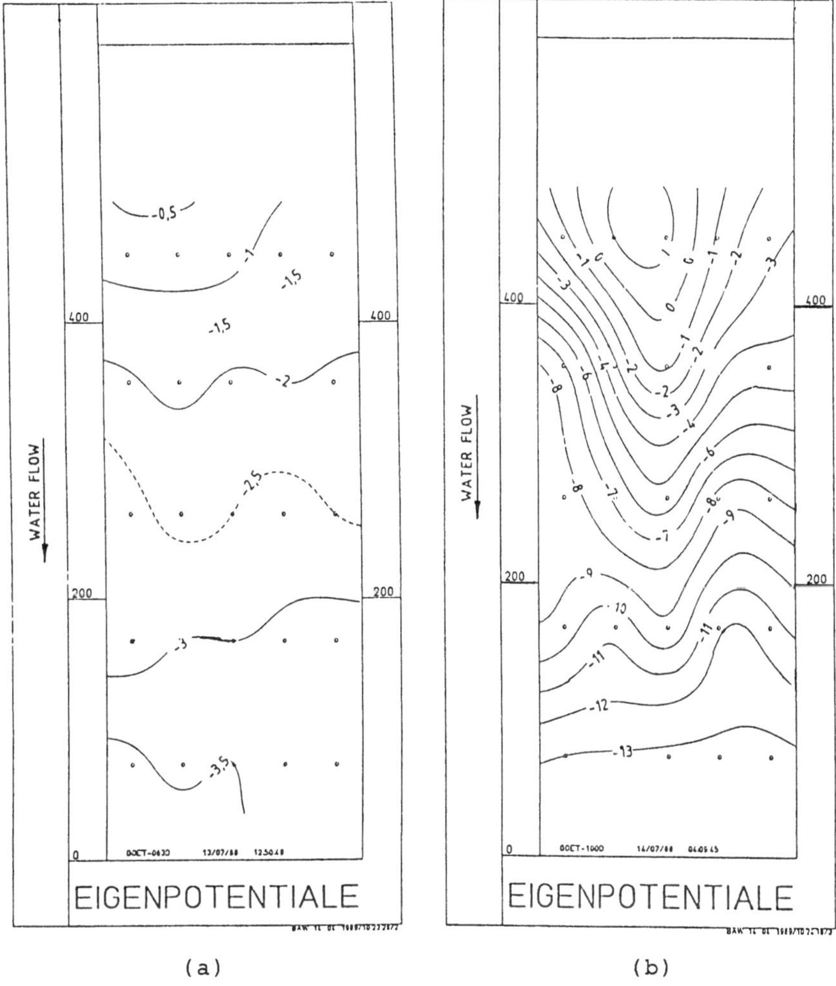

Fig. 19: Map of isolines of self potential values measured at a water level of (a) 25 cm and (b) 27.5 cm

As it is obvious from Fig. 19 (a) there is no differentiation of the isolines according to the flow direction at low water levels.

An increase of the water level, or even if water coming out of the unsaturated zone reaches the drain , leads to an orientation of the privileged streaming directions.

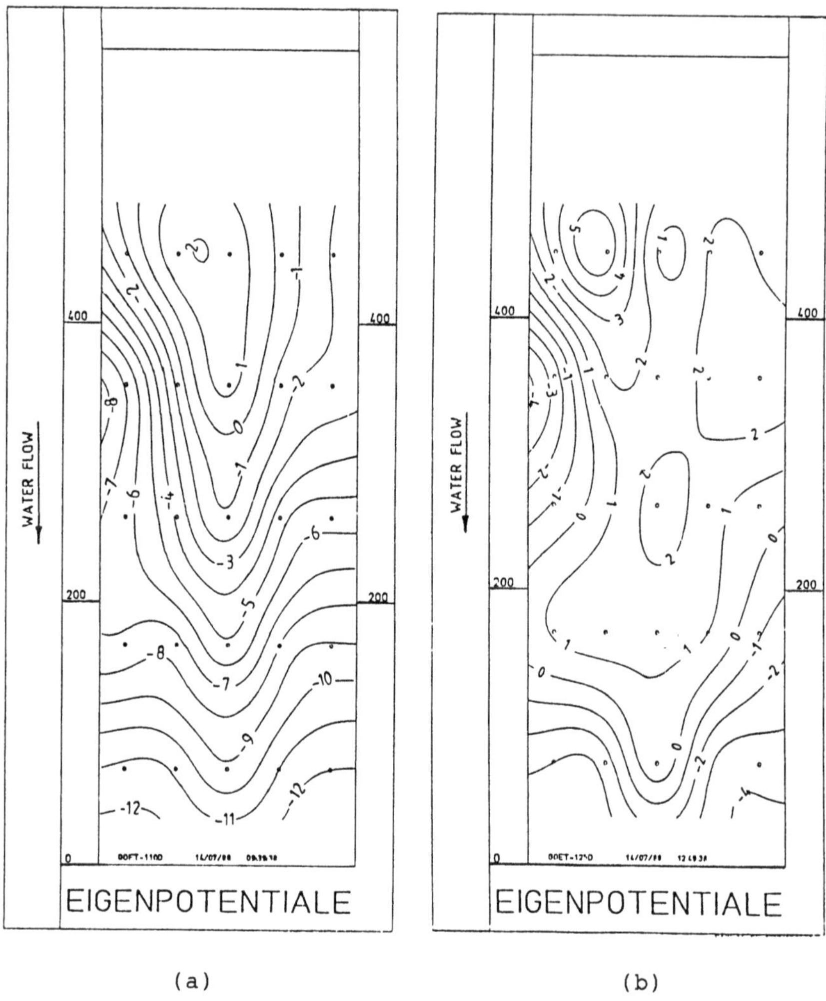

(a) (b)

Fig. 20: Map of isolines of self potential values measured at a water level of (a) 30 cm and (b) 50 cm

An increasing water level shows that the orientation of the isolines is defined as long as the seepage line or the unsaturated zone, respectively (Fig. 20 (a)), does not reach the drain of gravel (Fig. 20 (b)). As soon as the seepage line reaches or exceeds the drain, the orientation of the isolines according to the flow direction is disturbed. Finally, if the drain is flooded, the orientation is practically lost.

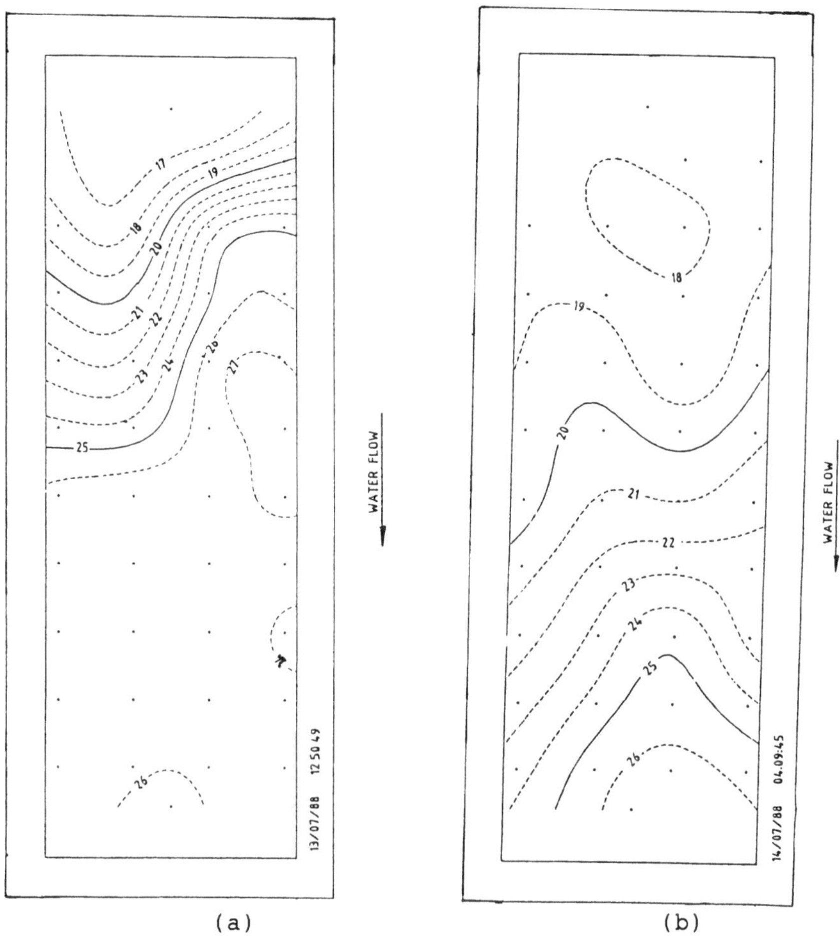

Fig. 21: Map of temperature isolines measured at a water level of (a) 25 cm and (b) 27.5 cm, measuring level I

Fig. 21 (a) and (b), respectively, show that the degree of orientation of the temperature isolines also increases with an ascending water level, and that the heat flow indicates itself by a marked tendency of orientation. A short distance to the drain or the reaching of the unsaturated zone of the gravel filter (Fig. 22 (a)), respectively, causes a reinforcement of this tendency of orientation.

Same as in the maps of self potential isolines it is obvious from Figs. 22 (a) and (b), that the degree of orientation of the temperature isolines tend to decrease with a water level reaching the drain.

It has to be examined in further experiments of how much importance convective and diffusive heat transfer phenomena respectively their interdependence are in this connection. It is known that there is almost exclusive a heat conductivity within the unsaturated zone (Lütkestratkötter 1977).

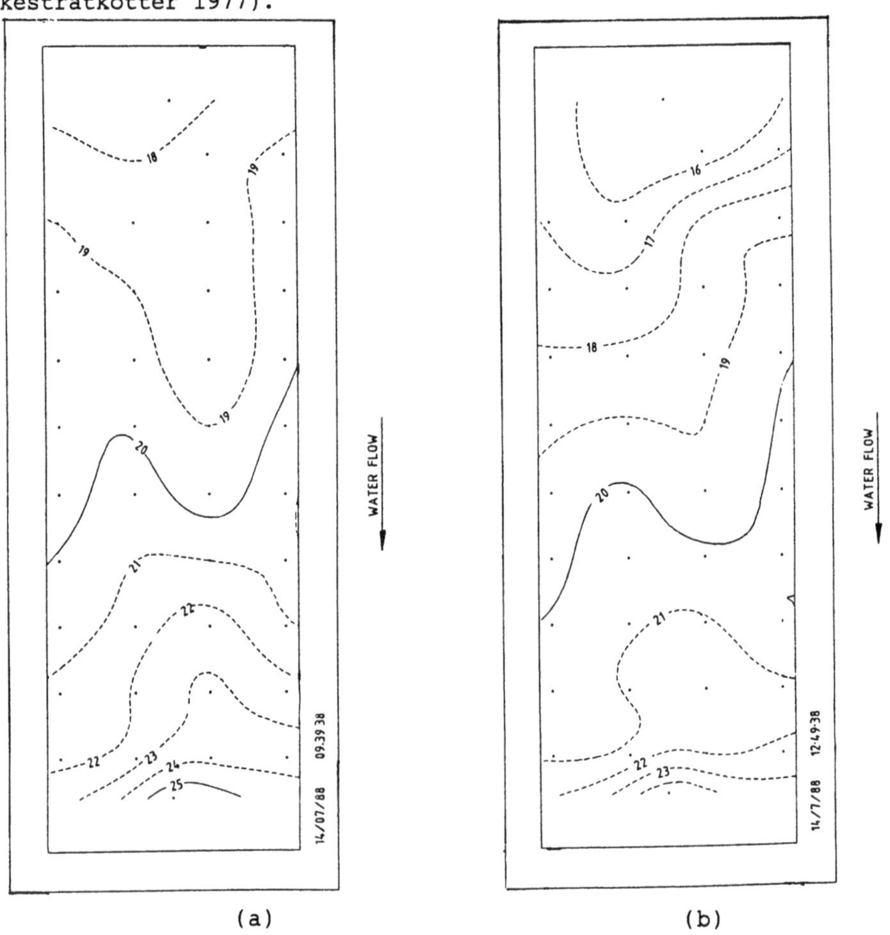

(a) (b)

Fig. 22: Map of temperature isolines measured at a water level of (a) 30 cm and (b) 50 cm, measuring level I

c) Field Example

In order to illustrate a change between model experiments and field measurements or a correlation between them, respectively, the results of the measurements at a leakage zone of the Rhine-dam are briefly presented in the following.

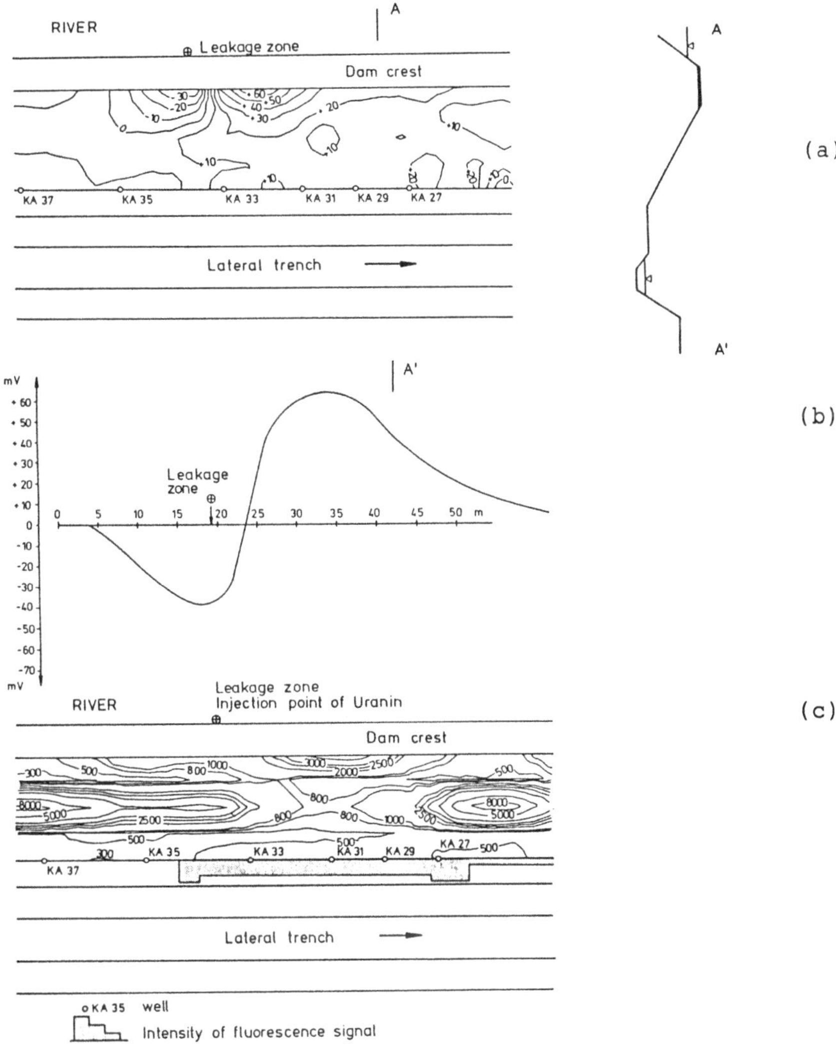

Fig. 23: Self potential data along a dam with a seepage zone:
(a) Self potential voltage isolines map (contour interval 10 mV)
(b) Self potential profile at the upper part of the dam slope
(c) Map of isoresistivity data (Wenner array) with the outlet zone of the tracing experiment

Figs. 23 (a) and (b) show that the leakage zone indicates itself in the self potential values with a significant dipole anomaly. The results of the self potential measurements are confirmed by the resistivity values. In the middle range of the dam slope - below the leakage - is a striking and marked resistivity minimum, which attributes to a saturation of the coarse-grained dam fill with water on the one hand and to an enrichment of the cohesive materials on the other hand.

It is obvious from Fig. 23 (c) that the outlet zone of the tracer is located at the lower part of dam slope in extension of the minimum of resistivity.

Results and Conclusions

The experiments, which had been carried out, have shown that a connection between self potentials, the hydraulic and the thermic field can be observed. Until now it was only possible to establish qualitative correlations for the dam model. A dependence of the self potential values on the distance between seepage line and surface or self potential electrodes, respectively (Merkler & Hötzl 1989).

A relatively prompt reaction of the electrical and the thermic field on even little changes in the hydraulic field could have been ascertained.

During the experiments on the dam-model one of the main problems was caused by the applied platinum electrodes. As platinum has no defined output potential, the measurements of the different experiments did not yield reproducible values. Certainly, the geometry of the dam body complicated the quantification of the correlation between the hydraulic field and the interdependent thermic and electric field, which had been strived for.

The experiments on the plain model confirmed the results on the dam-model. The assumption that there are quantifiable correlations between the three fields, mainly in the range of stationary hydraulic processes, has to be followed up.

In the hydraulically stationary domaine, that is to say, until the water level reaches the gravel filter, certain linear correlations have been established. That is obvious from Figs. 24, 25, 26 and 27.

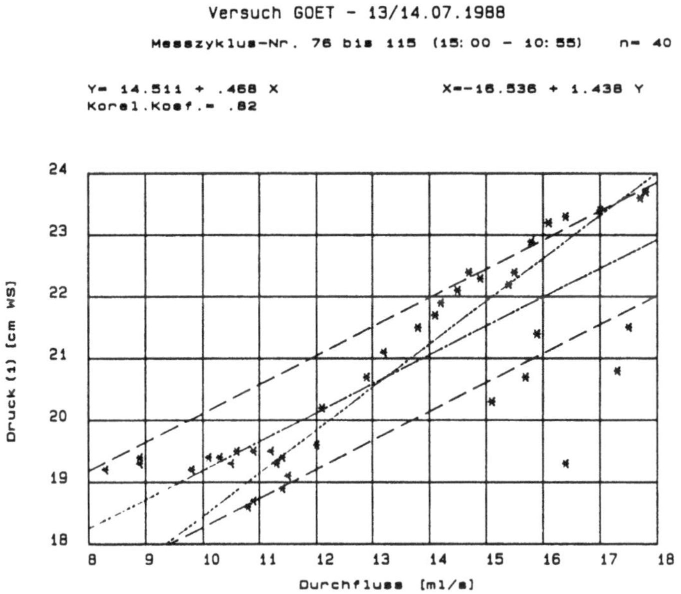

Versuch GOET - 13/14.07.1988

Messzyklus-Nr. 76 bis 115 (15:00 - 10:55) n= 40

Y= 14.511 + .468 X X=-16.536 + 1.438 Y
Korel.Koef.= .82

Druck (1) [cm WS]

24 23 22 21 20 19 18

8 9 10 11 12 13 14 15 16 17 18

Durchfluss [ml/s]

Fig. 24: Correlation between piezometer and quantity of water flow for a part of the experiment

Figs. 25 and 26 show a certain correlation between self potential and piezometer, or temperature and self potential, respectively.

Versuch GOET - 13/14.07.1988

Messzyklus-Nr. 76 bis 115 (15:00 - 10:56) n= 40

Y=-19.191 + .629 X X= 27.541 + 1.106 Y
Korel.Koef.= .834

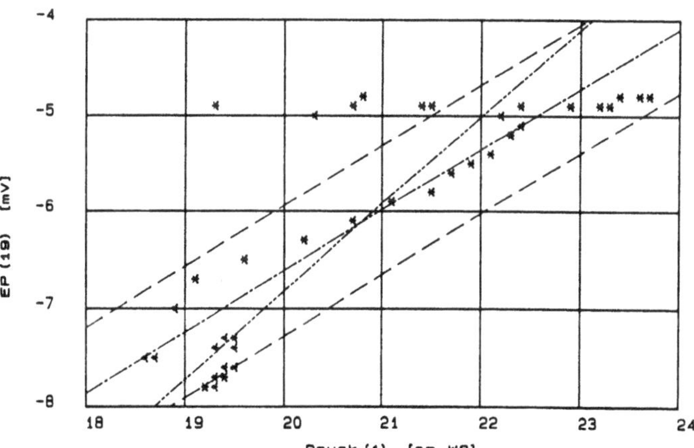

Fig. 25: Correlation between piezometer and self potential values for a part of the experiment

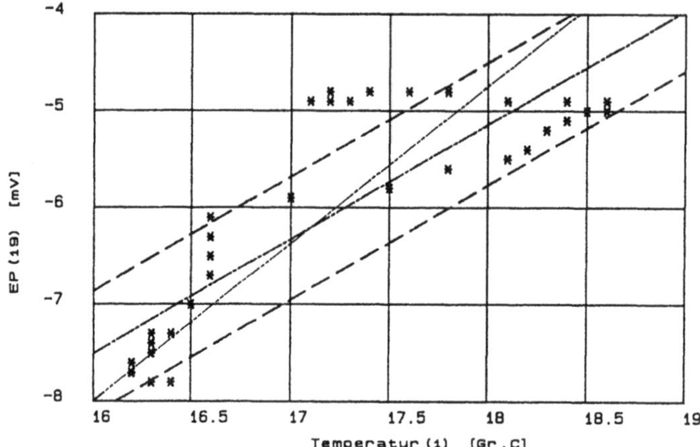

Fig. 26: Correlation between temperature and self potential values for a part of the experiment

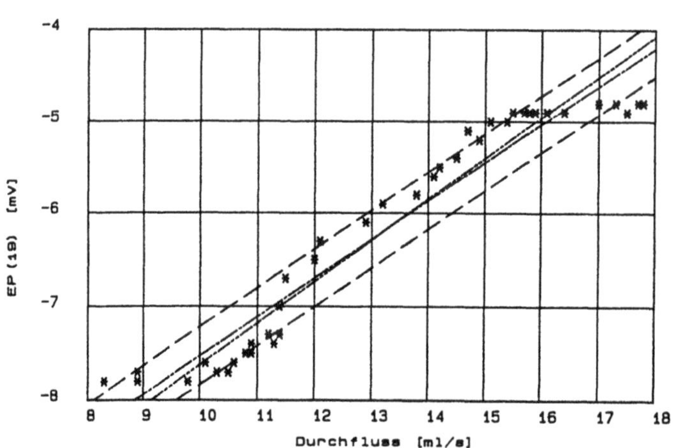

Fig. 27: Correlation between seepage quantity and self potential values for a part of the experiment

From this, mathematical-statistical linear correlations between the hydraulic values and the appertaining thermic and electric fields in the domaine of the laminar and stationary flow are derivable.

The influence of the drain filter on the flow processes and, as a result, on the electric and thermic field was illustrated in Figs. 19-22. It is to assume, that the differentiated or orientated image isolines of self potential and thermic values vanishes as soon as even the unsaturated zone reaches the drain filter. The orientation of the isolines according to the flow direction is obvious, until the water level reaches the filter. The mentionned differentiation of the isolines vanishes where the seepage line reaches the drain filter. This is caused by a change of stationary and laminar flow processes.

The field example confirms the correlation between a leakage zone and the appertaining streaming potentials on the one hand and the correlation between the leakage zone and the changes in the apparent specific resistivity in the domaine of the leakage zone on the other hand (Fig. 23). At the same time it should yield a confirmation of geophysical methods in application of tracers.

Fig. 28 shows the fit between the measured self potential values and the calculated in the domaine of the leakage zone (Wilt & Corwin 1988, Corwin & Ticken 1988).

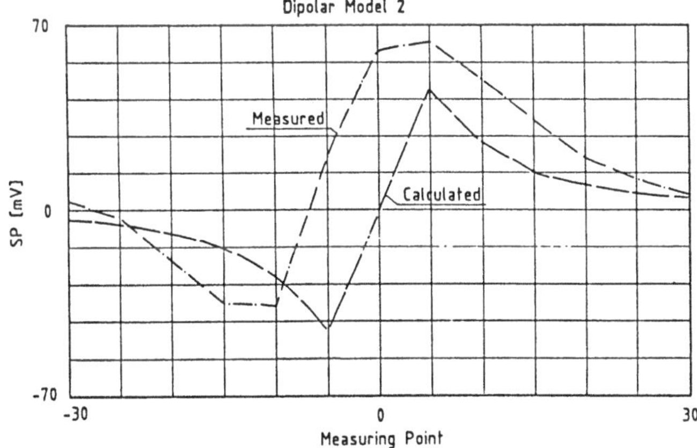

Fig. 28: Comparison between calculated and measured self potential values for the results given in Fig. 23

The example given refers to a model-calculation after CORWIN & TICKEN (1988) for a dipolar model. The assumed depth of the dipolar sheet source is between 2 and 10 m, the length is around 10 m. This example (see Wilt & Corwin 1988, too) is given as an additional evidence of certain quantitative correlations between streaming potentials and the calculated locality of the source.

For the future it is intended to realize experiments at the big channel, which are dependent on different ion solutions and concentrations, in order to examine the interdependence between the electrical, thermometrical and hydraulic field.

Acknowledgements

We are in a debt of gratitude to the Volkswagen Foundation for the generous financial support of the research project. We would also like to thank Mr. R.F. Corwin and Mr. M.J. Wilt for making available to us computer codes for the calculation of the self potential anomalies for models of leaky dams. For helpful translation and paperwork on this manuscript many thanks to Mrs. A. Sass.

Literature

ARMBRUSTER H, BRAUNS J, MAZUR W, MERKLER G P (1988) Effects of Leaks in Dams and Trials to Detect Leakages by Geophysical Means. Proc. Detection of Subsurface Flow Phenomena by Self Potential/Geoelectrical and Thermometrical Methods. Int.Symp. Karlsruhe, FRG, March 14-18, 1988

ARMBRUSTER H, MERKLER G P (1985) Geoelectric, Thermic Field Measurements for Dam Control.- Proc. of the XI Int. Conference of Soil Mechanics and Foundation Engineering, San Francisco, August 12-16, 1985

BOGOSLOVSKY W A, OGILVY A A (1970) Application of Geophysical Methods for Studying the Technical Status of Earth Dams. Geophys. Prosp. Vol. 18, pp. 758-773

CORWIN R F, TICKEN E J (1988) Development of Self Potential Interpretation Techniques for Seepage Detection, unpublished report, Harding Lawson Associates, USA

LÜTTKESTRATKÖTTER H (1977) Numerische Behandlung von Wärmeausbreitungsvorgängen in porösen Medien nach der Methode der Finiten Elemente. Institut für Wasserbau und Wasserwirtschaft der TH Aachen, FRG

MERKLER G P, HÖTZL H (1988) Model Experiments on a Small Test Channel. Empirical Correlations between Flow Potentials and the Hydraulic Field. Proc. Detection of Subsurface Flow Phenomena by Self Potential/ Geoelectrical and Thermometrical Methods. Int. Symp. Karlsruhe, FRG, March 14-18, 1988

MERKLER G P, BLINDE A, ARMBRUSTER H, DÖSCHER H D (1985) Field Investigations for the Assessment of Permeability and Identification of Leakages in Dams and Dam Foundation.- Proc. XV Congr. of Large Dams, Q 58 R7, pp. 125-141, Lausanne

SILL W R (1983) Self Potential Modelling from Primary Flows. Geophysics, Vol. 48, No. 1, pp. 78-86

WILT M J, CORWIN R F (1988) Numerical Modelling of Self Potential Anomalies due to Leaky Dams: Model and Field Examples. Proc. Detection of Subsurface Flow Phenomena by Self Potential/Geoelectrical and Thermometrical Methods. Int. Symp. Karlsruhe, FRG, March 14-18, 1988

Special Applications of Geoelectrical Measurements

GEOELECTRICAL MAPPING AND GROUNDWATER CONTAMINATION

Rainer Blum
Hess. Landesamt f. Bodenforschung, Leberberg 9-11, 62 Wiesbaden, FRG

Abstract

Specific electrical resistivity of near-surface materials is mainly controlled by the groundwater content and thus reacts extremely sensitive to any change in the ion content. Geoelectric mapping is a well-established, simple, and inexpensive technique for observing areal distributions of apparent specific electrical resistivities. These are a composite result of the true resistivities in the underground, and with some additional information the mapping of apparent resistivities can help to delineate low-resistivity groundwater contaminations, typically observed downstream from sanitary landfills and other waste sites. The presence of other good conductors close to the surface, mainly clays, is a serious noise source and has to be sorted out by supporting observations of conductivities in wells and geoelectric depth soundings. The method may be used to monitor the extent of groundwater contamination at a specific time as well as the change of a contamination plume with time, by carrying out repeated measurements. Examples for both are presented.

1 Introduction

Geoelectrical mapping is a well established and widely used method. Generally, the results will be too ambiguous, a drawback which in certain cases might be compensated by its operational simplicity. In this chapter only its usefulness with respect to monitoring groundwater contaminations and their subsurface movements is discussed.

Lecture Notes in Earth Sciences, Vol. 27
G.-P. Merkler et al. (Eds.)
Detection of Subsurface Flow Phenomena

2 Physical Basis

The physical basis is shown in Fig. 1, i.e. a plot of the amount of dissolved solids versus the specific electrical resistivity of the solution, valid at 25°C. Obviously the resistivity depends strongly upon the ion content; the higher the ion content, the lower is the resistivity. In a NaCl solution 10 g NaCl dissolved in 1 liter of water (lower graph) results in a resistivity of 0.5 Ωm (that is, a conductivity of 20000 µS/cm). The allowable content for drinking water is 2.5 g/l and corresponds to 2 Ωm. The upper graph gives the same relation for natural groundwater, containing a multitude of dissolved solids. It is just slightly shifted towards higher resistivities. Overall, however, the resistivity reacts in the same way, extremely sensitive to the ion content.

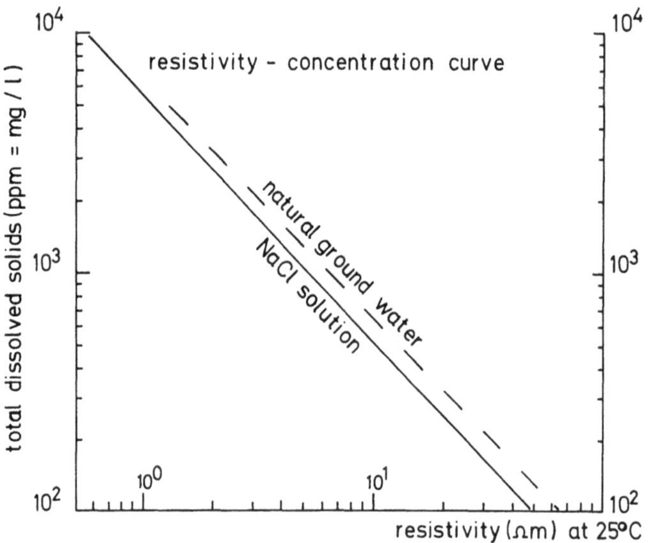

Fig.1 Relation between specific electrical resistivity and amount of dissolved solids for a solution

Actual groundwater resistivity can be observed only in a well. Geophysical surface measurements react upon the resistivity of the groundwater reservoir, the aquifer. The relation between these two resistivities, water and aquifer, may be described by Archie's law. Its original form is for example given by Asquith (1983). In a simplified version (e.g. Telford et al. 1976, p.775) the parameters porosity, cementation factor, and tortuosity are combined into a formation factor F resulting in the relation:

$$\rho = F \cdot \rho_w$$

where ρ = bulk resistivity of the water saturated aquifer, ρ_w = resistivity of the water. Typical values of aquifer formation factors lie between 4 and 7 (Deppermann et al.1961, p.724).

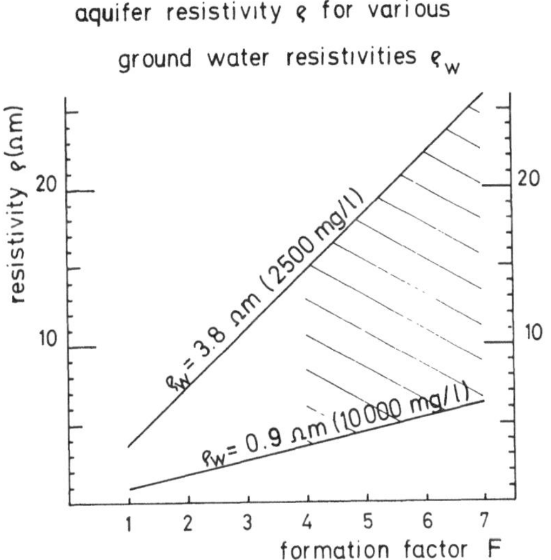

Fig. 2 Aquifer resistivities following Archie's law, hatched area contains range for realistic aquifers with water containing between 2500 and 10000 ppm dissolved solids

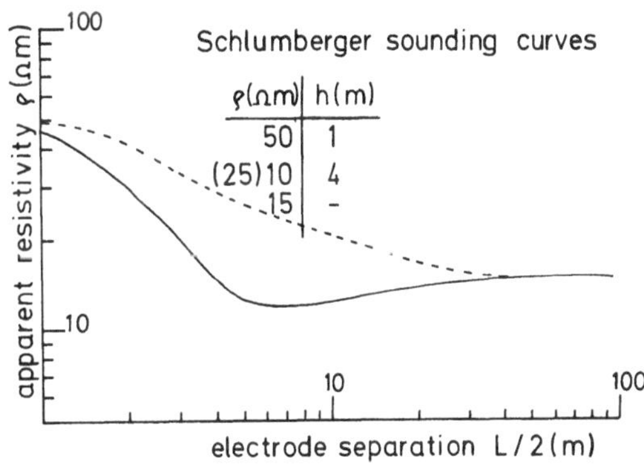

Fig. 3 Electrical sounding curves for equal structures with aquifer resistivity of 10 (solid line) and 25 Ωm (dashed line)

Figure 2 shows the range of possible aquifer resistivities for various water resistivities in a realistic aquifer. The lower boundary is valid for water with a resistivity of 0.9 Ωm, corresponding to about 10000 ppm dissolved solids at 10°C. The upper boundary describes groundwater with 2500 ppm (limit for drinking water quality). The shaded area, ranging from a formation factor F=4 to F=7, then contains possible aquifer resistivities. They vary between 5 and 25 Ωm. While 5 Ωm is an extremely low value which would attract the attention of any observer, resistivities between 15 and 25 Ωm are quite commonly found in clayey environments, for example directly below an aquifer. This makes clear that for moderate groundwater contaminations the effect in terms of lowering the aquifer resistivity does not have to be too distinct. Besides, the recognition of a lower aquifer resistivity by means of surface measurements might be a problem. They only yield apparent resitivities which are somehow an integral effect of the true resistivities under and around the electrode setup.

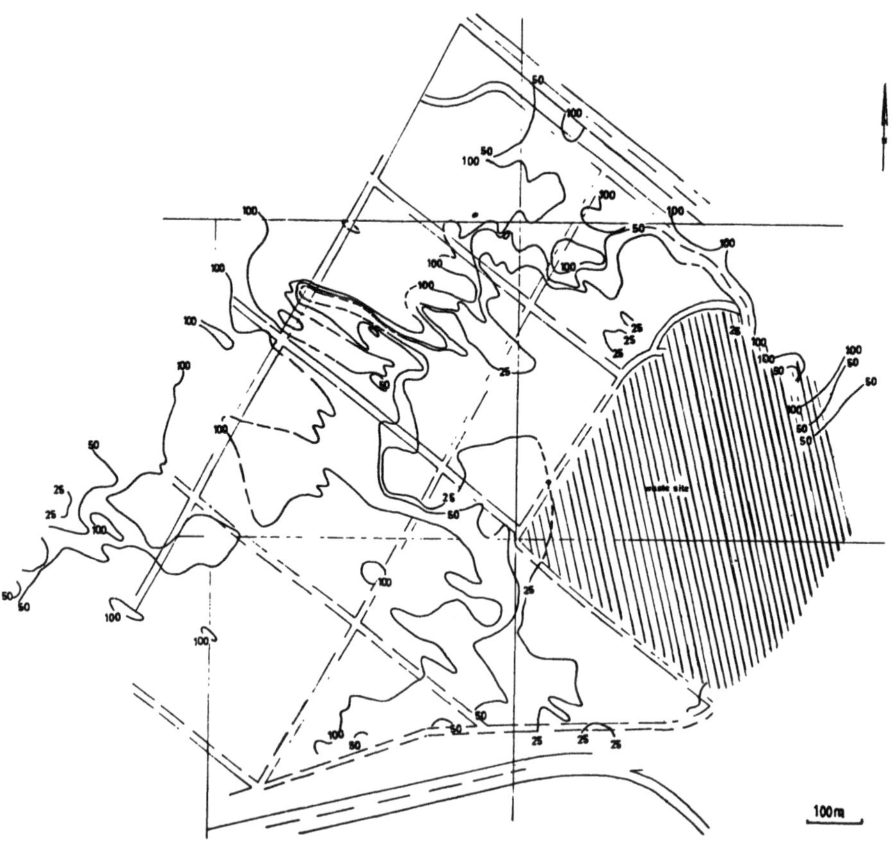

Fig. 4 Contour map of electrical mapping results at waste site

Geoelectric mapping, i.e. measuring the apparent electrical resistivity with a fixed electrode spacing for an array of field points, can only show variations of that apparent resistivity. It does not reveal the source of such a variation. The only way to determine the true resistivity-depth structure is by geoelectrical soundings where apparent resistivities at a fixed point are measured as a function of electrode spacing. Even then, however, one might not be able to distinguish between a variation in a layer's thickness or resistivity, if that layer is, for example, sandwiched between two high-resistivity layers. In such a case a high value in the underlying clay layer cannot be distinguished from a lowered groundwater resistivity. Such ambiguities have to be resolved by including borehole information, e.g. electrical logging data or depths of geological boundaries. A well interpreted sounding then in turn can help to design a successful field setup by the choice of a proper electrode spacing for a mapping campaign.

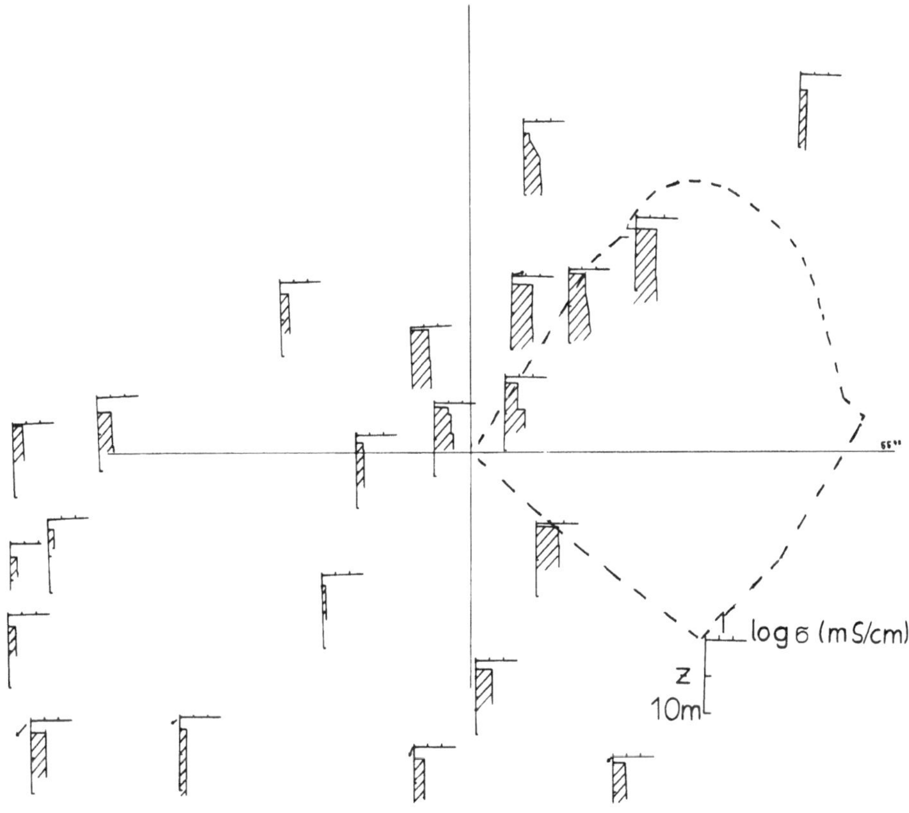

Fig. 5 Salinometer logs around waste site, same area as Fig. 4. Conductivity is plotted versus depth (1 corresponds to 10 mS/cm = 1 Ωm)

3 Examples

An example for such a geoelectrical mapping survey, calibrated by electrical soundings has been carried out around a sanitary landfill. The problem was to probe the surrounding of the waste dump in order to determine how far away a groundwater contamination could be observed and whether it escaped through buried channels to larger distances. A number of observation wells were available together with a detailed geological description.

In a first step geoelectric soundings were observed and interpreted using the geologic information. Typical results are shown in Fig. 3. Both sounding curves represent a simple three layer case, i.e. top soil above the aquifer, above clay. The broken line denotes an aquifer with a 25 Ωm resistivity. The full line results from an aquifer with 10 Ωm, indicating a contaminated water content. The third layer in both cases is formed by a clay with 15 Ωm. A comparison of both sounding curves shows that they differ most for a electrode spacing between 5 and 10 m. A mapping survey with a corresponding electrode spacing then should give optimal information about resistivity variations in the aquifer.

Such a mapping survey has been carried out and the result is presented in Fig. 4. The plot clearly shows low apparent resistivities below 25 Ωm mainly to the NW and W of the waste site. In the E and S, not included in the plot, the resistivity values are not influenced by the waste site. Obviously, the low resistivities follow more or less the groundwater flow, from SE to NW. However, they recover with increasing distance, somehow elongated towards the NW. But there too, the electrically visible contamination dies out. Overall, this gives a clear picture of the distribution of a contamination plume, with a minimum of effort. The field work, a couple of short sounding profiles included, was done by three men in 1 day.

The results are confirmed by logging data as presented in Fig. 5, showing salinometer logs for various wells around the dump site. The logarithm of the conductivity is plotted versus depth, the scale is marked on the right. In the area of the low apparent resistivities in the mapping picture (NW of the waste dump) very high conductivities, between 10 and 100 mS/cm, are observed. With increasing distance the conductivities return to normal values between 1 and 10 mS/cm (10 - 1 Ωm).

It is of passing interest how the conductivities return to normal values, beginning at the top and proceeding to the bottom, indicating some chemical precipitation process.

Nevertheless, it can be seen that in this case electrical mapping gives a rather reliable impression of the extent and shape of the contamination plume in the groundwater.

Another example from a different waste site is presented in Fig. 6a,b. Here the surroundings of a sanitary landfill in old gravel pits have been probed. The primary reason for electrically mapping the area was to delineate gravel and silt occurrences. The second idea was to additionally judge the results with respect to any leakage of contaminated water into the groundwater. Figure 6a shows the distribution of apparent resistivities from a first measuring campaign. The groundwater here flows from N to S and no contamination plume is seen on the resistivity picture. The only features visible are striking E-W, typical for the geologic setting.

A repeat of the measurements 6 years later (Fig. 6b) did not reveal any change.

Fig. 6a Contour lines for electrical mapping at waste site and b repeat 6 years later

Overall the values are slightly higher, probably caused by a variation in humidity. There is, however, still no sign for any leakage of contaminated water from the landfill.

4 Conclusions

Summarizing, it might be stated that geoelectrical mapping can serve as an efficient, low-cost tool for monitoring contamination plumes. Simple mapping alone may be sufficient for recognizing time variations or the rough outline of groundwater contamination. If further details are needed one should combine the method with additional exploration tools.

References

ABEM (ed) (1987) Instruction manual terrameter SAS systems. Bromma

Asquith G (1983) Basic well log analysis for geologists. AAPG, Tulsa, OK

Deppermann K, Flathe H, Hallenbach F, Homilius J (1961) Die geoelektrischen Verfahren der angewandten Geophysik. In: Bentz A (ed) Lehrbuch der angewandten Geologie. Enke, Stuttgart, pp.718-804

Telford W M, Geldart L P, Sherrif R E, Keys D A (1976) Applied geophysics. Cambridge Univ Press, London

Complex resistivity measurements on granites

G. Nover and G. Will

1.0 Introduction

The electrical properties of crustal rocks strongly depend on the chemistry of the natural formation waters in the pores, the pore volume available, and the temperature and pressure conditions (Keller and Frischknecht 1966; Schön 1983; Duba et al 1988). In rocks with low porosity (<around 5 vol. %) two processes contribute to the total current transport (σ_{rock}), conduction current (σ_{el}) and dissipation current ($\sigma_{surface}$).

$$\sigma_{rock}=\sigma_{el}+\sigma_{surface}$$

Both, the conduction current, due to the transport of charge carriers, e.g. cations and anions, and the dissipation current frequency dependent as a result of drifting of the ions, liquid viscosities, local forces and the interaction between the dipols of the formation water (e.g. H_2O) and the charges on inner surface of the pore structure (Hasted 1973; Olhoeft, G.R., 1986). The displacement current is associated with time changes in the distribution of bound charges, that are not free to drift in the elctrolyte (Anorgan and Madden 1977). An orientation of the dipols (Fuller and Ward 1970, Cammann 1973; Schultze 1986; Washburn 1982) can be observed in this electrochemical double layer (DL) (Schmickler, 1986). Such polarization effects are found for example at the boundary layers electrolyte-mineral grain (Fig. 1).

As result of the rock-fluid interactions, the resistivity of the core samples must be described as a complex quantity. If an ac field is applied to such a system, and the complex resistivity is measured over a wide frequency range, then different polarization effects can be recognized and determined (Amstrong and Fireman 1973; Mansfeld, 1981; Brauer and Piroth 1986; Will and Nover 1986; Duba et al 1988). In general the phase shift in the high frequency (0.01 - 100 kHz) region is due to the bulk polarization of the sample, and in the low frequency (5×10^{-5} - 10Hz) region mainly polarization effects with

Mineralogical Institute, University Bonn, Poppelsdorfer Schloß, 5300 Bonn 1

Lecture Notes in Earth Sciences, Vol. 27
G.-P. Merkler et al. (Eds.)
Detection of Subsurface Flow Phenomena

longer relaxation times which are due to surface polarization effects can be seen. This electrical response of rocks can be modelled by an

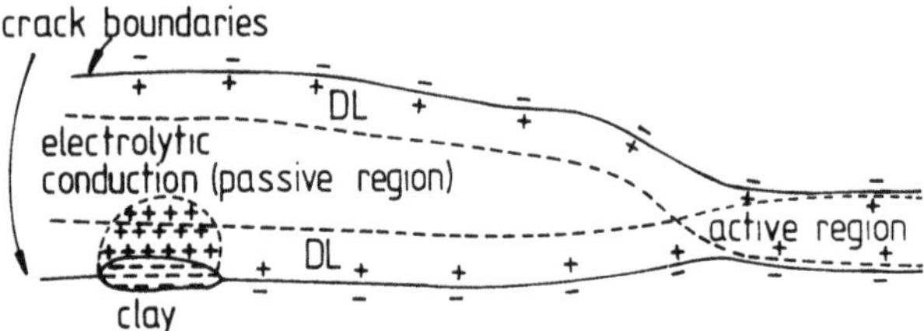

Fig. 1: Schematic representation of a flat crack in a crystalline rock. Shown are the regions of electrolytic conduction (passive region), and the DL on each side of the crack, where the conduction process is due to a dissipation current (active region, surface conductivity).

electric relaxation circuit consisting of a resistor and a capacitor in parallel (Jonscher 1978).

For the interpretation of the electrical resistivity data of core samples the knowledge of the petrophysical rock parameters porosity, permeability, density and inner surface is necessary. The zeta potential of the rock forming minerals must be measured as a function of the electrolyte type and concentration too, in order to get the thickness of the DL (Davies and Rideal 1961; Delahay 1965; Ahmed and Maksimov 1969; Hills 1970; Ney 1973) and an independently determined value for the capacity of the DL. We therefore have measured beside the petrophysical rock parameters both, zeta potential (ZP) and complex electrical resistivity (CR) of several core samples from the Falkenberg granite as a function of electrolyte type and concentration and the hydrostatic pressure conditions.

2.0 Experimental Investigation

2.1 Porosity

The porosity of the core samples was measured with a mercury intrusion method on samples 30 mm in diameter and 35 mm in height. For details see (Becker et al 1986). The measurement gives the gradual increase of the total pore volume with pressure, the radial distribution and the inner surface area of the sample.

2.2 Permeability

The permeability measurements were performed on cylindrical core samples of 30 mm in diameter and 15 mm in height in an apparatus that is described in Huenges, 1988. The permeability was determined with a pressure transient method. The pressure (P_1) on one side of the sample is kept constant and a lower pressure P_2 is on the other side of the sample. The pressure increase of P_2 a function of time is gives the permeability can be calculated from this data.

2.3 Complex electrical resistivity measurements

The frequency dependent complex electrical resistivity measurements were performed on different core samples with dimensions of 16 and 30 mm in diameter and 3 and 10 mm in height. An electrochemical cell (EG + G, K47) was used for the determination of the transfer resistance (R_b) and the capacity of the electrochemical double layer (C_{dl}) for different types and concentrations of electrolytes at 1 bar confining pressure. The measurments under higher pressure conditions (up to 500 bar) were performed in a pressure vessel (Nover et al 1984; 1987). (Fig. 2). These samples had a diameter of 30 mm and a thickness of approximately 10 mm.

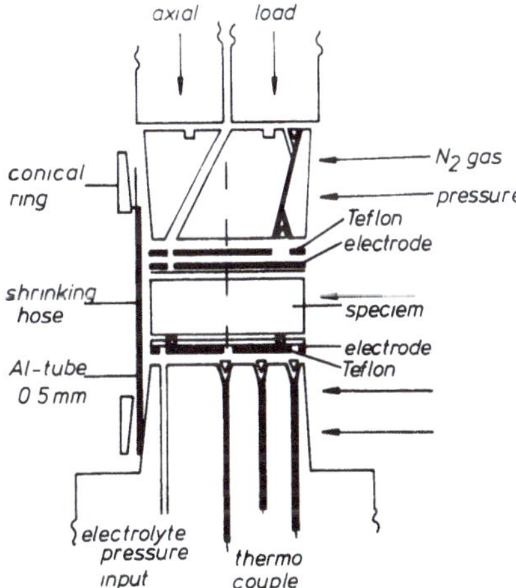

Fig. 2: Schematic drawing of the autoklave (for details, see text)

Pressure can be applied by different stress components: (1) uniaxial outside pressure P_{uni}; (2) perpendicular to the uniaxial load operates the so-called mantle pressure P_m, and (3) a spindel press allows to increase the pore pressure P_{el} up to 1.1 x P_m with P_m=P_{uni}. In this cell a two-electrode arrangement was used.The frequency dependent complex resistivity measurements were done in the frequency range between 10^{-5} Hz upto 10^{5} Hz using an AC impedance system (EG + G). Between 5 Hz and 10^{5} Hz a lock-in analyzer (LIA) was used, while at frequencies below 10 Hz a potentiostat with a Fast Fourier Transform (FFT) technique was used to measure the impedance (Fig. 3). M 178 is an electrometer. M 173 a potentiostat operating in the frequency range 10^{-5} up to 10 Hz using a Fast Fourier Technique, and M5206 is a lock in analyzer (LIA). M276 is an IEE 488 Interface, controlling via the Apple computer both, the Lock in Analyzer and the potentiostat. The connections between M173 and M5206 are for the signal inputs and the frequency oszillator.

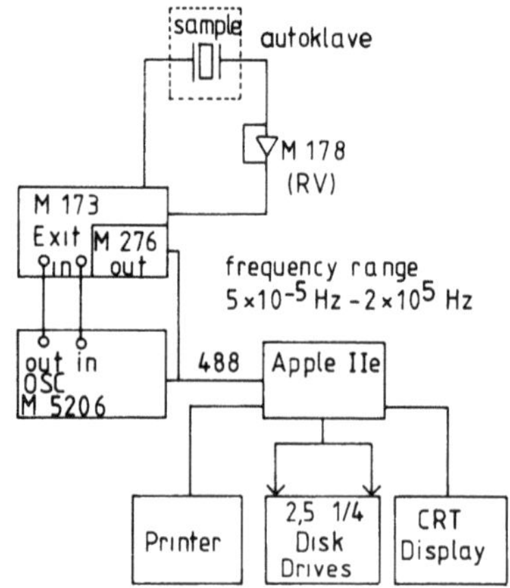

Fig. 3: Principle of the equipment for the measurement of the frequency dependent complex electrical resistivity of core samples

The complex input signal experiences an phase shift due to the electrical characteristics of the rock sample. If we describe the electrical phenomena in the frequency domain, then the input voltage is $E((x,\omega)$ where the vector x denots the space dependence, and $\omega = 2\pi f$. The electric field input $E(x,\omega)$ is seperated into two compounts, the real part (conduction $\sigma(\omega)$ of current $I_c(x,\omega)$ and the imaginary part (dissipation $\varepsilon(\omega)$ $I_d(i\varepsilon\ K(x,\omega))$. In this way the sample is described to be linear, homogeneous, isotropic and time invariant, but each of the electrical parameters is still permitted to be a complex function of frequency. The physical meaning of $\sigma^*(\omega)$ is the

representation of transport of free charge, but current density and electric field are not necessarily in phase. This means that the "free charges" do not follow instantaneously the variations of the applied electric field.
The physical meaning of $\varepsilon^*(\omega)$ is the describtion of various electric polarizations, which are complex values. Their changes in time contribute to the total observed current density. The complex admittance Y* of the sample can therefore be described by the sum of the real and imaginary part of the admittace:

$$1/Z^* = Y^* = Y' + iY''$$

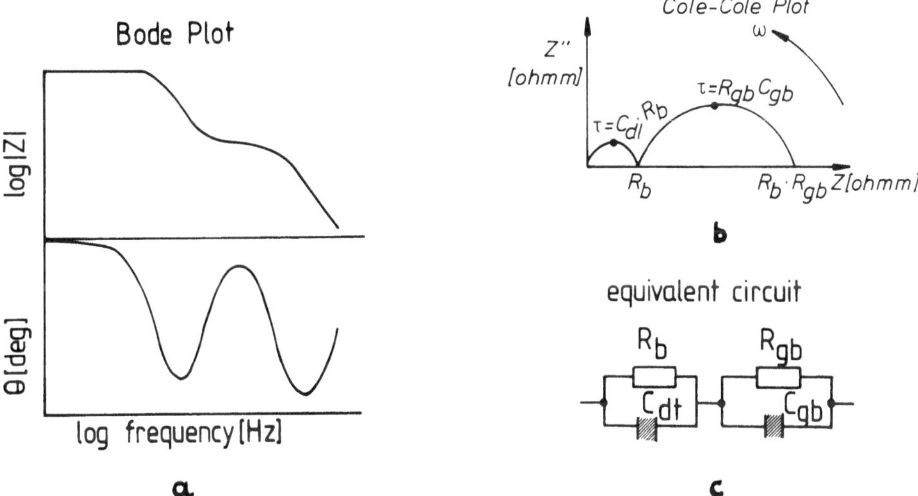

Fig. 4 a-c: Complex electrical resistivity data shown in a) a Bode plot where the absolute value of the impedance and the phase angle are plotted as a function of frequency, b) Cole-Cole plot where the real part is plotted versus the imaginary part of the impedance, and c) in terms of an equivalent circuit model consisting of an array of RC elements in series. The indices b= bulk, gb= grain boundary and dl= double layer

where Z*= the impedance and Y' and Y" are the real and imaginary part of the admittance. For the interpreation of the complex response of the sample simple equivalent circuit models are used (Fig. 4c), they allow the determination of the values of the resistor and the capacitor (Grahame 1952; (Mund 1986), for one relaxation process. If the relaxation times τi of the different polarization effects differ by about an order of magnitude, then we can expect well separated semicircles. In such an equivalent circuit model represent the parameters ρi, Ci and τ i (i= 1, 2, 3, ...) resistors, capacitors, and relaxation times for the different processes (Cole and Cole 1941; Mason et al 1974; Sheppard and Grant 1974; Jonscher 1978). The resistivity values can be read at the intersection of the semicircles with the real axis, while the relaxation times τi can be read at the top of the semicircle where $\omega_{max}=1/C \cdot R$ or $\omega \cdot \tau=1$.

$$\rho^*_{calc}= \sum_{i=1}^{n} (1/\rho i+ \tau i/\rho i \cot (n_i \pi/2)+\omega^n + i \tau_i/\rho_i{}^n)^{-1}$$

2.4 Zeta potential

The surface potential ψ_o (the wall potential) of the rock forming minerals (Fig. 1) is one of the electrochemical parameters that are responsible for the occurance of polarization processes in the interface solid/liquid. The structure of such liquid-solid interfaces have been investigated by Helmholtz 1879; Gouy 1910; Chapman 1913; Stern 1924; McCafferty and Zettelmoyer 1977). Different layers within the electrochemical double-layer (DL) can be distinguished. Of special interest are the Stern-layer (potential determing layer) in a rock-water system, and the diffuse part of the double layer. The thickness is approximately 80 nm for distilled water and 2 nm for a 2 molar electrolyte. Within the diffuse part of the DL the potential of the "Outer Sternplane" is the zeta potential (ZP), which can be determined by electrophoretic measurements. The electrophoretic method gives a proportionality between the surface charge of a particle and its velocity in an applied dc field when disperged in an electrolyt.

3.0 Results

3.1 Rock Status Determination

The results of the measurements of the petrophysical rock status parameters porosity, inner surface, density and permeability are compiled in tables 1 and 2.

Table 1: Rock status determination for some Falkenberg granite core samples from various depths (Nover et al 1984).

sample depth (m)	condition	porosity (vol. %)	inner surface (m^2/g)	density (g/cm^3)
Hb4a 25,40	Fresh	1.61	1.03	2.674
HB4a 50.00	Fresh	1.53	0.76	2.678
HB4 70.40	Hydrothermal	1.20	0.73	2.648
HB4 71.00	altered	1.59	1.04	2.644
HB4a 279.55	Fresh	1.33	1.01	2.712
HB4a 286.80	Fresh	1.46	0.94	2.691

Table 2: Permeability data for the Falkenberg granite samples at confining pressures of 60, 100, 200 and 400 bars.

sample depth	permeability µD at a pressure of (bar)			
	60	100	200	400
HB4a 25,40	-	116,0(3)	54,9(8)	-
HB4a 50,00	5,21(3)	4,93(2)	2,68(2)	1,81(2)
HB4 70,40	94,0(12)	38,0(5)	[110,0(2)	(cracked)]
HB4 71,00	-	44,0(5)	22,0(1)	-
HB4a 279,55	1,85(1)	1,65(2)	1,90(4)	-
HB4a 286,80	-	13,0(1)	2,9(7)	-

3.2 Zeta Potential Measurements

Increasing electrolyte concentration reduces the zeta potential. The strongest decrease of the ZP of more than 58 % (from - 59.3 to -25 mV) for an increase in concentration of NaCl from 0 g/l up to 0.03 M could be observed for quartz (Table 3). The decrease of the ZP for the fieldspar crystals in NaCl solutions is about 52 % and for mica we have observed almost no change of the value for the ZP in NaCl solutions. A similar behaviour could be detected for the decrease of the ZP for quartz in KCl solutions of different concentrations. In this case the decrease of the ZP is about 52 % for quartz in the concentration range 0 g/l up to 0.03 M, 49 % for feldspar and for mica we get a decrease of 39 %.

Table 3 : Zeta potential for the major minerals comprising the Falkenberg granite. Measurements were performed in distilled water and solutions of different molarity of NaCl and KCl. Thickness of the DL (1/x) according to Deby and Hückel 1923.

mineral	dist. water	0,006	0,017	0,03	(M NaCl)
Quartz	59,3	45,9	27,0	25,0	
fieldspar	49,3	45,9	37,6	23,9	
mica	32,6	32,3	33,8	28,7	
Quartz	59,3	44,1	32,9	25,6	(M KCl)
fieldspar	49,3	38,7	36,8	30,3	
mica	32,6	39,5	31,3	20,1	
1/x (nm)	80,0	6,3	3,7	2,7	

3.3 Complex Electrical Resistivity Measurements

The chemical and mineralogical homogeneous Falkenberg granite samples are marked by different porosities, pore radii distributions, inner surface areas and permeabilities. Of special interest was the measurement of the electrical data, using the same type of pore saturand, in order to detect deviations of the electrical response. These deviations should than be compared, and if possible, be correlated with the parameters of the rock status. In general the volume resistivity increases with decreasing porosity Tables 1, 4. The phase angle increases with increasing inner surface area. The time constant τ_{gb} decreases for the relaxation process with decreasing surface area.

Table 4: Bulk resistivity data, (R_b) bulk relaxation times (τ_b) of the Falkenberg samples. Pore filling were distilled water 0.5 M and 2.0 M NaCl. Parameters of the Archie equation a= 1.5, m= 1.58, ϱ_{el}= 1 x 10^5, 0.24, 0.05 ohm-m, porosity data are from Table 1. The time constant τ_b for the volume relaxation process was calculated from the data with distilled water as a pore fluid.

sample	25.40	50.00	70.40	71.00	279.55	286.80
R_{Archie} (10^4)	7.1	7.7	25	14	9.5	8.3
R_{bulk} [x10^4]	0.22	0.76	0.96	0.32	0.3	0.23
τ_b(sec) x 10^{-5}	7.9	4.2	6.3	0.32	-	18
sample	25.40	50.00	70.40	71.00	279.55	286.80
R0.5 M NaCl	306	127	-	217	-	-
	245	265	390	250	-	-
R2M NaCl	50	25	39	56	-	-
(ohm m)	51	55	81	52	-	-

Results of the resistivity measurements in distilled water with a resistivity of 1 x 10^5 ohmm on the Falkenberg core samples from 25.4, 50.00, 70.40, 71.00 m in depth are compiled in Fig. 5. The phase angle in frequency range 1 x 10^4 Hz up to 1 x 10^5 Hz shows two polarization processes. In the high frequency region decreases the phase with increasing porosity, and the time constant of this volume polarization process is decreased. In the low frequency region dominates the contribution of surface conductivity, due to the decrease in the crack width an interaction of the DL on each side of the crack. Figure 6 shows the results of the measurement on sample 70.4 m using distilled water, .5 M NaCl and 2 M NaCl as a pore electrolyte. In this case the pore geometry is fixed, but the thickness of the DL is decreased with increasing electrolyte concentration (Debeye and Hückel 1923). The rate of interaction between the DL's on each side of the crack is thereby reduced.

Table 5: Parameters of the equivalent circuit for Falkenberg sample HB 50.00 in NaCl and KCl solutions of different molarities.

NaCl	τ_b (sec)x10^{-5}	τ_{gb} (sec)x10^{-2}	R_b (ohm m)x10^{3}	R_{gb} (ohm m)x10^{2}
dist. water	4.1	2.9	7.6	-
.00005	4.1	2.8	5.4	3.9
.0001	1.5	2.7	3.8	1.9
.001	0.78	2.7	1.9	2.2
.01	0.44	0.74	0.82	3.5
.1	0.29	0.42	0.36	2.8
.5	0.22	0.10	0.13	1.4
KCl				
0.001	0.71	0.014	1.5	17.2
0.01	0.40	0.11	0.95	17.2
0.1	0.16	0.32	0.79	0.8

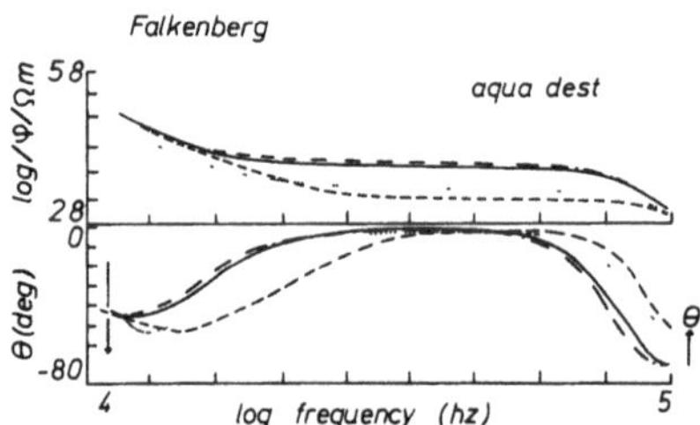

Fig. 5: Bode plots of the electrical data of the Falkenberg granite samples HB 25.4 (dashed line), 50.00 (dash and point), 70.4 (solid) and 71.00 (dotted line) measured with distilled water as a pore fluid

The decrease of the impedance as a function of the electrolyte conductivity covers more than 3 orders of magnitude and cannot be explained by the Archie equation (Table 4,5). The phase angle decreases in the high frequency part of the spectrum, and increases in the low frequency region with increasing electrolyte concentration. The thickness of the electrochemical DL is thereby decreased (Table 3)and the surface area involved in the interfacial polarizations increased. The relaxation time for the bulk effect of the sample is thereby decreased (Fig. 6). In the case of 2 m NaCl electrolyte pure ohmic conduction and no dissipation current is from approximately 100 kHz down to 0.5 Hz.

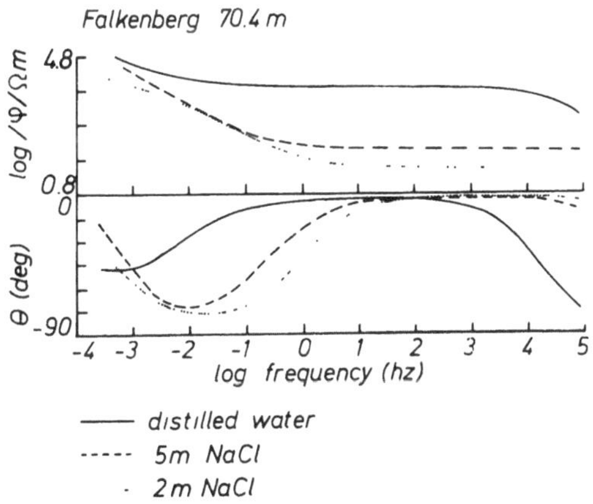

Fig. 6: Bode plots of the Falkenberg sample HB 70.4 with electrolytes of different molarity in the pore space of the sample

An increase of the confining pressure on core samples reduces the permeability due to a closing of the cracks. The electrochemical result is an overlapping and an interaction of the DL's on each side of a crack. Now two effects

a) surface conductivity and b) decreasing pore volume

are operating contrary. The closing of the cracks increases the rock resistivity, but the overlapping DL's increase the contribution of the surface conductivity with the result, that the impedance of the sample 70.40 (Fig. 7) is decreased up to a pressure of 150 bar. In this way increasing load produces an exchange reaction between the two double layers on each side of the crack. The electrical effect is a shift of the time constant of the relaxation process to shorter times, when the double layers begin to overlap. A further increase of the confining pressure leads to a rapid increase of the impedance caused by a decrease of the conduction paths. An increase of the pressure decreases the phase angle in the high frequency region, thus decreasing the relaxation time for the samples bulk polarization process. In the low frequency region one can observe an increase of the phase angle under increasing pressure conditions, too. The relaxation times due to surface effects are thereby increased. A further increase of the confining pressure leads to a rapid increase of the impedance caused by a closing of the conduction pathes in the pore system.

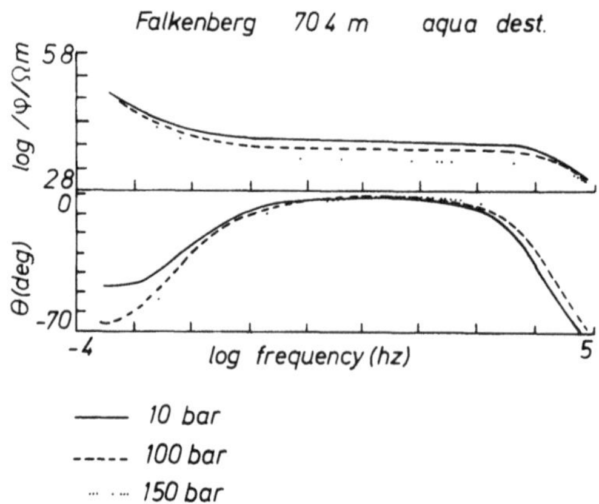

Fig. 7 Bode plot of the Falkenberg sample 70.40 with distilled water as a pore saturand, measured under different confining pressure conditions

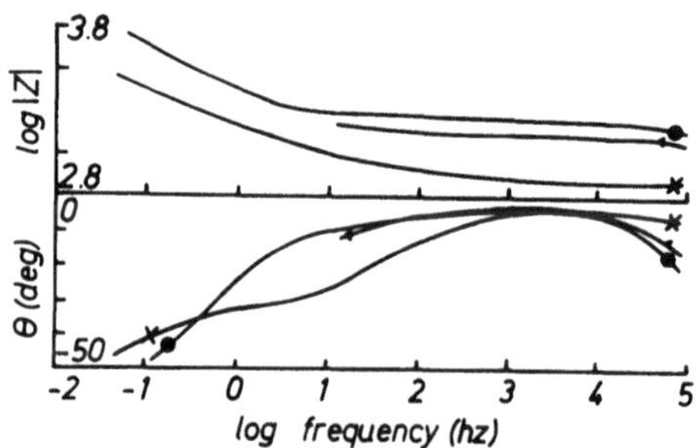

Fig. 8. Bode plot of the Falkenberg sample 25.40 m with a pore electrolyte of 10 g/l NaCl. (x 10, <300, 500 bar confining pressure)

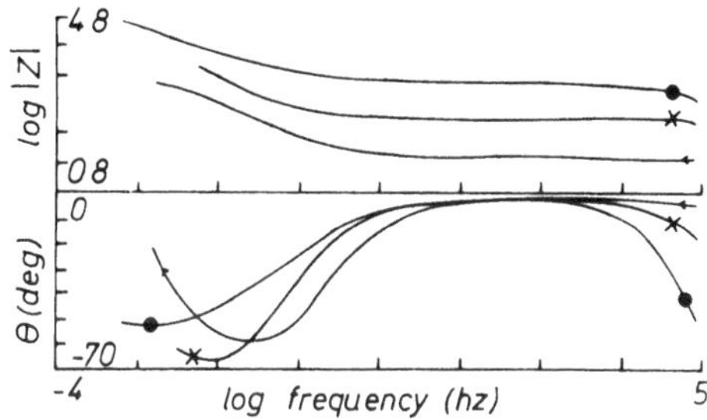

Fig. 9. Bode plot of the Falkenberg sample 25.40 m (● distilled water, ✗ 0.5 m NaCl, < 2 m NaCl) at a fixed pressure of 10 bar

4.0 Discussion

The conduction process in rocks with a narrow crack system is influenced by the degree of interaction of the double layers on each side of a crack. The time constant of the volume and grain boundary polarization is varied if either the characteristics of the electrolyte are changed, or if the geometry of the pore space is decreased by an external pressure. To seperate the bulk polarization effect from the surface polarization effects, both,the "thickness" of the electrochemical double-layer (DL), and the crack geometry were varied in our experiments.

Increasing the electrolyte concentration means to increase the contribution of the ohmic electrolytic conduction process. In this way the contribution of the surface conductivity is reduced, due to the decreasing thickness of the double layer. The bulk resistivity data are then comperable to the calculated "Archie" resistivities (Table 4) of the core sample.

The low frequency relaxation is shifted towards lower frequencies with shorter time constants for τ_{gb} with increasing salt concentration in the pore fluid (Table 4, 5).

An increase of an axial load perpendicular to the crack orientation leads to a closing of cracks. The time constant for polarizations at the inner surface of a rock sample (Fig. 7) depend on the surface area that is involved in the water/rock interfacial polarization. If we consider an electrolyte of low salinity, then the contribution of surface conductivity can t be neglected, because the thickness of the double layer is stronly increased in low electrolyte concentrations. The surface area involved for that particular polarization is thereby decreased, if the thickness of the DL exceeds the crack width. If the thickness of the immobile part of double layer is in the dimensions of narrow cracks in a rock, then we will get different viscosities if two neighboured double layers overlap (active and passive region in Fig.1). The mobilities of ions in these regions are different, and this can be measured either by frequency dependent complex resistivity measurements, due to the fact that the relaxation time of polarization processes are different, or by direct measurement of the mobilities of particles. In each of the two experiments such an effect is correlated with a shift of the capacity of the double layer. So one should compare the change of the capacitors from both experiments zeta-potential and complex resistivity (Table 6).

Table 6 Capacities for a plate capacitor (ZP) as calculated from the ZP measurements and complex resistivity data (CR). The factor of geometry for equal surface areas is not regarded.

molarity M KCl	capacity F(ZP)	capacity F(CR)
0.0011	-	8.1×10^{-8}
0.003	9×10^{-13}	-
0.006	1×10^{-12}	-
0.01	-	6.4×10^{-7}
0.017	2×10^{-12}	
0.034	3×10^{-12}	
0.1	-	3.8×10^{-6}

We can see that in both experiments the capacity decreases with increasing electrolyte concentration due to the fact of the decreasing thickness of the double layer.

The results discussed so far have shown that it is possible to distinguish by frequency dependent complex electrical resistivity measurements between the bulk polarization of the sample and the polarizations on the inner surface. Fluid flow phenomena are strongly influenced by the area of the solid/liquid interface in a pore system, and thereby different polarizations should influence the fluid flow characteristics. To improve this, further measurements were done, in which the electrical parameters and the water permeability were monitored at the same time. The decrease of the gas-permeability (Ar) in the pressure range 50-200 bar, for the sample HB 70.40 m is shown

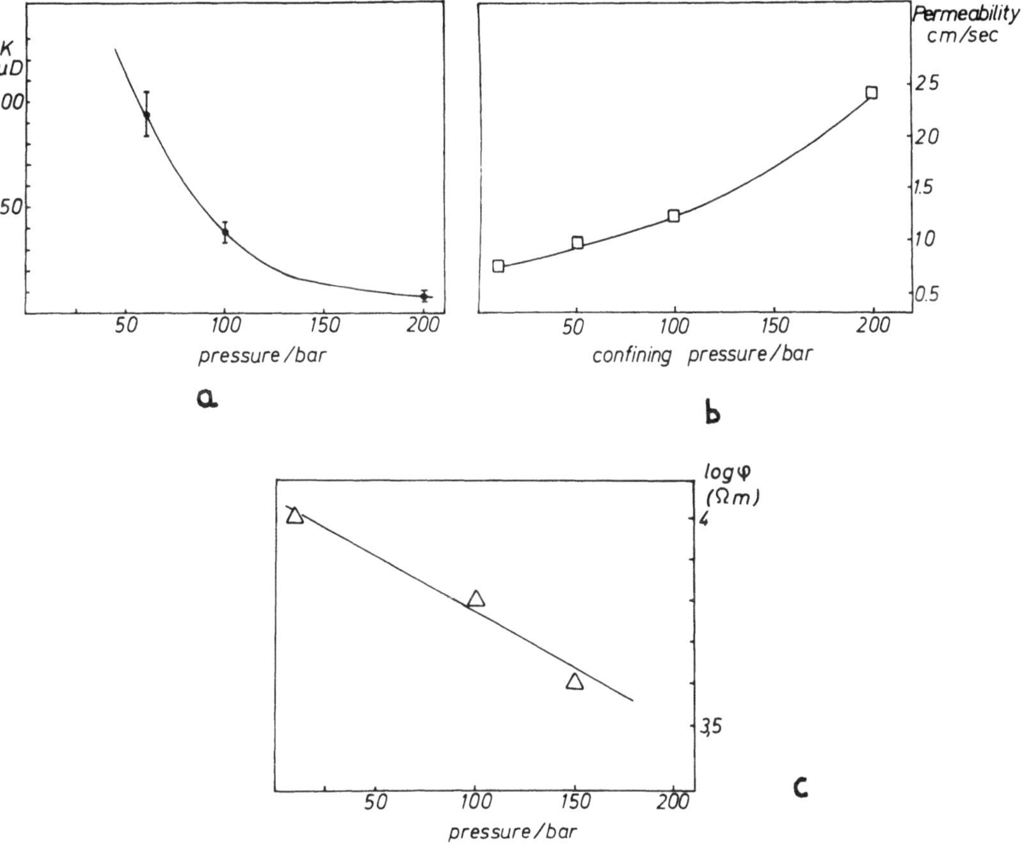

Fig. 10 a. The permeability of sample 70.40 m measured in the pressure range of 60-200 bar confining pressure with. b. The permeability for water under the pressures of c. Bulk resistivity of the sample in the same pressure range. A decrease of the resistivity due to surface conductivity can be observed up to pressures of 200 bar

in Fig.10a, Figure 10b shows the water-permeability (pressure gradient is 1.5 bar) for equal experimental conditions. The flow rate of water is increased by about a factor of 5 when the confining pressure is increased up to 200 bar. A further increase of the confining pressure then reduces the flow rate. This observation can be explained by a change of the mobility of water molecules in the pore structure when the DL's begin to overlap under increasing uniaxial load.
The impedance of the sample is decreased too Fig. 10c. A similar experiment performed with 2 M NaCl as a pore fluid shows a nearly

unchanged impedance and a decrease of the flow rate, due to the small thickness of the DL. This experimental observation is not a general result but due for the sample HB 70.40 m. The characteristics of this sample were a hydrothermal alteration process with the result of a nearly "polished" inner surface due to the solving of minerals. Thus the pore radii distribution shows a maximum in the range of more than 200 nm, which is more than two times the thickness of the DL.

Acknowledgement

This work was supported by the Deutsche Forschungsgemeinschaft (DFG), Bonn, West Germany, under grant Wi 349/23-3, which is gratefully acknowledged. We also wish to thank Prof. Dr. Al Duba (Lawrence Livermore Laboratories), and Prof. Dr. E. Hinze (Mineral. Inst. Universität Giessen) and Prof. Dr. K.-F. Seifert (Bonn) for helpful discussions.

References

Ahmed S, and Maksimov D (1969) Studies of the double layer on cassiterite and rutile. J Coll Interf Sci 29, 1: 97-104

Anorgan Y and Madden TR (1977) Induced polarization: a preliminary study of its chemical basis. Geophysics 42, 4: 788-803

Armstrong RD and Firman RE (1973) Impedance plane display of the adatom model for metal dissolution/deposition. Electronanal Chem Interf Electrochem 45: 257-266

Becker R, Lentz H, Hinze E, Nover G and Will G (1986) Ein Quecksilberporosimeter zur Charakterisierung mineralischer Stoffe. Berichte der Bunsenges Phys Chem 90: 833-838.

Brauer E und Piroth J (1986) Impedanzspektroskopie in der Elektrochemie. GIT Fachz Lab 6: 533-543

Cammann K (1973) Das Arbeiten mit ionenselektiven Elektroden. Springer, Berlin Heidelberg New York

Chapman DL (1913) Philas Mag 25 6: 475

Cole KS and Cole R (1941) Dispersion and absorption in dielectrics. J Chem Phys V9: 341-351

Davies JT and Rideal EK (1961) Interfacial phenomena. Academic Press New York London 56-107

Debeye P and Hückel E (1923) Phys Z 24: 185

Delahay P (1965) Evolution of ideas on the electrical double layer. Double layer and electrokinetics. John Wiley & Sons, New York

Duba Al, Huenges E, Nover G and Will G (1988) Impedance of Black Shale from Münsterland 1 Borehole: An Anomalously Good Conductor? submitted to Geophys J

Fuller BE and Ward SH (1970) Linear System Description of the Electrical Parameters of Rocks. IEEE Transact on Geosci Elect GE-8 1: 7-13

Gouy G (1910) J Phys 9 9: 457

Grahame DC (1952) Mathematical theory of the faradaic admittance. J electrochem Soc 99: 370-385

Hasted JB (1973) Water- a comprehensive treatise. Plenum New York vol. 1: pp. 255-458; vol 2: pp 405-458.

Helmholtz H (1879) Wied Ann 7: 337

Hills GJ (ed) (1970) Electrochemistry. Chem Soc, Durlington House London, pp 117-167

Huenges E (1988) Messung der Permeabilität von niedrigpermeablen Gesteinsproben unter Drücken bis 4 kbar und ihre Beziehung zu Kompressibilität porosität und elektrischem Widerstand. PhD Thes Univ Bonn

Jonscher AK (1978) Analysis of the alternating current properties of ionic conductors. J Math Sci 13: 553-562

Keller GV and Frischknecht FC (1966) Electrical methods in geophysical prospecting. Pergamon, New York

Mansfeld F (1981) Recording and analysis of AC impedance data for corrosion studies. Corrosion 36, 5: 301-308

Mason PR, Hasted JB and Moore L (1974) The use of statistical theorie in fitting equations to dielectric dispersion data. Advanc molec relax proc 6: 217-232

Mc Cafferty E und Zettelmoyer AC (1971) Adsorption of Water Vapor on α-Fe_2O_3. Disc Faraday Soc 52: 239

Mund K (1986) Untersuchung poröser Elektrodenstrukturen mit Impedanzmessungen. Dechema Monographien 102 Grundlagen von Elektrodenreaktionen VCH Verlagsgesellschaft

Ney P (1973) Zeta Potentiale. Springer Berlin Heidelberg New York

Nover G, Hinze E and Will G (1984) Elektrische Leitfähigkeitsmessungen an Gesteinen in Abhängigkeit von Druck, Temperatur und Gesteinsstatus. Forschungsbericht T84-279, Fachinformationszentrum, Karlsruhe

Nover G, Huenges E and Will G (1987) Messung der Frequenzabhängigkeit elektrischer Gesteinswiderstände unter in situ Bedingungen. Abschlußbericht zum Forschungsvorhaben 03E-6187-A, BMFT

Olhoeft GR (1986) Electrical properties of rocks and minerals. Short Course Notes, Golden Colorado

Pottel E (1973) In: Frank F (ed) Water- a comprehendive treatise. Vol 3. Plenum New York, pp 401-432

Schmickler W (1986) Die Doppelschicht in wässriger und nicht wässriger Lösung. In: Dechema Monographien 102 Grundlagen von Elektrodenreaktionen, VCH Verlagsgesellschaft

Schön J (1983) Petrophysik. Enke, Stuttgart

Schultze JW (ed.) (1980) Grundlagen von Elektrodenreaktionen. Vol 102 Verlag Chemie Weinheim

Sheppard R J and Grant E H (1974) Alternative interpretations of dielectric measurements with particular reference to polar liquids. Advances in Molecular Relaxation Processes 6 61-67

Stern O (1924) Zur Theorie der elektrolytischen Doppelschicht. Z Elektrochemie 30 508

Washburn J C (1982) Parameterization of spectral induced polarization data and Laboratory and in situ spectral induced polarization measurements: West Shasta copper-zinc district, Shasta, CA. Thesis, Colorado School of Mines, Golden Colerado

Will G and Nover G (1986) Measurment of the frequency dependence of the electrical conductivity and some other petro-physical parameters of core samples from the Konzen (West Germany) drill hole. Annales geophysicae 4 B2 173-182

The MIMAFO Direct Current Cross-Hole Method : A Support to Hydraulic Investigations

G. Pottecher and C. Poirmeur[1]

Abstract

A direct current geophysical method using a pole-pole array between two boreholes, named MIMAFO, can provide structural information in all kinds of rocks. Data interpretation is made possible by flexible 3D modelling software based on a surface integral algorithm.

Apparent resistivity data interpretation through model fitting yields the spatial distribution of electrical resistivity. Electrical field data interpretation yields connections between the two drillholes by the recognition of characteristic patterns.

A survey has been carried out on a site used for cross-hole hydraulic tests. The electrical connections match the hydraulic connections.

[1]Bureau de Recherches Géologiques et Minières,
Département Géophysique,
BP 6009, 45060 Orléans Cédex 02, France

Lecture Notes in Earth Sciences, Vol. 27
G.-P. Merkler et al. (Eds.)
Detection of Subsurface Flow Phenomena

1 Introduction

The hydraulic properties of hard rock at great depth determine widely the possibility or the impossibility of carrying out some applications like storage of radioactive waste and hot, dry rock geothermy. In both cases permeable areas must be localized, either to avoid them or to use them. They can be determined by single hole observations, directly with hydraulic tests or indirectly with geophysical logging, core inspection and water geochemistry. Cross-hole hydraulic and geophysical measurements outline the extension of the fractures in the volume, which is a fundamental factor in waste repository design and in geothermal exchange capabilities evaluation. The geophysical methods alone provide 2D or 3D geometric information and they may replace some hydraulic tests, if enough experience has been gained on a given site.

The MIMAFO method, a BRGM concept used in mining exploration, can determine permeable structures in hard rock as they are electrically more conductive than the surrounding medium. However, it will also detect conductive but dry structures such as intrusions or a fossile hydrothermal system, but may miss some permeable discontinuities.

After explaining the principles of MIMAFO, this chapter will demonstrate with a field study how it can contribute to hydraulic exploration.

2 The MIMAFO Method

This method consists in exploring the zone comprised between two boreholes with a large number of measurements, as is done in tomography.

The data acquisition uses a pole-pole array, namely a current injection between a moving electrode A in the injection hole and a fixed remote grounding B, while the potential is measured between a moving electrode M in the second hole and a remote reference N (Fig. 1).

The complete data set is represented as a potential map on a conventional X, Y diagram where X and Y are respectively the measurement and injection depths. Three other maps can be derived from the potential map (Fig. 2):

1. The X-gradient map is the electrical field E_m that could have been measured in the receiver hole instead of the potential;

2. The Y-gradient map is the electrical field E_i that would have been measured in the same manner but with exchanged injection and receiver holes (reciprocity);

3. The apparent resistivity map is computed with the formula:

$$\text{rhoa} = 4\pi \frac{V}{i} \cdot \left(\frac{1}{AM} + \frac{1}{A^*M}\right)^{-1},$$

where $\frac{V}{i}$ is the measured potential normalized by the injected current and A* is the electrical image of A. B and N can also be taken into account if more precision is needed.

These four maps undergo a quantitative or qualitative interpretation, possibly after some signal processing to enhance contrasts.

Interpretation differs from a tomographic inversion as the electric potential behaves nonlinearly with respect to the ground resistivity distribution. Therefore, interpretation is done in comparing the field data to computed 3D models. The currently used algorithm is based on the surface integral method (references). It can compute the potential produced by any point electrode array and assumes a layered half-space including limited inhomogeneities such as spheres, hexahedrons and joined hexahedrons. The mesh density on the surface of these bodies is optimized as a function of the distance to the electrode locations.

This approach could be extended to other inhomogeneous shapes in implementing the relevant meshing algorithms and to nonlayered host medium structures, provided that Green's function can be computed easily. Extended current sources would need another formulation of the equations to be modelled.

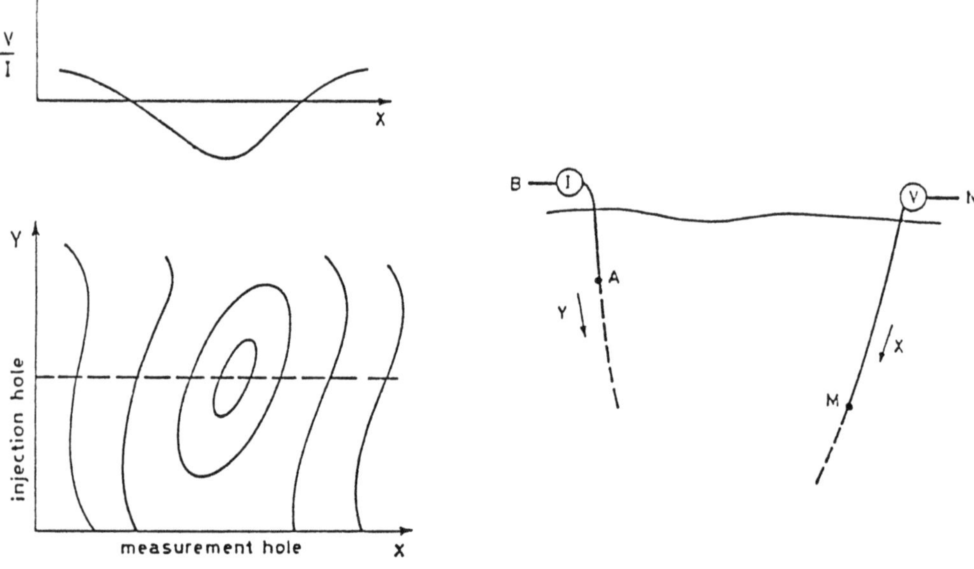

Fig. 1. Principle of operation and plot diagram for the MIMAFO method

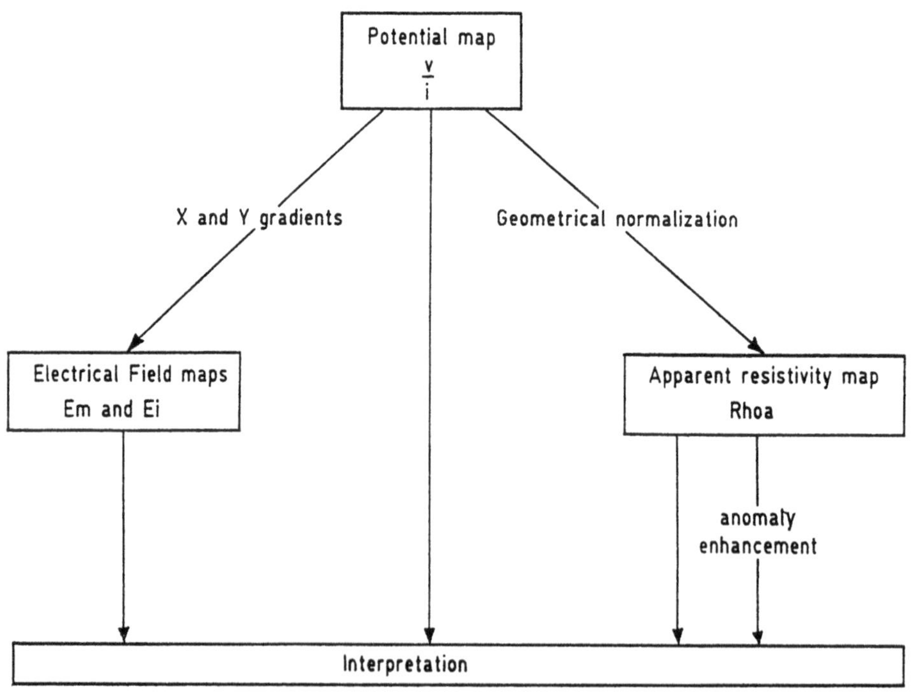

Fig. 2. MIMAFO data processing

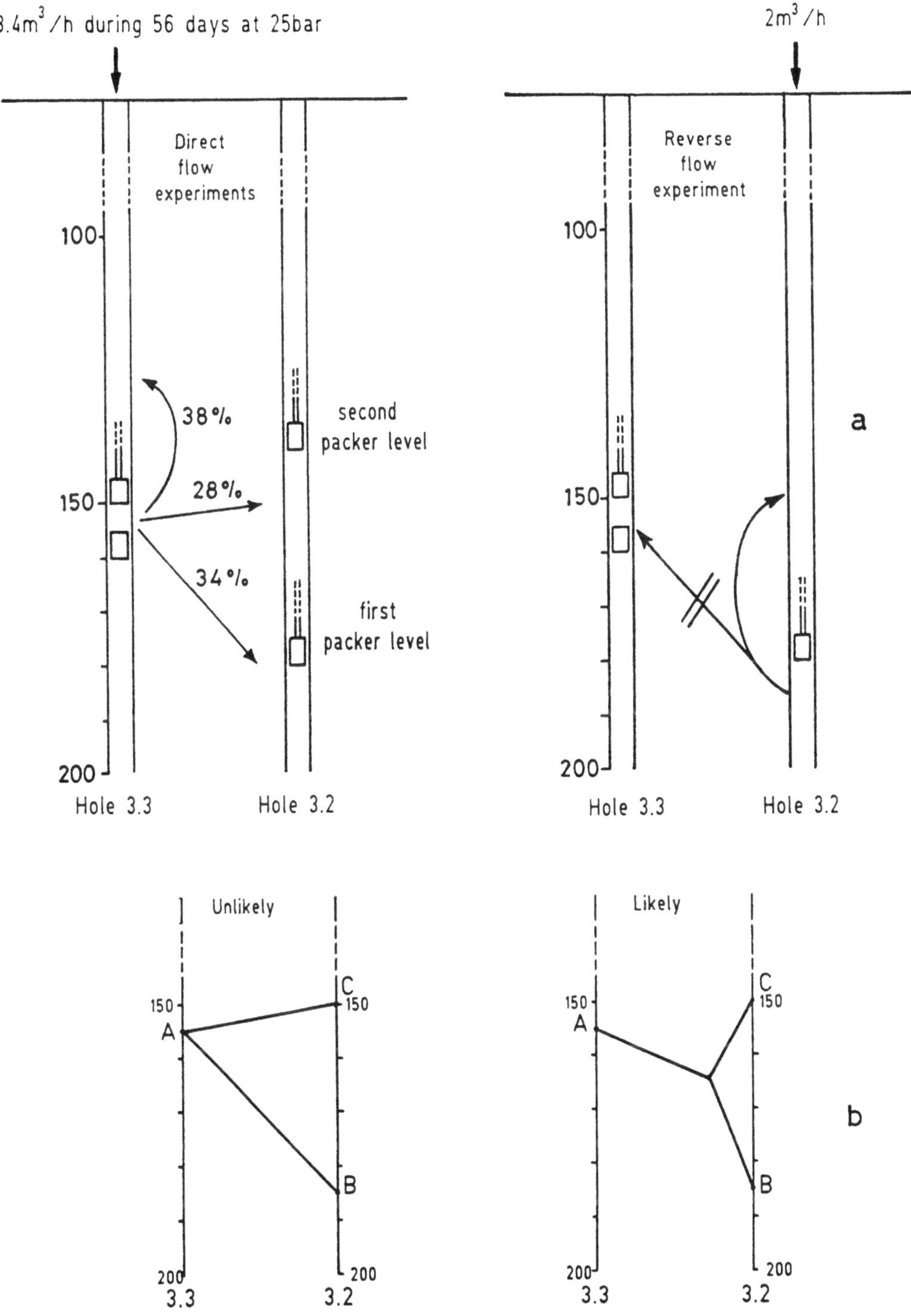

Fig. 3. Cross hole hydraulic test (INAG); possible hydraulic connections

MIMAFO has the drawbacks of all electrical methods: a lower resolution compared to some other geophysical methods, the nonlinearity of the inverse problem and relative inefficiency in the vicinity of metal bodies. On the other hand, it has a long range (more than 300 m with a 1 kW transmitter) and is a robust and quick process in the field. It investigates not only the zone comprised between the drillholes but also the volume around them. It can be used in all kinds of rocks, crystalline as well as sedimentary.

3 Hydraulic Tests at Le Mayet de Montagne (France)

The INAG (National Institute for Astronomy and Geophysics) carried out a small-scale "hot, dry rock" experiment at Le Mayet in the late 1970s. It aimed to determine the possible flow rates and the thermal exchange regime of water flowing between two boreholes. The holes named 3.2 and 3.3 were drilled into granite at a maximum depth of 200 m and 30 m apart. Videocamera and geophysical logging were used to determine the fractures intersecting them. A fracture at 186 m in 3.2 was widened by hydraulic fractionation and sustained with calibrated sand.

The first cross-hole experiment (Fig. 3a) was then carried out with water injection at 156 m in 3.3 between two packers; 3.4 m^3/h of water was injected during 56 days at 25 bar. The water was collected in 3.2 underneath a packer placed first at 180 m and later at 140 m. Water temperature measurements showed that the water entered 3.2 at two points only: at 186 m (36% of the injected water) and 150 m (28%). The remaining 38% flowed back into 3.3 over the injection packer. The water temperature evolution at the 186-m outflow could be fitted with a finite difference model in assuming a laminar flow within a rectangular fracture joining the two holes.

A second kind of experiment took place between the two previously described injections. Immediately after the injection with the packer at 180 m in 3.2, the pressure was released in 3.3 so that water flowed out through the injection chamber A, in the reverse direction as previously. When water was injected into 3.2 through B, it did not affect the outflow in A, but all of it came out in 3.2.

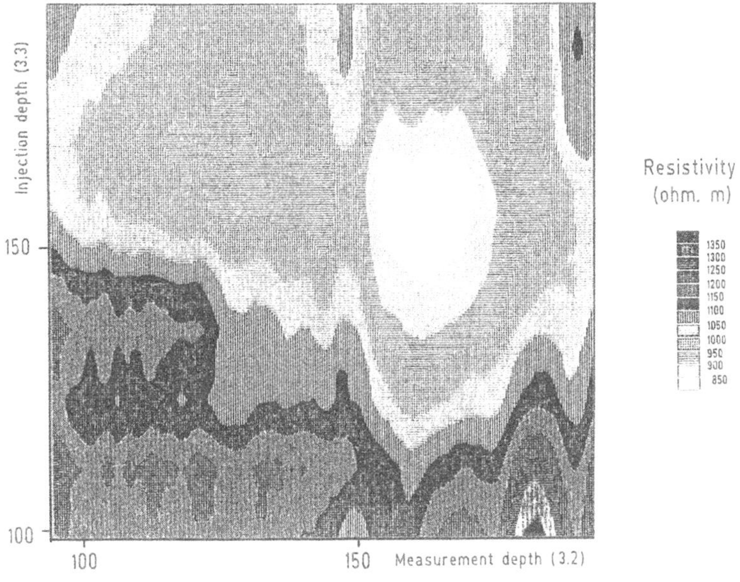

Fig. 4. Apparent resistivity data (Le Mayet)

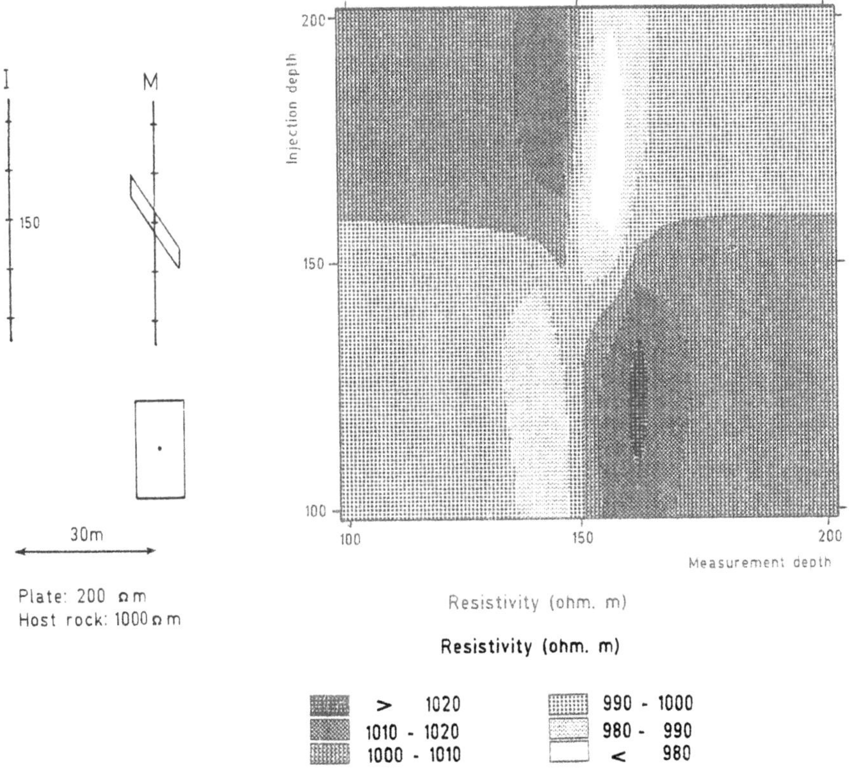

Fig. 5. Apparent resistivity map for a simple plate model

Concerning the flow paths, this suggests that B and C are two branches of a single channel starting in A, in opposition to a two-channel hypothesis (Fig. 5b).

4 MIMAFO Measurements and Interpretation

Electrical cross-hole data were acquired in the same pair of boreholes in fall 1987. Thus, 34 injections were carried out in 3.3 between 92 and 192 m, at a 2-m step between 130 and 172 m and a 4-m step elsewhere. The potential was measured in 3.2 from 94 to 194 m at a 0.5-m step. The remote electrodes B and N were grounded 1.5 km away from the site in opposite directions. Outstanding data quality has been obtained despite a power line running on the site.

The apparent resistivity map (Fig. 4) shows smooth and well--defined anomalies of various wavelengths, the most prominent being a flat minimum at depth (165 m, 155 m) and elongated maxima at (148 m,-), (-,112 m) and (-,145 m). It is interpreted by fitting computed models to the field data. Experience helps to find the correct structures, but many computations are necessary in each survey to understand the combined influence of the ground surface, of the borehole spacing and of nonlinearity.

We assume that this data set can be interpreted with conductive plates only, representing fracture zones. Figure 5 shows the basic quadripolar anomaly generated by a plate intersected by the measurement hole. As a rule, a resistivity maximum occurs when conductive material is between the A and M electrodes and a minimum when it is behind or lateral to them. The anomaly pattern depends on the plate size, dip, centering relative to the borehole and conductance value (thickness/resistivity). Only the 2D dip in the plane of the holes is important, the 3D dip component is a secondary parameter.

Combining two structures in a model "sums" the respective anomalies if the structures remain distant from each other (Fig. 6). This result eases the interpretation of anomalies parallel to the X or Y axis in terms of small structures close to the injection or measurement hole respectively. High pass filters can be used at this point to enhance the

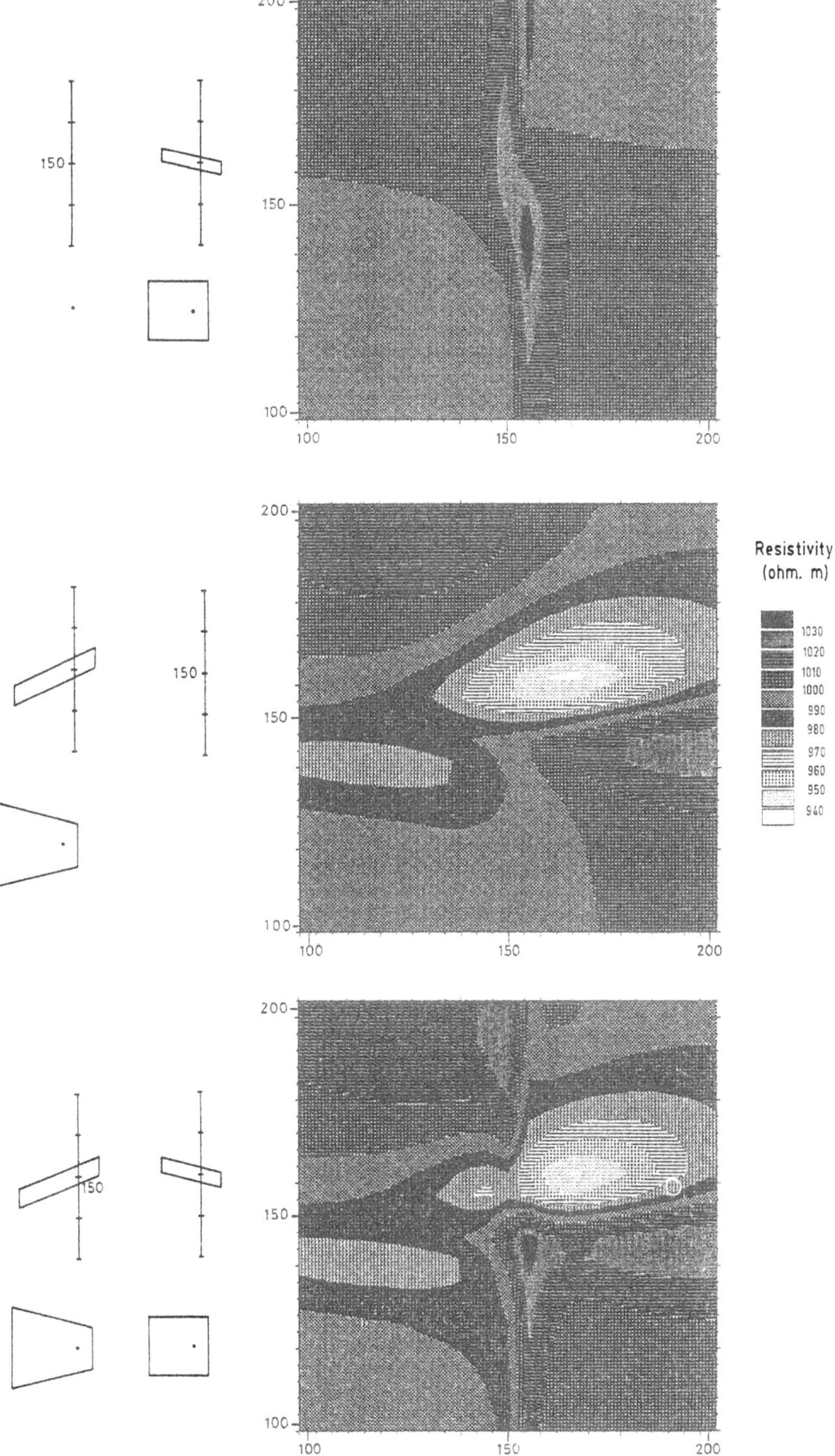

Fig. 6. Linear combination of resistivity responses, in the case of noncoupled structures

MIMAFO field data - Le Mayet de Montagne

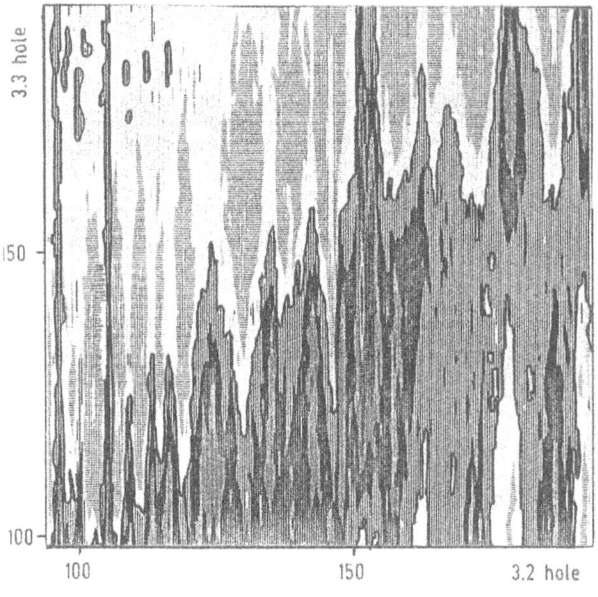

Electrical field Em
with injection in 3.3
Computed with
a 1m span filter
(mV/A. m)

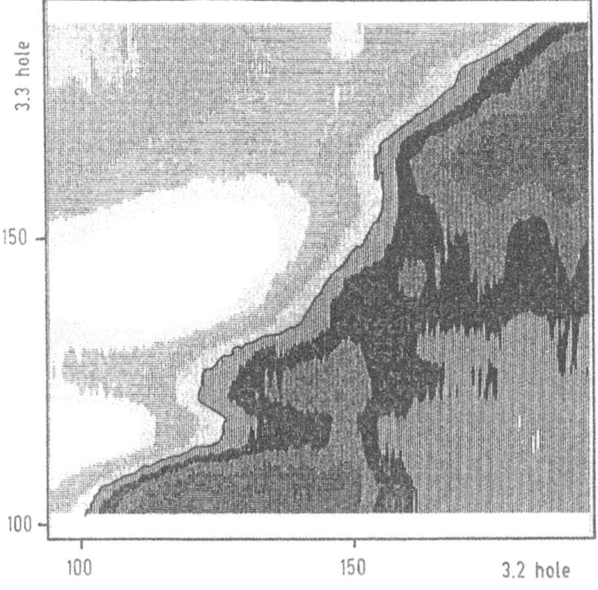

Electrical field Ei
with injection in 3.2
Computed by reciprocity
with an 8m span filter
(mV/A. m)

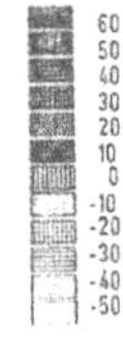

Fig. 7. Electrical field data

anomalies in a qualitative interpretation phase. The second derivatives of the apparent resistivity in the X and Y directions proved to be useful.

Coupling must be taken into account if large structures are to be determined. Once the main features of the map have been fitted, smaller details can be adjusted, providing information on structures not intersected by the boreholes. In the end, apparent resistivity interpretation yields the distribution of resistivity in space. Some 3D information will be obtained if contraints are put on the model, e.g. the structure flatness in this case. At Le Mayet, the resistivity map interpretation was stopped after obtaining the intersection depth of the main structures, their 2D dip and their centering relative to the boreholes.

As expected from the theory, the electrical field (Fig. 7a, b) is positive when the measurement point is above the injection point and inversely, and it tends to 0 when the electrodes move away from each other.

Interpretation is still supported by modelling but to a limited amount as it is qualitative. Figure 8a, b, c shows the three basic features:

1. An undulation of the 0 isoline occurs where the borehole crosses a conductive structure (a, b and c);

2. A sharp dissymmetry between the positive and negative areas indicates an upward or downward dip (b);

3. A local field intensity maximum indicates an electrical connection between the boreholes.

Direct map reading provides all information, making this procedure very efficient.

Some features could not be interpreted on the E_m map (Fig. 7a): e.g. the local field inversions in the lower right corner. A steep connection between the two holes or 3D effects might be the answer.

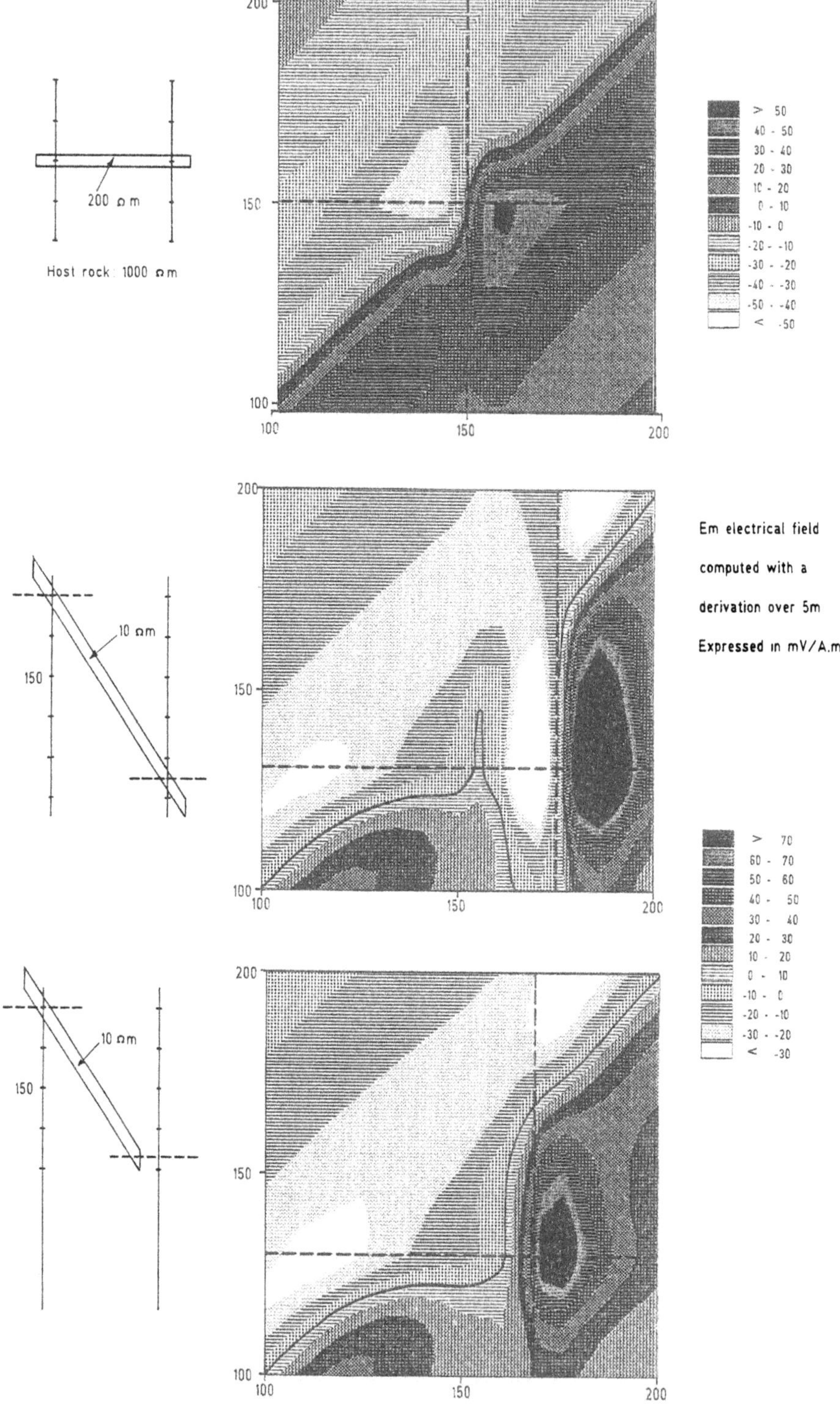

Fig. 8. Electrical field model computations

5 Comparison of results

Figure 9 presents all information gathered with MIMAFO and cross-hole hydraulics.

Thirty intersected structures could be positioned in both holes and an indication on their 2D dip is available in most cases. Moreoven 21 electrical connections could be established with good confidence (the lines on the figure show the existence of connections, not their geometry). In addition, the main extension of major structures is given, either between the boreholes or outside.

The water injection interval A includes the top of an oblique structure that is probably connected with two structures joining the B outflow area (broken line). Nothing appears in B itself, although the outflow depth has been determined with 1-m precision. A possible explanation is that a fracture complex surrounds B and links it to the other hole, and that B has been washed out of its conductive filling (clay ?) by the high water speeds.

The C outflow appears to be indirectly connected with A, as had also been inferred from the reverse hydraulic experiment. It is located inside one of the major MIMAFO structures.

Comparison with the 16" resistivity yields a partial agreement only between the two electrical methods, which is due to their very different types of investigation.

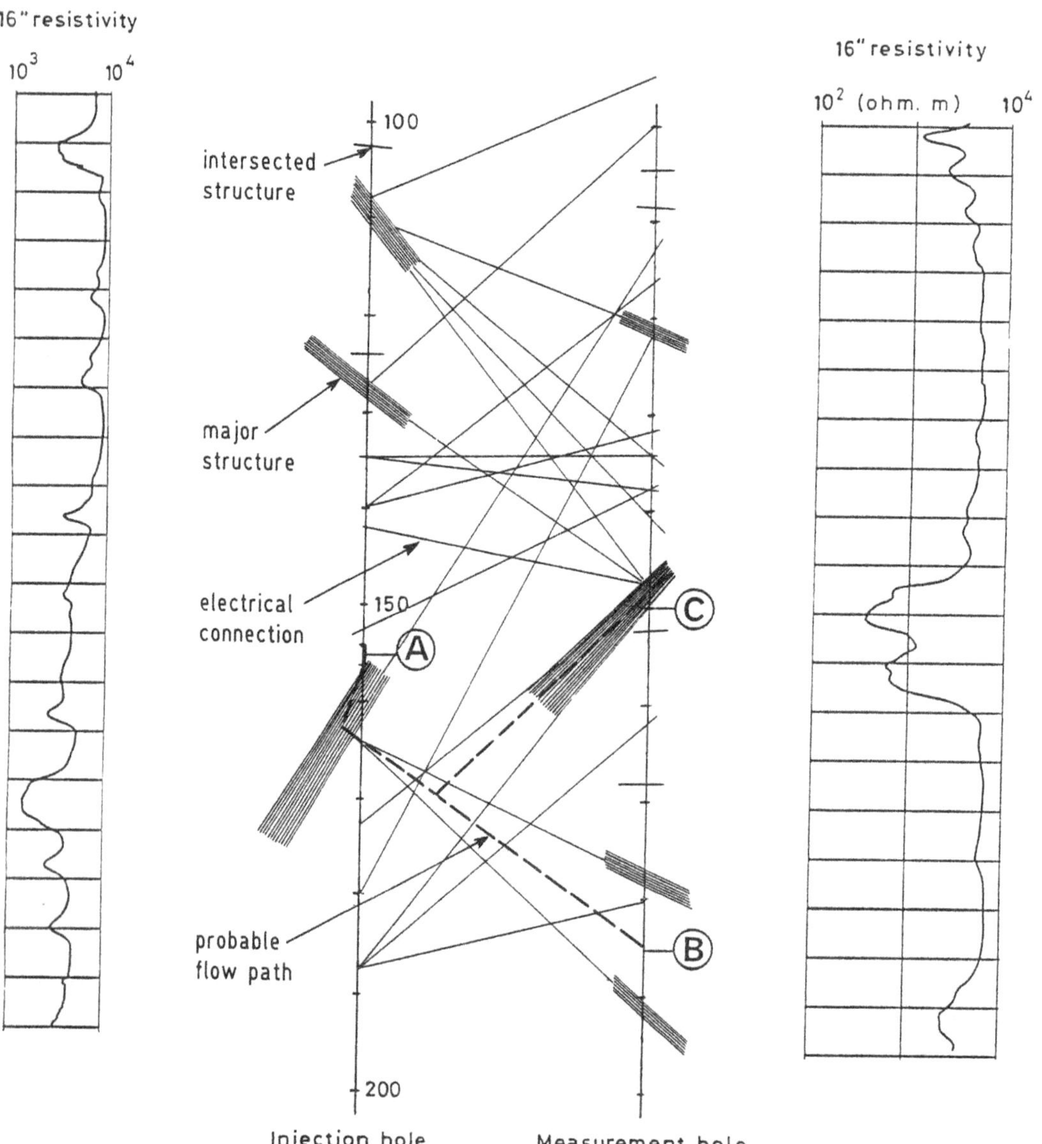

Fig. 9. Interpretation results

6 Conclusions

This case study shows how the structures of hard rock can be determined with MIMAFO between and around two boreholes. The results seem relevant to fluid circulation phenomena and may help the planning and interpretation of cross-hole hydraulics.

On other sites, the electrical and hydraulical conductivities might be less obviously related, however, structural information will be obtained.

In geothermal exploration, MIMAFO can help to locate the injection points and to choose an interpretation model. In waste repository design, it can point out short- and long-range rock structures and show disturbed zones that have to be ruled out for safety reasons.

Other applications will certainly be discovered in the future for this simple and general purpose method, complementary to more refined techniques such as radar and underground seismics.

Acknowledgements. We are very grateful to Mr. Cornet of INAG for permission to use the Mayet boreholes and for complementary information on the hydraulic tests.

References

BARTHES V, VASSEUR G (1978) Use of DC electrical sounding for the detection of a conducting heterogeneity buried in a stratified medium. Sem Geothermal Energy, Brussels, Dec 1977. Commiss Eur Commun, EUR S 920, V I, 309-319, Luxembourg

HOSANSKI J-M (1980) Contribution à l'étude des transferts thermiques en milieu fissuré. Thèse Doct Ing, Univ Pierre et Marie Curie - Paris VI et Ecole Nat Sup Mines, Paris

POIRMEUR C (1986) Modélisation tridimensionnelle en courant continu (méthode et applications à la prospection géophysique). Thèse, Univ Sci Tech Languedoc

POIRMEUR C, VASSEUR G (1988) Three dimensional modeling of a hole-to-hole electrical method; application to the interpretation of a field survey. Geophysics 53, 3: 402-414

Geoelectrical Measurements at the Salt Mine Asse to Observe an Underground Barrier Construction

U. Yaramanci and D. Flach [1)]

Abstract

In the salt mine Asse (FRG), the sealing, and other civil engineering characteristics of an underground barrier construction for a nuclear waste disposal site will be tested. One important aspect of the barrier construction experiment is to determine the brine migration properties in and around the barrier. To achieve this, direct current geoelectrical measurements were planned and partly conducted. Twenty-one boreholes with some 200 fixed elctrodes were used for the measurements, for which a fully computerized measurement system had to be developed. So far, preliminary measurements indicate a fairly homogeneous rock salt with 10^6 Ωm. Calculations with finite difference resistivity modelling on simulated models using different current configurations show that the brine migration can be easily detected. These calculations also help to choose an optimal measurement configuration.

[1)] Gesellschaft für Strahlen- und Umweltforschung, Institut für Tieflagerung, Theodor-Heuss-Str. 4, 3300 Braunschweig, FRG

Lecture Notes in Earth Sciences, Vol. 27
G.-P. Merkler et al. (Eds.)
Detection of Subsurface Flow Phenomena

1 Introduction

For the sealing of a nuclear waste disposal site, barriers at different levels and of different types are necessary. These barriers have to restrain eventual leakage of gases and fluids to and from the site and consequently restrain radioactive contamination of the biosphere. In order to study various aspects and properties of a barrier construction, a project was launched in the salt mine Asse (FRG) to build such a barrier and various experiments were designed for the investigation. Along with experiments concerning civil engineering aspects, many experiments were designed to study the sealing properties of the barrier. For this purpose a chamber with salt fill at one side of the barrier construction will be injected with gas and later with brine at high pressure. The aim of this test is to study the migration of the brine and gas in and around the barrier, particularly in the loosening zone.

Geoelectric methods are well-suited to study the diffusion, migration and leakage of brine in and around the barrier. Furthermore, using geoelectric methods, rock properties of the loosening zone around the barrier gallery can be studied, provided they notably influence the resistivity. A major advantage of geophysical methods is that structures can be studied with remote sensors, i. e. without interferring in the structure, which is especially important for a barrier sealing a radioactive site. Therefore, further investigations will be also carried out with seismic tomography, i. e. seismic sounding of the structure at different distances and frequencies, and microseismological observations will be conducted.

2 Test Site and Measurement Technique

The Asse Salt Mine is located 30 km south-east of Braunschweig. Formerly, it was used for commercial production of rock salt. However, since 1965 it has been used mainly for research purposes in order to study the problems of nuclear waste disposal. Asse is a subdued ridge elongated in the north-west direction about 8 km. It is about 3 km wide in the north-east direction. Rock salt deposited ca. 240 million years ago in the Zechstein has moved due to its plasticity and low density into the core of the present Asse structure. The rock salt has thereby broken and penetrated the flat-layered lower and middle sandstone regions (1750 to 2050 m) and has unified with the salt of the upper sandstone from the south-west flank. The main salt anticline is composed of rock salt of the Stassfurt and Leine series. Within the body of salt, in which the former mine Asse is located, there is a main anticline with a flat diving axis in the south-east direction. A depression separates a second-order, small-sized anticline to the south. The flat layered base of the Zechstein formation is about 2200 m deep and the top of the salt reaches a depth of 300 m.

The barrier construction to be investigated is located at a depth of 925 m. There are two barriers planned in two separate galleries (Fig. 1). The barrier in the

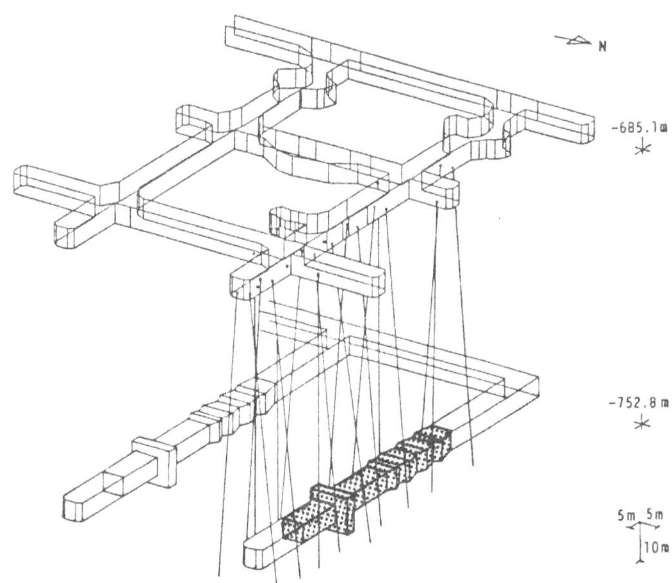

Fig. 1. Boreholes with fixed electrodes and planned barrier construction.

north gallery will be investigated using various geophysical, geomechanical and other methods. Sixtyseven m above the barrier gallery level there is an upper gallery system with two main galleries almost exactly over the barrier galleries and three cross-cutting galleries. The upper gallery system is used mainly for borehole instrumentation and to accomodate data collection systems. The barriers have an average cross-section of 8 x 8 m and are 60 m long with a 17 m long chamber at the end. This latter chamber will be filled with salt filler. It is connected to the upper level gallery by a borehole which will be used to pump the high pressure gas and later the high pressure brine into the chamber.

For direct current geoelectrical measurements, 21 boreholes were drilled from the upper level gallery (Fig. 1). They were placed pairwise to define a measurement section cross-cutting the barrier. There are nine such measurement sections. In each section the two boreholes of 74 m depth lead in approximately 4 m distance from the barrier. This distance is great enough that the boreholes do not influence the civil engineering characteristics of the barrier and are yet small enough that small voltages can be used for current injection. In every borehole there are ten separate electrodes, the deepest being 73 m, and each having 2-m spacing.

The geometry of the electrode array is designed to allow a great variety of measurement configurations. In order to fix the electrodes in the boreholes, a special concrete mixture containing salt was used. The usage of fixed electrodes is necessary since repeated measurements will be made over the span of a few years in order to investigate the change in apparent resistivity due to the loosening zone and brine migration. The boreholes should be sealed wherever possible in any case, in order to avoid possible artificial pathways, in which the brine could migrate.

The measurement system (Fig. 2) consists mainly of the geoelectric instrument with an AC-DC power supply converter, a multiplex box and a computer to control the

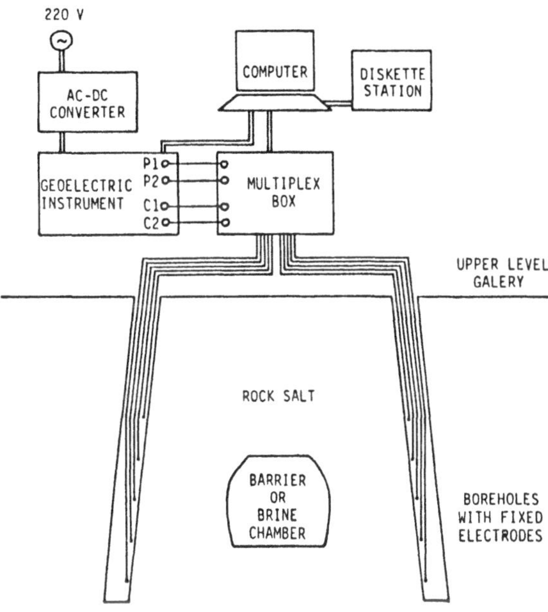

Fig. 2. Schematic of the geoelectric measurement system.

system and conduct the measurements. A computer is essential to regulate the measurements due to the great number of possible electrode configurations. Furthermore, the routine observation of a barrier and a nuclear waste disposal site must run with the highest possible automation. The computer controlling the measurement systems has three outputs, via standard serial interfaces, to the multiplex box, to the geoelectric instrument and to an external storage device, like a diskette or magnetic tape station. There are special programming statements designed and worked into the standard programming language which sets up necessary signals at the interfaces to drive the connected devices. In a measurement cycle, the first command provides the desired electrode connections by furnishing code numbers of current and potential electrodes. The second command is sent to the geoelectric instrument and contains the time parameters for the square wave current, number of square waves for stacking and the injection command. With the third command the result of the measurement with accompanying peripheral data, which are first recorded in the memory of the geoelectric instrument, will be sent over to the computer.

Preliminary processing, like calculating apparent resistivities, comparison to earlier measurements, in order to detect time variances in the resistivity, and mapping of the data can be done immediately. For further processing, the data can be transferred to more powerful computers via diskette, magnetic tape or telephone. It is possible to run a continuous measurement program by designing a proper computer program. At present there are few geoelectric instruments offered commercially which can be driven by a computer, so that only the multiplex box with special relays had to be designed and produced. The AC/DC converter for the power supply had to be modified also in order to be driven by the computer.

3 Measurements

The gallery for the barrier construction is as yet unbuilt, so the measurements taken were done in pure rock salt. These test measurements serve the purpose of an in situ investigation of the order of primary voltages needed to produce measurable current in salt and also to study the effect of contact resistance of the electrodes. In general, over a distance of 20 m, currents of an order of a few mA can be injected using approximately 80 V. In a special survey we were able to study the drying process of the concrete in the boreholes and consequently the change of the contact resistances of the electrodes. Within the first 30 days, the resistivity first increased rapidly and then slowly (Fig. 3). From then on it has remained relatively constant.

To determine the general resistivity, measurements were conducted using two boreholes of a measurement section. Many different current electrode configurations were used and for every current configuration all possible potential pairs were measured. Since there was no systematic behavior in the apparent resistivity to indicate a possible inhomogeneity in and around the boreholes, the results are displayed in a histogram (Fig. 4). It can therefore be safely said that the investigation area is fairly homogeneous and deviations are

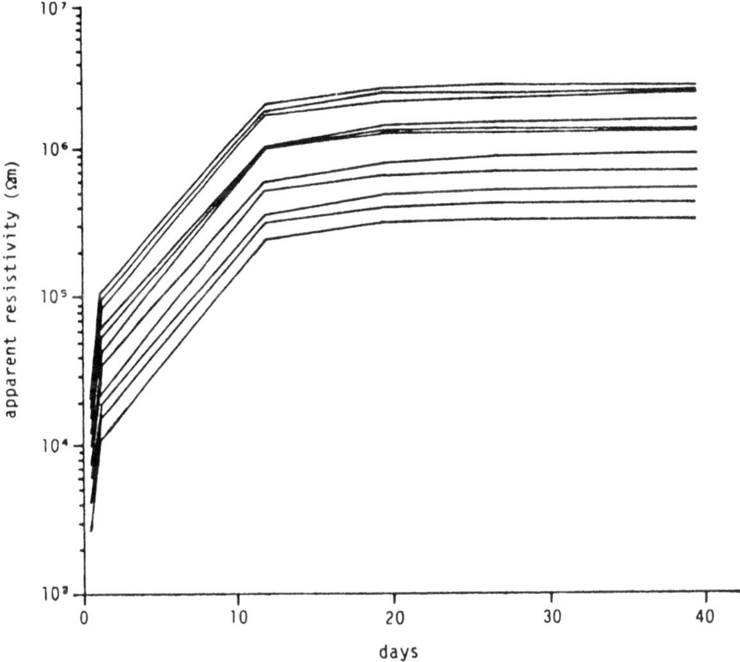

Fig. 3. Change of the apparent resistivity on different electrode pairs in a borehole sealed with special concrete.

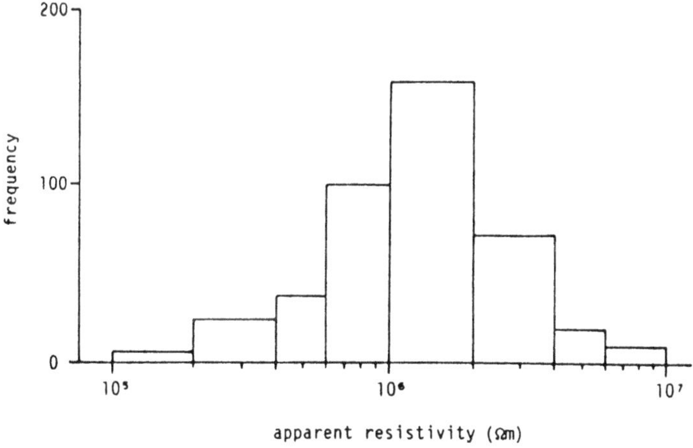

Fig. 4. Histogram of measured apparent resistivities in rock salt of Asse.

mainly due to the measurement errors. According to this measurement, in situ rock salt resistivity was determined to be 1.5×10^6 Ωm, with a relatively small deviation. This value agrees well with measurements conducted earlier at the gallery wall (Kessels et al. 1985) and is also in agreement with values cited by Matula (1981).

4 Models

In order to study the effect of the change in resistivity on the measurements due to brine in the chamber and brine migrations in the environment, the potential fields and apparent resistivity sections for a large number of models and current configurations were calculated. Further attention was directed to studying electrode configurations which are most sensitive to the change in resistivity for optimal current injection, thereby optimizing the measurements.

In calculating apparent resistivities and designing electrode configurations, the infinities in the geometric factor must be taken into consideration (Daniels 1977). Geometric factor is here defined as:

$$G = \frac{1}{\frac{1}{r_{11}} - \frac{1}{r_{12}} - \frac{1}{r_{21}} + \frac{1}{r_{22}}}, \tag{1}$$

where r_{ij} is the distance of the i-th potential electrode to the j-th current electrode, and apparent resistivity is calculated with the well-known formula:

$$\rho = \frac{4 \cdot \pi}{I} \cdot \Delta V \cdot G, \tag{2}$$

where ΔV is the potential difference at the potential electrodes and I is the current injected into the ground by the current electrodes.

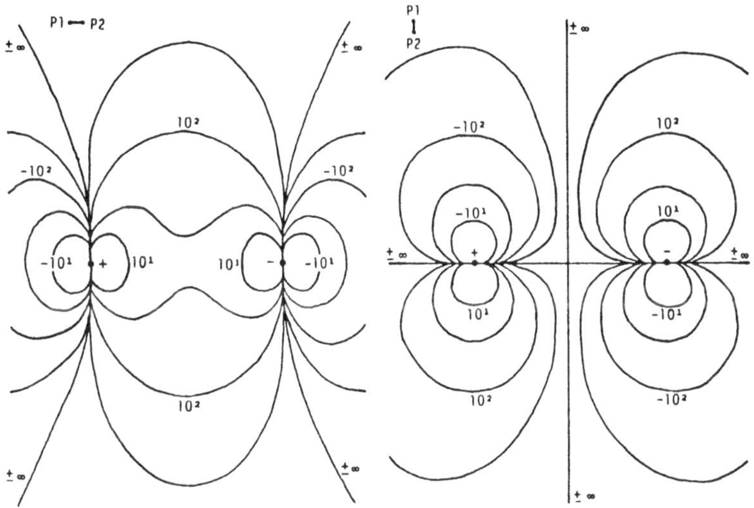

Fig. 5. Geometric factor for a bipole source with horizontal and vertical orientation of the potential receiver. The current poles are 16 m apart und potential bipole length is 1 m.

As is apparent from Eq. (1) and as shown in Fig. 5, for bipole source with a vertical and horizontal orientation of the potential electrode pairs, the geometric factor becomes great, even infinite in certain areas, which corresponds to areas of very small, or even zero potential difference. Consequently, measurements of the potential difference in and around these areas will lead, due to the arithmetic instabilities, to erroneous apparent resistivities. A current electrode configuration must therefore be chosen that does not have these infinities of the geometric factor in the area of interest. In fact, it is preferable if the area of interest has low geometric factors, for which, due to the high potential difference in the area, the measurements are more accurate.

Model calculations are conducted using a two-dimensional, finite-difference iterative procedure (Mufti 1976; Dey and Morrison 1979). The barrier structure to be studied and the location of main measurement sections are well-suited for two-dimensional modelling. The inclusion of the third dimension in the parallel direction of the barrier would not induce an improvement in the models here. Intensive tests were carried out to investigate the numerical aspects of the finite-difference resistivity modelling, concerning the grid size, boundary size, number of iterations, acceleration of the iteration and the initial guess. On the basis of these test calculations, a model frame was chosen with a 39 x 39 nodal point matrix, 1 m equidistant grid spacing and two rows at the borders with 10 and 100 m width respectively. The boundaries are far enough from the center section of the grid, where the current electrodes are located, so that the potential at all outermost boundaries can be set to zero.

Along with modelling of the potential field, apparent resistivity sections were calculated using these potential fields with potential differences of neighbouring points in the grid (vertical or horizontal) in order to study its resemblance to the original resistivity model. These maps were also used to study the apparent resistivities at the boreholes, which in the future will be measured explicitly. The contrasts in apparent resistivity to the original model and the contrast of apparent resistivities in the same borehole for different models should produce clues on the resolution of different electrode configurations for the measurements.

There are certainly many more possible current configurations, and among the many calculated, the four shown are typical for the principle types of configurations. These are current bipoles, in the same or different boreholes, and for the latter, horizontal and diagonal orientations of the current bipole. Monopole current injection configurations have not been studied at this point, because, in practice, one pole of the current must be placed very far away, thus the voltages needed to produce a measurable current are very high (due to the high resistivity of rock salt) and cannot be realized technically and, lastly, due to the mining safety regulations.

In Fig. 6, the potential fields calculated for one group of models are

Fig. 6. Potential fields for simulated brine migration models for different current injection positions.

presented. In these models, the chamber filled with brine and the brine migration are simulated. In Model I the potential distribution in a homogeneous medium of rock salt with 10^6 Ωm is calculated. This model serves as a reference model and is also used to normalize the other models. In Model II the chamber filled with brine was simulated. The resistivity of the chamber filling is set at 10^2 Ωm. This value is in agreement with the results of a survey in which the relationship between apparent resistivity and water content for different factors of cementation was studied (Kessels et al. 1985) for rock salt in the Asse. In Model III the migration of the high pressure brine into the loosening zone is simulated, where in a shell of two grid spacing units corresponding to 2 m, a gradual migration of the brine is assumed. The first shell is set up with 10^3 Ωm and the second shell with 10^4 Ωm. It is obvious that there are uncertainties in the resistivities assumed, particularly that of the loosening zone with migrated brine. Also a finer grid size would be appropriate to simulate the brine migrated loosening zone, which would unfortunately increase the computing time and cost considerably. However, the assumed orders are correct and the models produce a good qualitative overview of the process that is being simulated. For accurate quantitative results, one has to work with finer models, which was not the goal at this particular stage.

The equipotential lines in Fig. 6 are not equidistant, but in a logarithmic scaling in order to demonstrate the deformation of the potential field more clearly, i. e. in order to avoid too many tightly packed equipotential lines at higher potentials, especially around the current electrodes. The contours are at $a = \log |\Phi|$ with Φ being the potential, and are indicated with (a). Negative potentials are therefore indicated with a minus sign, -(a). As expected in Fig. 6, the common feature of the potential is that in Model II the potential lines are pushed away from the conductive structure, as compared to the potential lines of Model I. In Model III, the conductive zone being even larger, the potential lines are deformed even more. The resistivity contrast of the barrier structure to the surrounding rock salt is very high ($\sim 10^4$ to

10^2 times) so that within the barrier structure the change of the potential is very small, in fact it is practically constant and for certain models it is zero. Therefore, within the structure, there is virtually no change in potential from Model II to Model III. Another important feature is that by including the conductive zone in Model II and enlarging it in Model III, the potential lines get closer to the current injection points. This means that the injection voltage needed to produce a fixed current (here 1 mA for all models and current configurations) gets smaller, as was expected since the total resistance gets smaller by introducing a conductive zone.

In Fig. 7, apparent resistivity sections are shown as calculated from the potential field distribution in Fig. 6. The contours have been drawn for $\rho = 10^x$ where $x = \ldots,2,3,4,\ldots$. The appearance of the apparent resistivity section depends generally on the orientation and length of the potential bipole. The apparent resistivity section shown in Fig. 7 was calculated for a vertical bipole orientation in order to correlate the values in the sections later on to the measurements in the vertical boreholes. The length of the potential bipole is the same as the grid size, i. e. 1 m. Apparent resistivity sections were also calculated for horizontally oriented potential bipoles. These are generally similar to the vertical bipole sections, except in areas with critical geometric factors where the two differ considerably.

The distribution of the geometric factor has a great influence on the apparent resistivity section. In Model I (the model with homogeneous resistivity), the effect of the geometric factor can be seen clearly, and this effect matches the +/- infinite lines shown in Fig. 5. In fact, it is advisable to always calculate the apparent resistivity section for a homogeneous model in order to interpret anomalies induced by the geometric factor. Using this as a reference model, anomalies of high apparent resistivities for Models II and III can clearly be identified as arising from the geometric factor. In the first two

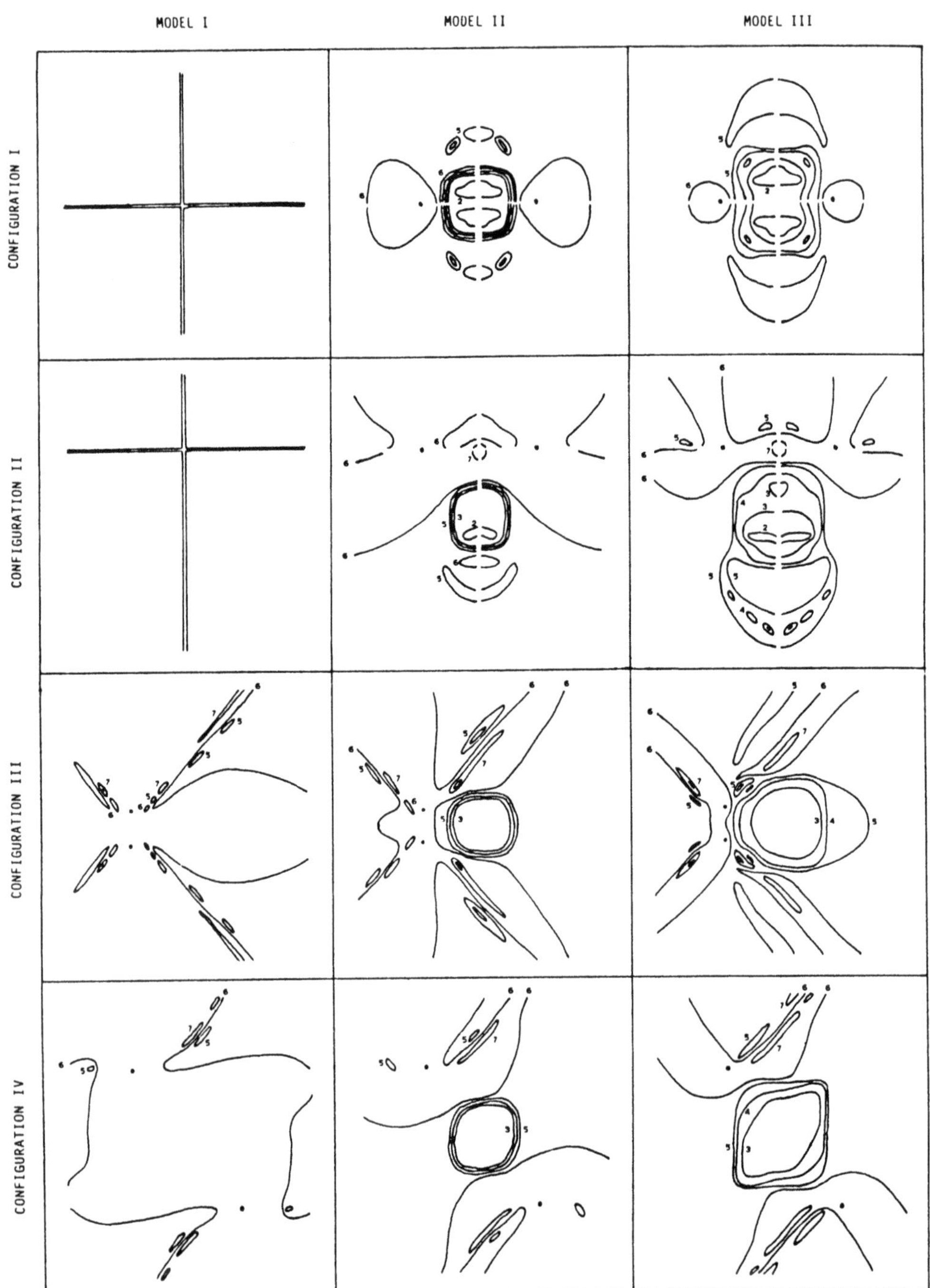

Fig. 7. Apparent resistivities for potential fields in Fig. 6.

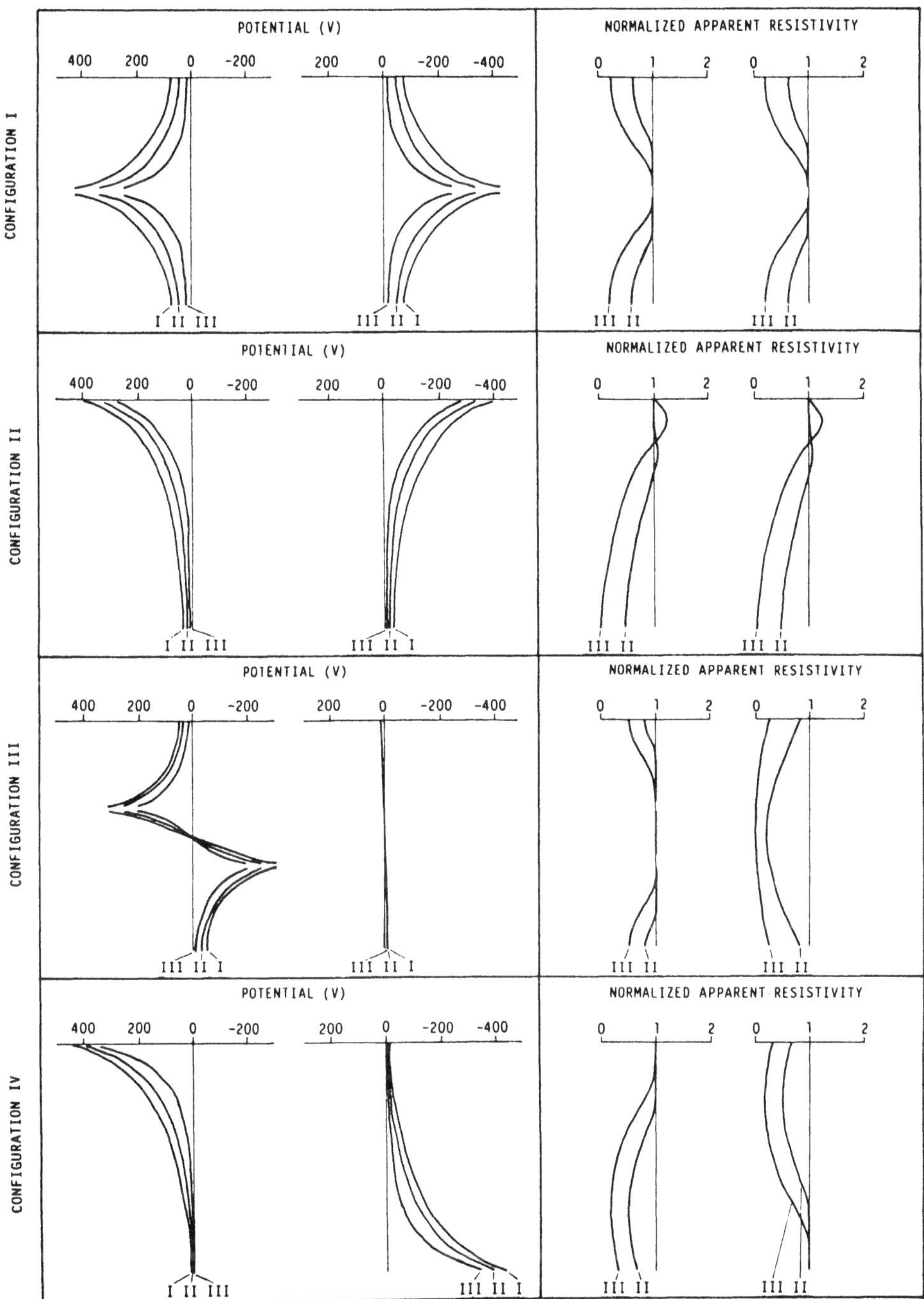

Fig. 8. Potentials and normalized apparent resistivities from Figs. 6 and 7 along two vertically placed boreholes around the chamber.

current configurations with horizontally oriented current bipoles, the anomaly due to the geometric factor is confined to only two lines. In the other two configurations, the anomaly covers a larger area and it follows the infinite lines of the geometric factor, but the detailed appearance is dependent on the grid size and length of the potential bipole.

The brine-filled chamber with 10^2 Ωm resistivity is well-defined and has relatively sharp contours. This is especially apparent in configuration I, where current electrodes are very close to the structure and, consequently, most of the current flows through the structure. The contours in Model III are less sharp, but the effect of the brine-migrated shell can be seen clearly. The chamber geometry is distorted, and generally elongated perpendicular to the current bipole axis and suppressed in the direction of the current bipole axis.

Since only measurements in the boreholes are possible, values of potential and apparent resistivities for different models and configurations along the boreholes were plotted. These can be directly compared to the measurements. The position of the boreholes and the portion for which the results are plotted in Fig. 8 are indicated in Fig. 6. Apparent resistivity values are normalized by the resistivity of the homogeneous model, i. e. 10^6 Ωm, so that the line at the number '1' in Fig. 8 corresponds directly to the homogeneous model. Due to the vertical symmetry of configurations I and II, the results in both boreholes are identical, and for configuration IV, the results for the second borehole are upside down from the first borehole due to the diagonal symmetry.

The change in the potential from Model I to Model II can be as much as 70 %, which does not, however, indicate the range of change in the apparent resistivity. Apparent resistivity is proportional to the potential difference. Therefore, the change of the potential along the direction of the potential bipole axis (here the borehole) is essential, and even more so is the change in the geometric factor along the borehole.

As expected, normalized apparent resistivities around the current source are equal to one, i. e. apparent resistivities are equal to the true resistivities, since current sources for all configurations and models are located in the homogeneous part of the medium. By moving away from the current source, the differences between the models become greater. Thus, in practice, it is preferable to locate the current electrodes away from the structure, but not so far away that the potential differences are no longer confidently measurable with a degree of accuracy. The changes in apparent resistivity from Model I to II might be as great as 50 times and from Model II to III as great as 10 times.

5 Conclusions

In order to study the sealing properties of an underground barrier construction for a nuclear waste disposal site, the use of direct current geoelectrical methods are suggested. This can be done only with a great number of electrode configurations and, consequently, with a computerized measuring system. First test results indicate a homogeneous rock salt in the construction area with a resistivity of 10^6 Ωm. Modelling of brine migration shows that it can be easily detected, but care must be taken in choosing the electrode configurations, especially considering the peculiarities of the geometric factor in certain areas.

Ackwowledgements. We are grateful to Mr. Pelz for his help in the measurements, and maintaining and designing of the geoelectrical system. Thanks are also due to Mr. Cooperrider for his valuable help in computing models. The project is supported by the Federal Ministry for Research and Technology (BMFT) under project no. 316-5691-KWA 5604 7.

References

Daniels J J (1977) Three-dimensional resistivity and induced polarization modelling using buried electrodes. Geophysics 42: 1006-1019

Dey A, Morrison H F (1979) Resistivity modelling for arbitrarily shaped two-dimensional structures. Geophys Prosp 27: 106-136

Kessels W, Flentge I, Kolditz H (1985) DC geoelectric sounding to determine water content in the salt mine Asse (FRG). Geophys Prosp 33: 436-446

Matula R A (1981) Electrical and magnetic properties. In: Physical properties data for rock salt. Nat Bur Standards Monogr 167, U S Gove Print Off, Washington, D C 20402

Mufti I R (1976) Finite-difference resistivity modelling for arbitrarily shaped two-dimensional structures. Geophysics 41: 62-78

GEOELECTRICAL MEASUREMENTS AT THE KTB LOCATION

HELMUTH WINTER, MIGUEL ARROYO, VOLKER HAAK,
JOHANNES STOLL and RICHARD VOGT

Institut für Meteorologie und Geophysik, Universität Frankfurt,
Feldbergstraße 47,D-6000 Frankfurt a.M. 1, F.R.G.

Introduction

A super deep drilling project is currently undertaken in the region Oberpfalz in the eastern part of the Federal Republic of Germany, called "Kontinentale Tiefbohrung" (KTB). The purpose of the project is to answer fundamental questions in geophysics and geology, such as: the nature of conductors and seismic reflectors in the crust, origin of ore deposits, mechanisms of crustal tectonics, and others. There are two drilling sites projected within a distance of 200 m: A pre-drilling hole with a target depth of 5,000 m and a main drilling hole with a target depth of 14,000 m (Fig.1).

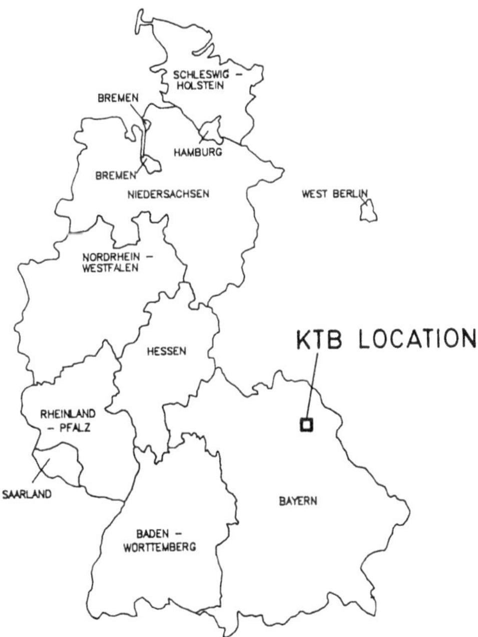

Figure 1: Federal Republic of Germany with KTB location

Lecture Notes in Earth Sciences, Vol. 27
G.-P. Merkler et al. (Eds.)
Detection of Subsurface Flow Phenomena

From March to June 1987 intensive studies of self-potential, VLFR and DC-resistivity were performed on a 1-km² area around the pre- and main drilling site. The measurements were made by the University of Frankfurt and by the Free University of Berlin. The aim was to secure geoelectrical data in that area as long as it was accessible before the installation of the drilling equipment. This study concentrates on an overview presentation of the large amount of data, which have been collected so far. However, we intend to perform additional data acquisition.

Measuring procedure and first results

The self-potential field procedure was performed by measuring the potential difference between a base electrode outside the area and 1100 points inside the 1-km² area. The point distance was 25 m. Unpolarizable copper-copper sulfate electrodes were used. To avoid telluric influences the cable length was limited to 200 m. By calculating the potential in two different loops the potential values at each point differed by about 20 mV only, which was sufficiently accurate according to the large potential differences observed in this area. The measurements revealed a large-scale anomaly of values down to -500 mV (Fig.2). The peaks of the anomaly are located around sites of graphitic material at the earth's surface.

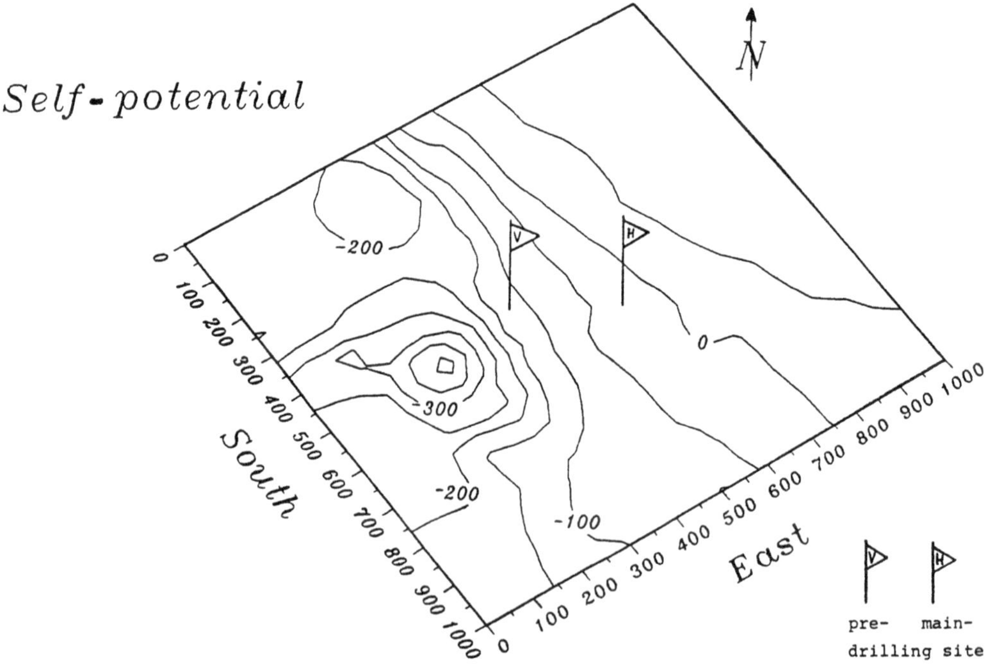

Figure 2: Map of self-potential

VLFR measurements at about 800 points display the apparent resistivity ϱ^* of the upper 20-50 m. The most important result is a clear division of the drilling location into two areas, one in the west and southwest with low resistivities around 10 Ωm, the other further east with high resistivities up to 3000 Ωm. Figure 3 presents a perspective view of isolines of the apparent resistivity.

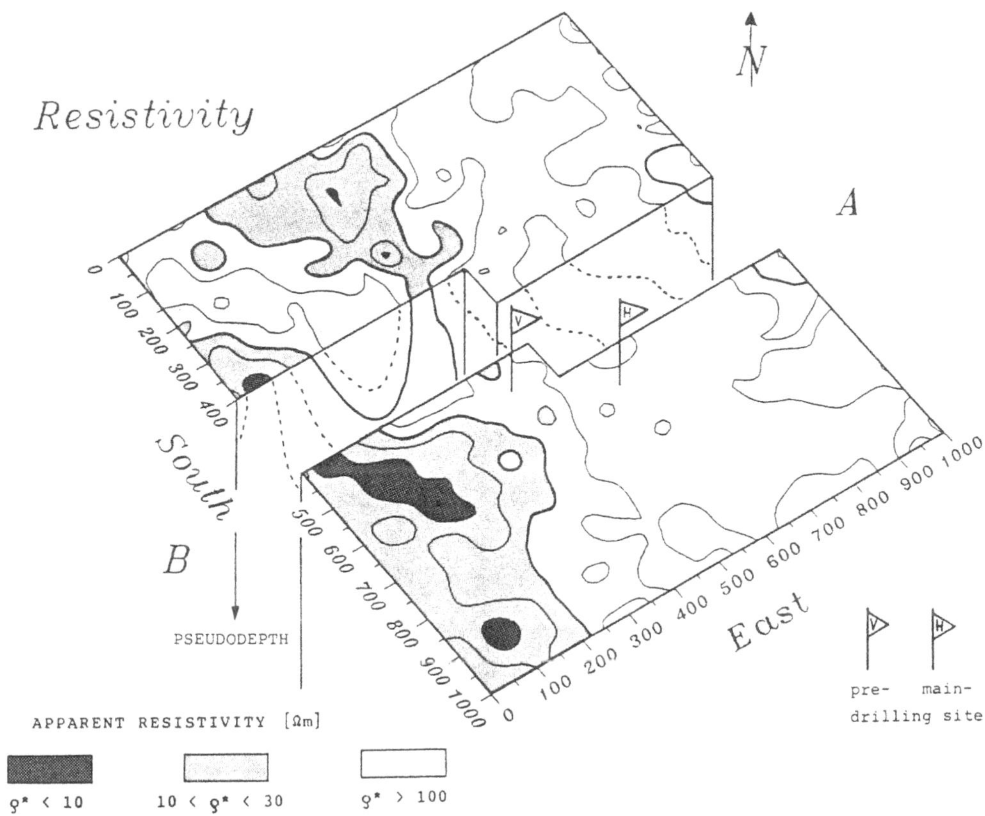

Figure 3: Map of VLF resistivity

Several pole-dipole profilings across the area in different directions with electrode separations up to 300 m confirmed the lateral resistivity variations obtained by VLFR. The results of one pole-dipole profile are presented in Fig.3 in the usual pseudo-section manner in the vertical direction.

At 14 sites Schlumberger sounding curves were measured. These measurements have been extended, with regard to the strong lateral resistivity variation, by measuring the complete tensor. The processing of the tensorial resistivity is still a topic of research and therefore still in progress.

In order to demonstrate the effect of strong lateral resistivity variation the sounding at the presite was performed according to the Hummel configuration ("Half-Schlumberger"). The measured curves are presented in Fig.4: Curve A represents the apparent resistivity measured with one electrode moving towards point A, curve B with the other electrode moving towards B. The profile AB agrees with the cross-section of Fig.3. One can clearly see the strong effect of lateral resistivity variation by expanding electrode separation. A Schlumberger sounding at the same site yielded the mean of both Hummel curves.

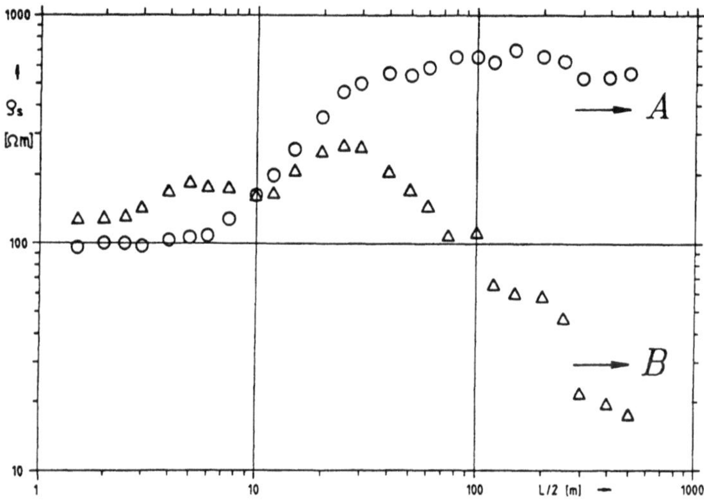

Figure 4: Hummel sounding curves of the predrilling site. Electrodes are moving towards A (o) and B (▲), see Fig. 3

Conclusion

The geoelectrical measurements in the neighbourhood of both drilling sites revealed a large variation of the resistivity from the southwest towards the northeast. It seems that this may be explained by the existence of considerable amounts of graphite down to great depths, which was indeed confirmed by the results in the predrilling hole with several graphitic "layers" down to 2000 m.

Geothermical Measurements

SOME PROBLEMS CONCERNING THE MEASUREMENT OF NEAR-SURFACE STREAMING POTENTIALS AS WELL AS OF SURFACE TEMPERATURES FOR THE DETECTION OF NEAR-SURFACE FLOW PHENOMENA

Militzer, H.[2] & Oelsner, Chr.[1]

The detection and control of flow phenomena of near-surface water in rocks or buildings is of interest for the following reasons:

1. The establishment and operation of hydrotechnical and hydraulic structures and on an increasing scale also for waste disposal;

2. The use of stored, natural water reservoirs;

3. The control of the disturbances of the natural water regime by mining or building operations.

Referring to such tasks it is necessary not only to investigate the grounds of the planed building but to extend the investigations also to the building itself. From the geophysical point of view, methods are used which give, on the one hand, direct information on streaming water (for example, by measuring the streaming (filtration) potential) and on the other hand, methods in which the parameters determined are strongly influenced by the water content (for example, geoelectrical or near-surface seismic methods or neutron methods). In practice, however, there are only a few examples which provided unequivocal results after the measurement of only one single parameter with respect to the existence and the moving force of streaming water near limitation surface of local areas which mainly showed descendent or ascendent water flow, respectively, with regards to the direction and the velocity of water movement. In most cases it is necessary not only to use different geophysical methods, but also to cooperate with neighbouring disciplines, e.g. engineering geology, geotechnics, hydrology and construction engineering. Such cooperation is the only way to master more and more the increasing difficulties of geological surveys for building operations and to fulfill all the increasing demands re-

[2] Bertolt-Brecht-Str.9, GDR-9200 Freiberg
[1] Bergakademie Freiberg, GDR-9200 Freiberg

Lecture Notes in Earth Sciences, Vol. 27
G.-P. Merkler et al. (Eds.)
Detection of Subsurface Flow Phenomena

garding a technically undisturbed operation of the buildings or the system building site.

Undoubtedly it is much more economical to anticipate problematic situations in engineering and geosciences during the projection and evaluation of the building site. Unexpected situations and the necessity of site stabilization, foundation corrections etc. often delay building operations and only increase the cost.

For the establishment of hydraulic structures a combination of geoelectrics and shallow seismics has proved best in controlling leakages.

An example is illustrated in Fig. 1. Corresponding to the existing possibilities, the dam was seismically sounded in a horizontal and vertical direction. The velocities of the longitudinal waves determined, above all, reflect an anomaly in porosity and structure, which might cause a failure or leakage hazard. The additionally determined distribution of streaming potential shows an anomalous behaviour in the middle part of the dam, for instance, the velocity. Obviously it is caused by leakages.

Problems discussed in the frame of this symposium "Detection of Near-Surface Flow Phenomena" necessitate the contemporary use of different geophysical methods. A first step in this direction is the combination of geoelectric and geothermal methods. Geoelectric and geothermal investigations at the surface or within boreholes for the detection of flow phenomena in a permeable medium belong to the "classic" geophysical methods. The first methods to be mentioned are different variations of resistivity mapping by four-point configurations, the method mise à la masse (mapping equipotential lines); further, tracer measurements, and more recently, electromagnetic high frequency methods (ground penetration radar); and, last but not least, self-potential measurements for mapping streaming potentials. The importance of geothermics is mainly related to the development of sensors sensitive to infrared (ir) radiation and to the possibility of remote sensing by such equipment. Later, the possibilities and limitations of these two methods and the advantages of combining both will be discussed.

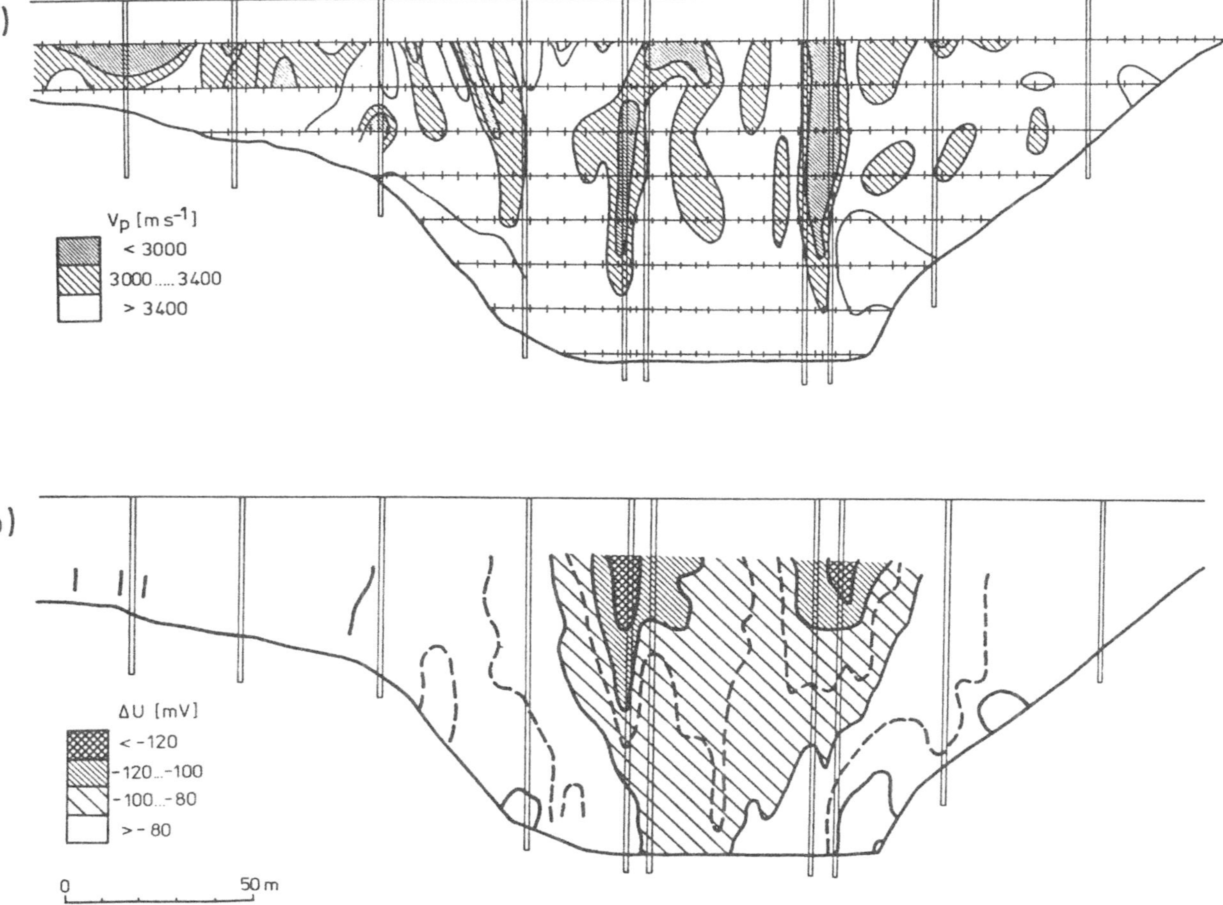

Fig. 1: Seismic and geoelectrical measurements for determining the water permeability of a barrage (measured by Hattenhauer, 1979; Militzer et al., 1986).

It is well known that electrical self-potentials in the sense of streaming potentials set in, if a fluid is forced with pressure difference through a capillary tube. A very important phenomena is the fact that the electrochemical double layer existing at the wall of the capillary splits. Regarding the energetic situation, the result is a potential jump (Fig. 2).

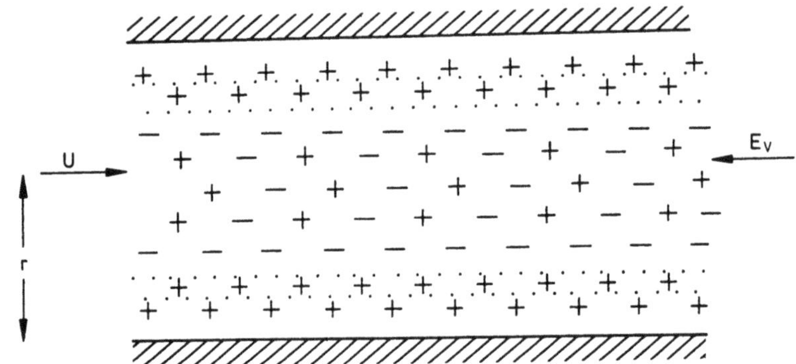

Fig. 2: Streaming of a fluid in a capillary tube.
U-Direction of fluid movement; E_V-field of filtration
(the dotted line denotes fluid movement off the range)

The simple idea of two neighbouring layers with positive and negative ions (first developed by Helmholtz) was later developed further, especially with regards to the acceptance of an ion atmosphere inside the fluid as a consequence of thermically excited movements. Thus it is possible to divide the Galvani voltage G into a fixed and diffused part:

$$G = G_{fix} + G_{diffuse}. \qquad (1)$$

The diffused part is identical with the electrokinetic part, the so-called ζ-potential. It mainly influences the amplitude of electrokinetic effects.

The charges taken along by the movement of the fluid inside produce a potential difference $\Delta\zeta$ between the ends of the capillary. This potential jump wants to be equalized by travelling ions. In the course of time the streaming potential becomes constant and conforms to the law of Hagen-Poiseulle:

$$\Delta\zeta = \frac{\zeta\rho\varepsilon}{4\pi\eta} \Delta p. \qquad (2)$$

Thus, the field of the streaming potential is directly proportional to the dielectric constant ε, to the specific electric resisitivity ρ of the fluid, to the ζ-potential and to the pressure difference Δp; it is inversely proportional to the viscosity η. In the mentioned form, Eq. (2) does not depend on the geometric parameters of the capillary. The latter are in evidence only, if we introduce the average velocity $\bar{v}$ of the fluid instead of the pressure. Regarding the above mentioned law of filtration (2) $\bar{v}$ becomes:

$$v = \frac{r_0^2 \Delta p}{8 \eta} \; ; \qquad (3)$$

$$\Delta\zeta = \frac{2\zeta\rho\varepsilon}{\pi r_0^2} \bar{v} \qquad (4)$$

These relations mean that changes in intensity of streaming fields are produced by the following:

1. Jumps in the ζ-potential; they always become effective, if the mineral composition of the solid material and chemical quality change;

2. Variations of the specific electrical resistivity of the subsurface water; mostly caused by variations of mineralization (e.g. rainwater containing 5-15 mg/l dissolved constituents (containing about 50 % NaCl): $\rho \cong 3x10-10^3$ Ωm; surface resp. near-surface water: $\rho \cong 10^{-1}-10^{-2}$ Ωm)

3. Enlarging the velocity of the water in porous media; however if increasing velocity is not responsible for increasing pressure, but for increasing porosity, we cannot expect any increase of the field intensity due to $\Delta\zeta \cong 1/\pi r_0^2$.

The resulting effects are of a manifold nature and it is very difficult to separate them. Furthermore, it must be noted that the intensity of the filtration fields not only depends on the properties of the filtration milieu, but also on those of the surrounding rocks and all other structures where the filtration process takes place. Alpin and Grünbaum were concerned with this question and they concluded that it is possible to measure filtration fields on the surface only if the filtration horizons are not situated too deeply and if the rocks lying over and under the filtration stratum have a high electrical resistivity. As a rough rule, therefore, we may assume that diminuation of the field intensity is proportional to the increasing thickness of the overlying burdens.

The direction of electrical streaming fileds is related to the direction of the movement of subsurface water. We can observe an increasing potential in the direction of the filtration. Consequently, descendent water is marked by a negative potential distribution and ascendent water causes a positive potential.

In summary the favourable conditions for the formation of streaming potentials are:

1. Intensive pressure decreased inside the filtration horizon;

2. High resistivity of subsurface water (little mineralization);

3. Water-saturated filtration horizon;

4. High specific electrical resistivity of the horizon over- and underlying the filtration stratum;

5. Small porous structure of the rocks with high water permeability;

6. High streaming velocity of the subsurface water.

What is the problem concerning the intensity of streaming fields? Normally, the gradient of the filtration potential amounts to less than 10 up to more than 100 mV/km. High amplitudes are observed especially in mountainous regions and in river valleys. Generally, there are no difficulties in determining the potential distribution on the earth's surface, provided a reference and a moved electrode, as in self-potential measurements is used. It is well known that the equipment consists of a voltmeter with high resistivity, two non-polarizing electrodes and a cable with only one wire; it allows a guaranteed accuracy of ± 0.1 mV for a single point without any problems. Difficulties first arise, if long-term observations are necessary to control special situations, e.g. leakages in dams, or if the results of measurements are to be quantitatively interpreted.

In quantitative interpretations, e.g. determination of the flow velocities or the exact quantity of flow-through, the following factors have to be take into consideration: some superimposed and masking effects; such as the influence of temperature, variations of the mineralization, influence of different porosity and also the absence of valid theory. At the present time the valid theory is based only on the behaviour of a fluid with a single capillary. For a generalization, which includes complicated in-situ situations, different results have not been sufficiently proved. Up to now it has not been possible to answer all the important questions with regards to a valid generalization and to support the theoretical results by laboratory tests. Such questions relate to the influence of grain distribution, grain size, solubility of mineral components, etc.

It is not yet possible to separate these effects with sufficient accuracy under normal conditions; quantitative interpretations can not be given at present.
Is there a possible step by step solution?
Undoubtedly the variations in mineralization simulate variations in streaming potential. But they are not only a question of chemistry, but also of the temperature of the environment. Therefore, as a first step, it is especially necessary to distinguish these effects, i.e. streaming potential/mineralization/temperature.

This requires the simultaneous measurement of temperature, variations of the specific electric resistivity caused by variations of the mineralization of the pore content and the streaming potential and at the same site. This could be realized by means of a "combination probe" as shown in Fig. 3.

Additionally, it is necessary to continue methodical lab studies which were mainly initiated by Ogilvy and Bogoslovsky, but also by Armbruster, Merkler and others. This is important in order to confirm first results concerning the relationships of streaming velocities with pressure variations. Besides, it is very important to obtain information with regards to the manifold petrophysical influences on the origin and the intensity of the streaming potential. The utility of geothermal measurements to prove streaming events, such as leakages in dams or dikes, has been well known for a long time. Note that the first detections of leakages in dams by means of geothermics were by Kappelmeyer (1957).

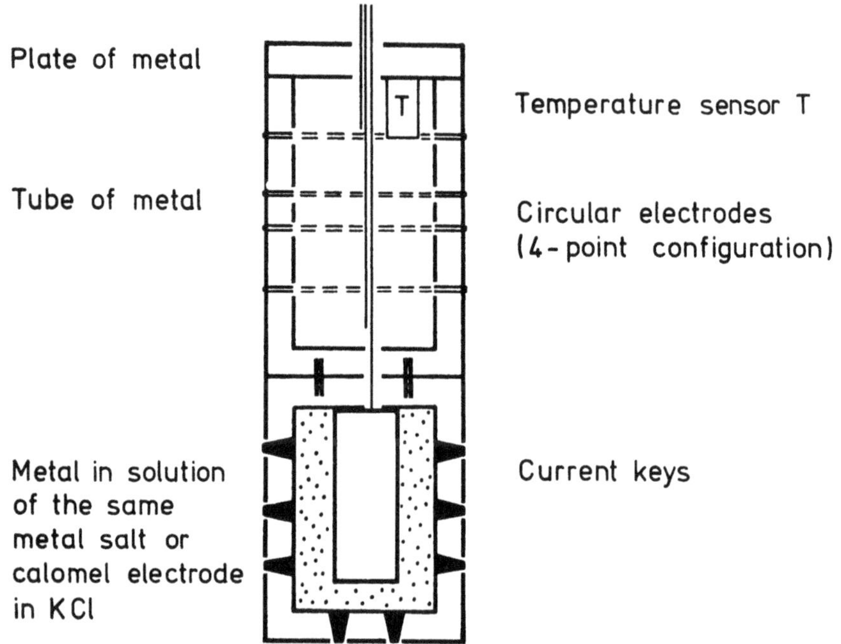

Fig. 3: Principle of a "combination tube" for stationary long-term observations of temperature, apparent specific electrical resistivity and streaming potential.

Often geothermal measurements seem to be more successful than other geophysical potential methods, because in many cases the geothermal effect is enlarged by the superposition of effects of thermal conduction and thermal convection. In these cases knowledge of petrophysical parameters of the medium are needed to solve the direct and inverse task. These parameters are the thermal conductivity λ, the thermal diffusivity k and the specific heat c. Whereas the thermal conductivity is a measure of any body to conduct heat, thermal diffusivity acts in a transient thermal process. It is well known that these three parameters are connected by:

$$\lambda = k\ c\ d \quad (\mathrm{Wm^{-1}\ K^{-1}}), \tag{5}$$

where d is the density.

Of considerable interest for practical studies in sedimentary rocks are some important relationships between thermal conductivity of the rocks, its porosity, density and water content. Many authors have studied this problems (see Fig. 4).

Thermal conductivity of the air only amounts to $\lambda_{air} = 0.027\ Wm^{-1}K^{-1}$ at 20 °C, i.e. about 1 % of the thermal conductivity of the rocks. Therefore, air-filled pores in sedimentary rocks strongly reduce its conductivity. The thermal conductivity of water-filled pores amounts to $\lambda_{water} = 0.6\ W\ m^{-1}\ K^{-1}$ at 20 °C. That means approximately 25 % of the matrix conductivity only, but it is much higher than that of the pores which are air filled.

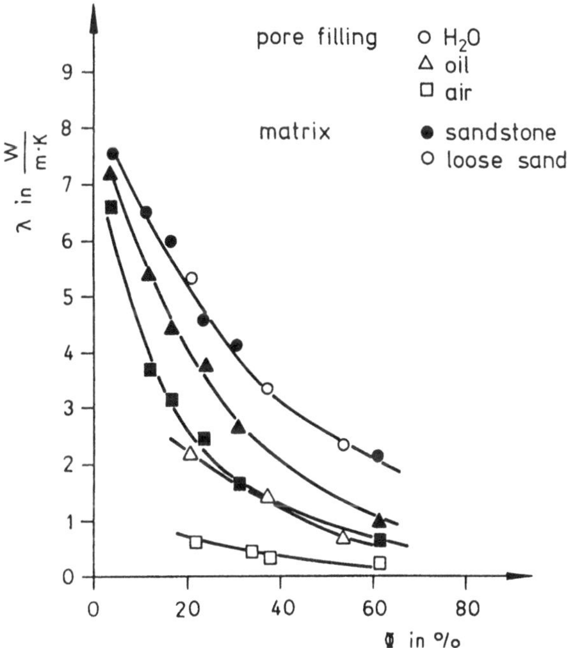

Fig. 4: Heat conductivity of sedimentary rocks as a function of porosity and pore content (Woodside; Messmer, 1961).

Due to the connection between the thermal conductivity and the thermal diffusivity, according to:

$$k = \lambda / c\ d \quad (m^2 s^{-1}) \tag{6}$$

variations of λ also transfer to k. For example, we may derive from data of Kanamori et al. (1968), that

$$\partial k / \partial T \sim \partial T. \tag{7}$$

Furthermore, we may also prove easily that:

$$\partial k \cdot / \partial \phi \sim \partial \lambda / \partial \phi. \quad (8)$$

Regarding the temperature anomaly ΔT of a moved point source, we may form new concepts about heat conduction effects of hydrodynamic processes.

The solution for a heat source of intensity Q (W) moving in an infinite medium with the velocity v parallel to the x-axis at a distance r is represented by:

$$T_{st} = \frac{Q}{4r\pi\lambda} e^{-\left(\frac{v(r-x)}{2k}\right)}, \quad (9)$$

where $r^2 = x^2 + y^2 + z^2$.

In Fig. 5, Eq. (9) is shown for different values of v/2k and the case z = 1, y = 0. The effect of levelling out at higher streaming velocities can be seen clearly.

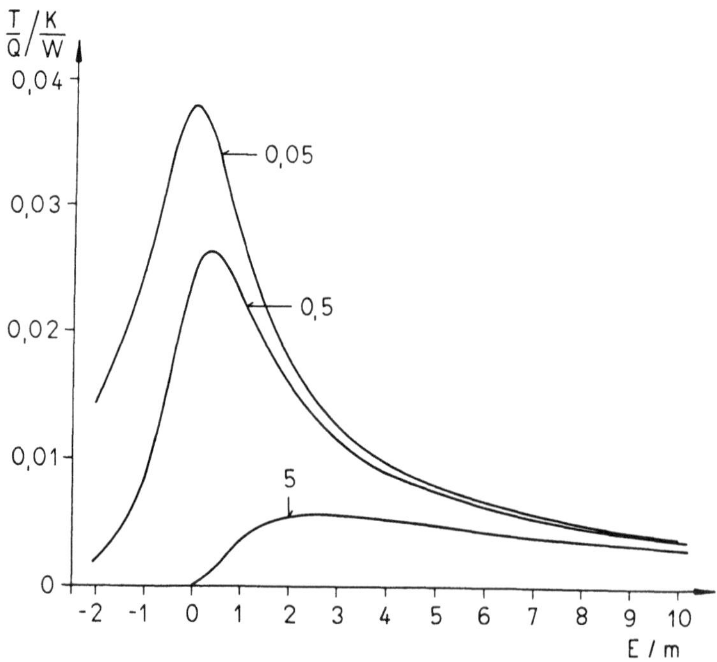

Fig. 5: Temperature caused by a moved point source; parameter: v/2k.

Leaks to be localized by means of temperature measurements are often related to non-stationary events (e.g. by switching pipes on or off, sudden leakages). In such cases we have to find a solution to the corresponding transient heat conduction process. Using the same symbols and units as in Eq. (9),

$$T_{inst}=\frac{Q}{8r\pi\lambda}\left[e^{-\frac{v(r-x)}{2k}}\cdot \operatorname{erfc}\left(\frac{r-vt}{2(kt)^{1/2}}\right)+e^{\frac{v(r+x)}{2k}}\cdot \operatorname{erfc}\left(\frac{r+vt}{2(kt)^{1/2}}\right)\right]. \quad (10)$$

In Fig. 6 the curves of the temperature buildup are compiled ($x = y = 0$ and $z = 1$) for three velocities ($v = 1x10^{-6}$, 2×10^{-6}, 3×10^{-6} m/s).

Looking at the smallest velocity it may be noted that this buildup is completed after 100 days. The duration to reach 90 % of the maximum amplitude at the curves 1, 2, 3 are 29, 12 and 10 days. In the estimation of the effect of such a source we have to pay attention to the fact that approximately 70% of the total time is necessary to increase the "last" 10% of the amplitude.

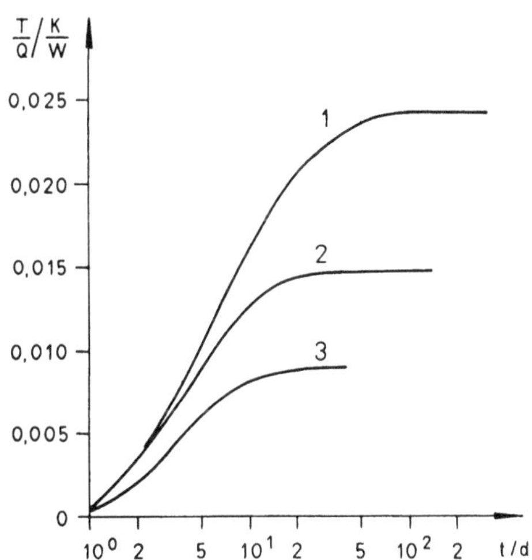

Fig. 6: Temperature buildup caused by a moved point source (depth 1) at the point $x = y = 0$.
Curve 1: $v = 1x10^{-6}$ m/s = 8.6 cm/day;
Curve 2: $v = 2x10^{-6}$ m/s = 17,3 cm/day;
Curve 3: $v = 3x10^{-6}$ m/s = 25.9 cm/day.
with $k = 10^{-6} m^2/s$ and $\lambda = 2$ W/mK

Under consideration of the chosen parameters, Eq. (10) leads to the following relationship between velocity and time:

$$\lg (v/v_o) = - 3/4 \lg (t/t_o) + 1/3, \quad (11)$$

where t_o is the time necessary to reach a definite value of temperature increase and v_o is the streaming velocity.

With regard to the time-depth relationship the quadratic connection is generally valid; that means an enlargement of the depth of the source by the factor a is followed by an increase of the time of the temperature buildup by a^2.

Figure 7 shows the application of the equation of a moved point source after infinite duration of action (stationary case). Compiled are the results of temperature measurements made in connection with an in-situ experiment for heat storage. Into a near-surface aquifer (depth of the upper brim was 12 m) warm water was injected and pumped back thereafter. The experiment is described in detail by Seipt (1987). Before rejection the temperature was measured along the radial profiles (starting at the injection hole) at a depth of 50 cm. The results along one profile are shown in Fig. 7.

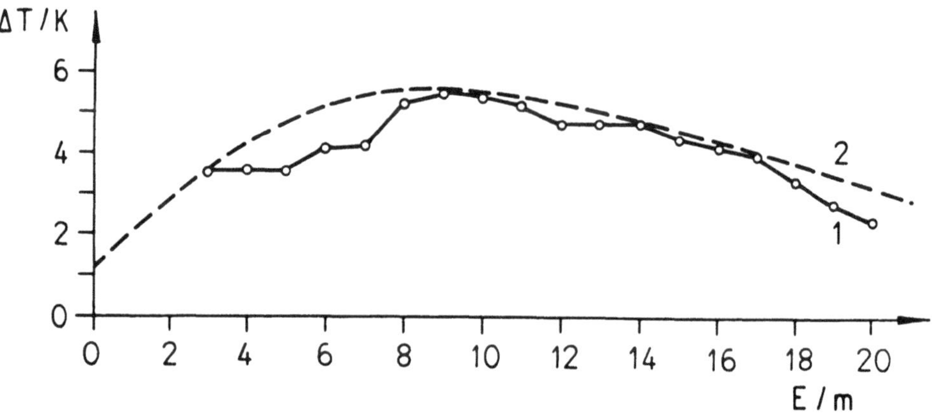

Fig. 7: Anomaly of temperature caused by injection of warm water into a near-surface aquifer. 1-Theoretical course of temperature; 2-measured course of temperature.

The accuracy of the device was 0.1 K only. The measurements were taken only between 3 and 20 m of the profile due to technical reasons. The parameters for the relatively good adaption of the theoretical values of the measured curve, especially concerning the position of the maximum and the course at > 9 m, are $z = 12$ m, $\lambda = 2$ W/m K, $v/2k = 0.1$ m. This value of $v/2k$ means for $k = 1x10^{-6}$ m^2/s a streaming velovity of 1.7 cm/day and for $k = 5x10^{-7}$ m^2/s respectively $v = 3.4$ cm/day. These values are quite realistic. Deviations between the measured and theoretical curve of the profile are caused by near-subsurface noise non-stationarity (end of injection).

In spite of the good results in investigation of streaming effects by geothermal measurements using contact thermometers (contact sensors) they are often less effective and not to be realized under all conditions, for example on the air side of a barrage. A remarkable increase of the effectivity of geothermal measurements is to be obtained by contactless temperature measurements, that is, the measurement of radiation temperatures.

The high measuring progress is a particular advantage, as it is characteristic for all the remote sensing methods. From the methodical point of view we have to distinguish between imaging and non-imaging methods.

Imaging ir-geothermics were used at the Institute of Applied Geophysics, Mining Academy Freiberg (GDR) with regard to application by using commercial equipment. This method seemed to be useful especially for detecting:

1. Leakages in barrage, dams and dikes;

2. Water-impermeable horizons in slopes;

3. Wet spots characterizing geological faults;

4. Smoulder burnings and for observation of dumps, which are in danger of becoming enflamed;

5. Loose rocks and loose parcels in underground minings.

Concerning non-imaging geothermics we carried out extensive preparatory work in all theoretical, methodological and technical aspects for approximately 20 years. During the last years we have had good cooperation with the Institute of Geophysics at the Mining University Leoben (Austria) with respect to further technical and methodological developments.

It is well known that the radiation balance S is reflected in the measurements. In the daytime some factors of S disturb the measurements.

For the time between sunrise and sunset S holds with

$$S = \varepsilon\sigma T^4 + I + H + G - R \pm W_{L,S}. \qquad (12)$$

where $\varepsilon\sigma T^4$ is the radiation of earth's surface (ε- coefficient of emission; σ- Boltzmann constant; T - temperature of the surface); I - direct sun radiation; H- diffuse sky radiation; R- reflected sun radiation; G- atmospheric counterradiation; $W_{L,S}$- latent or sensitive heat.
I, H and R are shortwave radiations and disappear after sunset. S is then reduced to:

$$S = \varepsilon\sigma T^4 + G \pm W_{S,L}. \qquad (13)$$

If the time of the measurements is short in comparison with the time of changing metereological situations, G may be assumed to be constant. If situations are avoided which cause variations of $W_{L,S}$, e.g. rain or dew, the measured radiation temperature becomes only a function of the coefficient of emission and of the soil temperature.

The coefficient of emission of natural soil ranges between 0.90 and 0.95, mainly to 0.95 (Watson 1975). Small coefficients of emission cause relatively high heating, due to the small radiation, but also to lower radiation temperatures. Both effects cancel out each other.

At the boundary soil/air the surface temperature is influenced by the air temperature due to the heat exchange. Aigner (1984) demonstrated that temporal variations of surface temperature as well as of air temperature may be reduced by measuring the difference of both the temperatures.

Hence, it follows that ir-radiation temperatures of the earth's surface may be measured in order to obtain geological information, if:

1. We measure between sunset and sunrise; and

2. We measure ir-radiation temperature of the earth's surface and the air simultaneously.

After having produced the first ir-thermometer (pyrometer) for low temperatures which was able to simultaneously measure the temperature of the object and of the air, we developed the conception for equipment; in the meantime it is produced by an Austrian firm. The block diagram of the device is shown in Fig. 8. The device consists of two parts, the sensor and the unit for measuring and data processing. The latter is responsible for indication, storage and processing of the data.

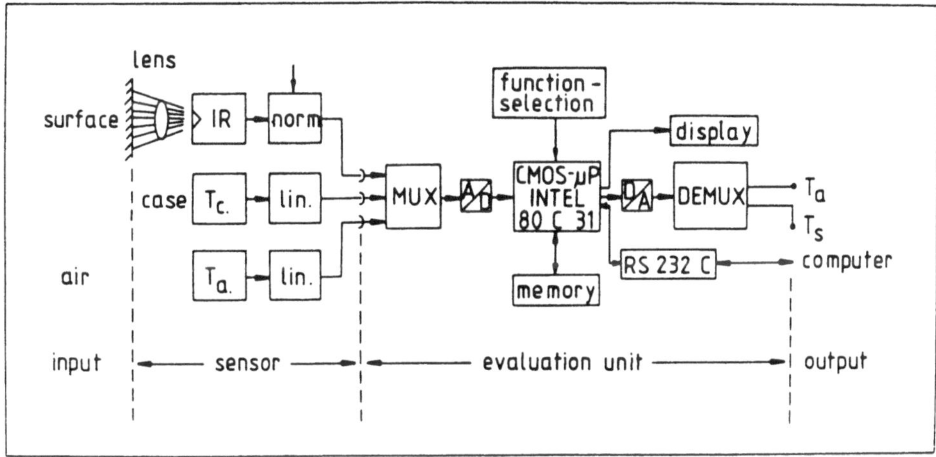

Fig. 8: Block diagram of a high-sensitivity ir-device (developed in cooperation with the Institute of Applied Geophysics and Mining University Leoben (Austria) and Institute of Applied Geophysics, Mining Academy, Freiberg (GDR).

The detector has a sensitivity of 20 V/W and a KRS-5 entrance window with a filter for 6-15 m wavelengths (range of permeability). The aperture is 90° and may be reduced to 5° by an optional lens. The temperature of the sensor is measured by means of a thermistor and is corrected automatically. All functions are checked by a CMOS microprocessor. With respect to the data of the sensor the microprocessor calculates the temperature of the sensor, the air temperature and the temperature of the object on the base of the preselected coefficient of emission. Furthermore, it controls the indicators on the display, corresponding to the preselected functions (temperature of object and air or rather the difference), the data transfer into or out of the memory, corresponding to the preselected mode of work, and filtering of the stored data by three possible low and band pass operators. 2048 values of the object and air temperatures may be fed into the memory. An interface RS-232C allows the direct data transfer to different computer systems. In Fig. 9 an impression of the measuring system is given; it may also be used like a two-canal data logger. Triggering of the device, i.e. transfer of the joining signal into the memory, may be done externally (any time or distance mark) or internally by unit of buttons to be pressed at the measuring and processing unit or at the handle of the ir-detector.

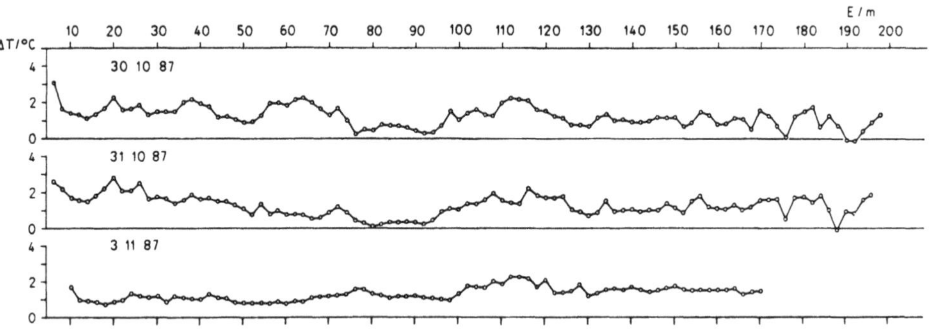

Fig. 9: View of the device

In table 1 the technical data are given (look at p. 17).

Figure 10 demonstrates an example of ir-measurements accomplished along a profile. The object of the study was the range of a heating line connecting two buildings within an industrial terrain. Standing time of the heating pipe was 35 years; it was in a narrow canal and it was not possible to walk along inside. Extensive losses of the heat carrier indicated at least one leak. Ir-surface temperature measurements were carried out to localize leakages on 30 Oct., 31 Oct. and 3 Nov. under different conditions. On 30. Oct 12 h before the measurements (time of the measurements: 5^{45} - 6^{20}) circulation of the heat carrier was started. After taking the measurements, the heating was switched off. Under these conditions the measurements were accomplished on 30 Oct. On 3 Nov. the conditions were again comparable with those on 30 Oct. In spite of an unfavourable, wet earth surface, resulting from rain before the measurements (5^{45} - 6^{45}), they were carried out due to methodologic reasons.

Table 1: Technical datas of the high-sensitivity ir-device shown in Fig. 9

SPECIFICATIONS

Measuring range:	
Object temperature (ϑ_T)	ϑ_C^* plus ± 15°C
Air temperature (ϑ_A)	-20 to + 43°C
Data logger (channels 1 & 2)	± 2.047 V
Absolute error:	
Object temperature (ε=1)	±(0.3 + 0.03 ($\vartheta_T - \vartheta_C^*$)) °C
Air temperature (0 to 43 ° C)	± 0.3 °C
Data logger	±(0.15% of display + 1 digit)
Reproducibility: (ϑ_C^*= ± 3° C, ϑ_T=const., ε=1)	0.03° C
Resolution: Both object and air temp. (ε=1)	0.01° C
Data logger	1 mV
Coefficient of emission:	0.30-1.00 (selectable)
Aperture angle of IR-sensor:	90°
Optional: Attached optical system	5°C
Analog output for both	
object and air temperature:	50 mV/ °C
Filter characteristics selectable:	
Low pass	0.1/0.2/0.5 Nyquist
Band pass	0.1-0.2/0.1-0.5/ 0.2-0.5 Nyquist
Storage capacity:	
Both object and air temperature	2048 measured values each
Power supply: NiCd battery	6 V/2 Ah
Time of data protection when replacing the battery:	min. 10 min
Load controller:	
Supply voltage	220 V ~ , 50-60 Hz resp. 12V-
Charging time	approx. 14 hrs.
Weight:	
Sensor	approx. 10 N
Evaluation unit	approx. 20 N

*... Housing temperature of IR-sensor

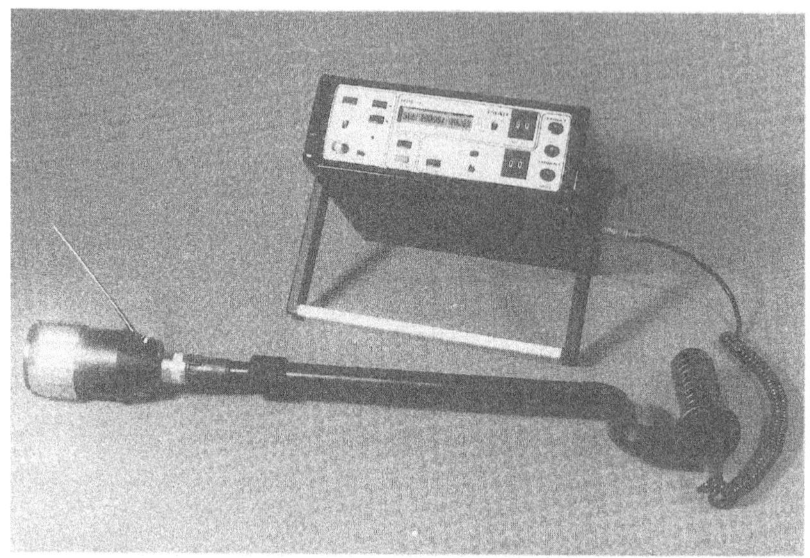

Fig. 10: Ir-surface temperature near a defective heating line.

Of course, a direct comparison with the results of the previous day was not possible.

The measured curves (Fig. 10) represent the average of three (Oct 30, Nov. 3) or two measurements. On 30/31 Oct. the mean deviation between 6-52 m, 76-100 m, 126-200 m equaled ± 0.2 °C, and between 52-76 m or 100-126 m it equaled ±0.5 °C. On Nov. 3 the deviations did not exceed ± 0.2 °C on all parts of the profile.

Regarding the measured curves the low decrease of temperature on both sides of the buildings limiting the area in a direction toward the middle of the profile is evident. Some single anomalies are superimposed by this "regional" trend. They are especially clear in the results of Oct. 30. The parts of the profile between 16-28 m, 34-44 m, 52-76 m and 94-124 m are to be regarded as anomalous. On Oct. 30 the indications became more developed between 16-28 m. The amplitude of the following disturbance decreased, but it widened to 52 m. The range between 44-52 m is distinguished by a hidden balance barrow of the heating pipe, i.e. here it is mainly outside of the profile. Both indications and the following between 52-76 m are interpreted as leaks. The positive temperature anomalies established over all three days, between 94-126 m, were only slightly influenced by different heating conditions. Therefore, they must be related to the different conditions of the foundation subsoil.

On Nov. 3 the measurements were completed at 168 m. Due to the metereological conditions, the grass surface prevented any ingenious results until the end of the profile.

This example shows that it is possible to obtain information on streaming effects very quickly by means of ir-radiation measurements. On the other hand, it becomes clear that there is no possibility to interpret the anomalies of surface temperature measurements alone in a positive manner. Unambiguous interpretation could be achieved by a combination of both methods, however, a simultaneous application is not possible.

In conclusion, the detection of near-surface streaming effects is possible by combined applications of geophysical methods as successfully demonstrated by some examples. Nevertheless, further theoretical, technical and methodological studies are necessary in order to add to our knowledge up to now and to quantify the proven effects.

References

Alpin, L.M. (1971): Praktičeskie raboty po teorii: Nedra, Moskva.

Armbruster, H.; Merkler, G.-P. (1983): Measurement of subsoil flow phenomena by thermic and geoelectric method.- Bull. IAEG No. 26-27: 135-142; Paris.

Kanamori, H.; Fujii, N.; Mitzutani, H. (1968): Thermal diffusivity measurements of rock forming minerals from 300° to 1100° K.- J. Geophys. Res., 73: 595-605.

Kappelmeyer, O. (1957): The use of near surface temperature measurements for discovering anomalies due to causes at depth.- Geoph. Prosp. 5, 3: 239-258; The Hague.

Militzer, H.; Schön, J.; Stötzner, U. (1986): Angewandte Geophysik im Ingenieur- und Bergbau, 2. überarbeitete und erweiterte Auflage.- VEB Deutscher Verlag für Grundstoffindustrie Leipzig, 419 pp.

Ogilvy, A.A.; Ayed, M.A.; Bogoslovsky, V.A. (1969): Geophysical studies of water leakages from reservoirs.- Geophys. Prosp. 17, 1: 36-62; The Hague.

Watson, K. (1975): Geologic application of thermal infrared images.- Proc. IEEE 63, 1: 128-137; Washington.

Woodside, M.; Messmer, R. (1961): Thermal conductivity of porous media I: unconsolidated sands, II consolidated rocks.- J. Appl. Phys. 32, 9: 1688-1699; Lancaster.

TEMPERATURE DISTURBANCE IN A DAM DUE TO LEAKAGE

C. Venetis
Rijkshogeschool,Groningen, The Netherlands

Abstract

Thermal disturbances may be viewed as natural tracers for detection and localization of leakage. Measurements taken at the model dam in the Bundesanstalt für Wasserbau (Karlsruhe) are very encouraging.

The sealing layer of the dam is assumed to have a leaking fissure. The problem is to estimate the temperature disturbance in the saturated sand of the dam in the vincinity of the fissure due to a temperature difference between reservoir water and the interior of the dam. Heat is transferred by means of the leaking water.

A long semicylindrical fissure and a circular hole are considered. In both cases radial flow with appropriate boundary conditions makes the problem analytically tractable. Conduction, convection and dispersion are considered. The problem is time-dependent. Numerial computations give rise to some general conclusions.

1 Introduction

Within the framework of the project "Quantifizierung von termischen und Eigenpotentialmessungen bei Fließvorgängen im Untergrund" (Armbruster, Hötzl and Merkler in the Poster Session) the formation and development in time of a thermal disturbance in a dam is of particular interest. Thermal disturbances are excellent tracers and can be used to predict the presence and approximate extent of a fissure in the sealing layer of the dam.

In what follows water of a different temperature enters the saturated sand of the dam via a crack or fissure and the temperature

Lecture Notes in Earth Sciences, Vol. 27
G.-P. Merkler et al. (Eds.)
Detection of Subsurface Flow Phenomena

distribution which is created in the originally uniform temperature field is analytically predicted. The problem is strongly formalized. The fissure is assumed to be semicylindrical (two-dimensional flow) or hemispherical (point source).

Conditions in the vicinity of the entering water are considered. The model is used to compute the time necessary for the temperature disturbance to reach the groundwater table. A finite differences solution would include the whole field of temperature; the assump-tion of realistic boundary conditions is the main difficulty in this approach. The problem has been modelled in the model dam of the Bundesanstalt für Wasserbau and we all await with great interest the elaboration and evaluation of the experimental measurements.

2 Flow through a fissure

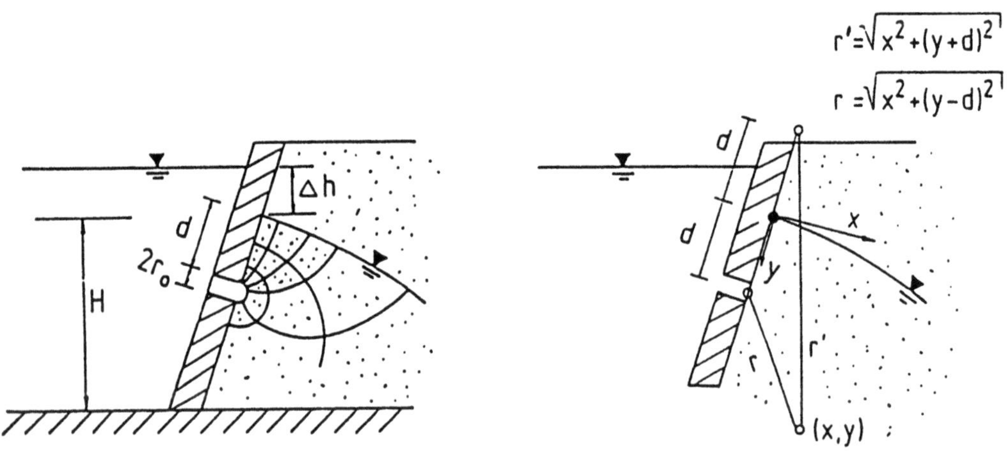

Fig. 1: Seepage due to fissure

A two-dimensional fissure may be viewed as a horizontal injection drain and the problem can be analyzed via the source-sink model, the imaginary sink being situated symmetrically with respect to the groundwater table (Fig. 1).

The results of this treatment are:

$$Q = \frac{\Delta h \pi K}{\ln(2d/r_0)} \qquad\qquad h - H = \frac{\Delta h}{\ln(2d/r_0)} \ln(r'/r)$$

$$V = \frac{\Delta h K}{\ln(2d/r_0)} \quad \frac{1}{r} \qquad (r>r_0) \qquad (1)$$

The definition of the symbols is given in Fig. 1; h is the piezometric head measured, e.g. from the bottom of the reservoir. The last equation expressing V(r) is the usual approximation which retains radial symmetrie. Q is the relevant recharge.

A crack in the sealing layer modelled as a circular hole generates, to a great extent, hemispherical flow because the ratio r_0/d may be assumed to be small. The hemispherical flow model produces the results:

$$Q = 2\pi K r_0 \Delta h \qquad h-H = \Delta h\left[1-r_0\left(\frac{1}{r_0} - \frac{1}{r}\right)\right]$$

$$V = K r_0 \Delta h \frac{1}{r^2} \qquad (2)$$

For the definition of the relevant symbols see Fig. 1. K is the Darcy permeability; h is the varying piezometric head. The pore velosity V_p is $V_p = V/n_e$ where n_e is the effective porosity; V is the Darcy velocity.

In order to obtain realistic numerical results the following values were assumed for the computations that follow:

$K = 7.5 \times 10^{-4}$ m/s;
$\Delta h = 1.5$ m;
$d = 2.5$ m;
$r_0 = 1$ cm;
$n_e = 18\%$.

The resulting pore velocities are:

$$V_p = 10^{-3}\frac{1}{r} \text{ m/s} \quad \text{two-dimensional flow;}$$

$$V_p = 6.25 \times 10^{-5}\frac{1}{r^2} \text{ m/s} \quad \text{hemispherical flow.} \qquad (3)$$

3 Radial Heat Transfer

If the water leaking from a crack in the sealing layer is of a different temperature than the assumed homogeneous medium, into which it flows, then heat transfer takes place in the vicinity of the crack in a radial fashion. The fissure constitutes a line or a point source (two-dimensional or spherical spreading) and a time-dependent temperature disturbance occurs in the medium, assumed to be extended and originally at uniform temperature. The mechanism of heat transfer is convection, conduction and dispersion and the equation governing radial transfer maybe be written in the form (see Appendix):

Two-dimensional (cylindrical) radial spreading:

$$\frac{\partial T}{\partial t} = (\alpha + \mathfrak{D}) \frac{\partial^2 T}{\partial r^2} + \left(\frac{\alpha}{r} - V_T\right) \frac{\partial T}{\partial r} \qquad (4)$$

Spherical spreading:

$$\frac{\partial T}{\partial t} = (\alpha + \mathfrak{D}) \frac{\partial^2 T}{\partial r^2} + \left(\frac{2\alpha}{r} - V_T\right) \frac{\partial T}{\partial r} \qquad (5)$$

The parameters involved are:

α : Thermal diffusivity.
For practical groundwater computations values between 10^{-7} and 10^{-6} m^2/s are appropriate.

$\mathfrak{D}$: Dispersion coefficient. It accounts for the effect of dispersion (see Appendix).
$\mathfrak{D} = l\ V_p$ with l as characteristic length expressing the dispersive properties of the aquifer; V_p is the pore velocity. Values of l reported in the literature range between 0.01 and 0.5 m (Kobus 1980).

V_T = The thermal front velocity. This velocity is smaller than the pore velocity because of heat exchange between the fluid and the solid matrix. According to Kobus (1980):

$$V_T = \frac{n_e \rho_W c_W}{\rho_c} V_P \approx 0.5\ V_P,$$

where ρc is density x specific heat of the saturated soil and $\rho_w \rho_c$ is the analogous product for water.

For the example treated here the following parameter values were chosen:

$\alpha = 0.7 \times 10^{-6}$ m²/s;

$l = 0.1$ m;

$V_T = 0.5\ V_P$ (V_P is given by $V_P = \dot{V}/n_e$)

With these values Eq. (4) assumes the form:

$$\frac{\partial T}{\partial t} = 0.2\ (5 \times 10^{-4})\ \frac{1}{r}\ \frac{\partial^2 T}{\partial r^2} - (5 \times 10^{-4})\ \frac{1}{r}\ \frac{\partial T}{\partial r} \qquad (6)$$

Conduction is found to be, in this case, unimportant compared to dispersion, because of the relatively high velocities of flow within the range of practical significance (e.g., $r < 4$ m). Also, when dispersion is taken to be a factor of 10 smaller ($l = 0.01$ m) the role of conduction (for $r < 4$ m) remains unimportant. This is not so for the hemispherical model where the velocities of flow are appreciably lower, although also in this case the influence of conduction is not overwhelming.

Hoopes and Harleman (1965) give an approximate solution of the radial diffusion-convection equation. This solution is reported (Raimondi et al., 1959) to perform very satisfactorily.

Since the equations are similar and the boundary conditions are analogous we can use the above mentioned solution for our problem of two-dimensional radial heat transfer. Denoting by T the temperature increase in the dam, which we wish to determine, and by T_o the temperature difference at the source, we may write the approximate solution of Eq. (6) as:

$$\frac{T}{T_0} = \frac{1}{2}(1 - \text{erf}\ \phi);$$

$$\phi = \frac{r^2/2 - 5\times10^{-4}t}{\sqrt{(4/3) \times 0,2\ r^3}} \qquad (7)$$

where erf ϕ is the error function of the argument ϕ.
Values of erf ϕ have been tabulated (see, for instance, Carslaw and Jaeger, 1959, etc.).

We continue the numerical example by computing the approximate time in which a temperature difference carried by leaking water is expected to reach the groundwater table where it could be detected.
The travel time of the heat front can be easily computed:

$$t_{fr} = \int_{r_0}^{r} V_T^{-1}(r)\ dr. \qquad (8)$$

For cylindrical radial flow with $V_T = V/2n_e$, we find [Eq. (1)]:

$$t_{fr} = \frac{n_e}{A}(r^2 - r_0^2)\ , \qquad A = \frac{\Delta hk}{\ln(2d)/r_0}\ ,$$

and this expression gives for r = 2.5 m about 1.7 h. For this duration Eq. (7) gives a 50 % temperature increase (at r = 2.5 m). From Eq. (7) we conclude further that a 90 % temperature increase would take about 3 h to occur. If the dispersion coefficient is reduced to one-tenth of the assumed value, the time needed for a 90 % temperature increase is about 2 h.

For the case of spherical flow we have to adjust the r-dependent coefficients of the differential equation via approximation, in order to express them in a form analoguous to that of cylindrical flow [Eq. (6)]. For the range of variation of r of practical interest (e.g. $r < 4$ m) the approximated coefficients lead to the solution:

$$\frac{T}{T_0} = \frac{1}{2}(1 - \text{erf}\,\phi);$$

$$\phi = \frac{r^2/2 - 1.9\text{x}10^{-5}t}{\sqrt{(4/3) \text{ x } 0,22\, r^3}} \quad . \tag{9}$$

The travel time of the heat front computed via Eqs. (2) and (8) is 46 h. Equation (9) gives for this time interval a 50 % temperature increase. A 90 % increase occurs at t = 75 h. If the assumed dispersion coefficient is taken to be a factor of 10 or smaller (l = 0.01), then the 90 % increase occurs at 54 h.

In conclusion, we can say, in general, that for the spreading of a temperature disturbance in a leaking dam, the role of dispersion is by far more important than that of conduction. Depending on the assumed dispersion coefficient (between 0.01 and 0.1 m) the travel time (of the thermal front) is about 60 to 85 % of the time needed for a 90 % temperature increase. This is valid for the vicinity of the sealing layer both for a line and a point leakage source. The travel time, of course, is a factor 10 higher in the case of a point source.

Appendix: Equations of Radial Heat Transfer

a. Conduction only

$$\frac{DT}{Dt} = \frac{1}{2}\frac{\partial}{\partial r}\left(\alpha r \frac{\partial T}{\partial r}\right) = \alpha\left(\frac{\partial^2 T}{\partial r^2} + \frac{1}{r}\frac{\partial T}{\partial r}\right) \longrightarrow \text{cylindrical source}$$

(A1)

$$\frac{DT}{Dt} = \frac{1}{r^2}\left[\frac{\partial}{\partial r}\left(\alpha r^2 \frac{\partial T}{\partial r}\right)\right] = \alpha\left(\frac{\partial^2 T}{\partial r^2} + \frac{2}{r}\frac{\partial T}{\partial r}\right) \rightarrow \text{spherical source}$$

where DT/Dt is the so-called substantial derivative.

b. Dispersion only

The equation of hydrodynamic dispersion refers to mass transfer and is written in terms of concentration c as

$$\frac{Dc}{Dt} = \frac{1}{r}\left[\frac{\partial}{\partial r}\left(r\mathcal{D}\frac{\partial c}{\partial r}\right)\right] \longrightarrow \text{cylindrical source}$$

(A2)

$$\frac{Dc}{Dt} = \frac{1}{r^2}\left[\frac{\partial}{\partial r}\left(r\mathcal{D}\frac{\partial c}{\partial r}\right)\right] \longrightarrow \text{spherical source}$$

In our case c may be viewed as the concentration of water with the higher temperature. Designating T_o as the temperature difference at the source and T the temperature increase at point r and time t, we have obviously $c = T/T_o$.

Further, we can write

$$r\mathcal{D} = rl\,|V_p| = rl\,\frac{A}{r} = lA \longrightarrow \text{cylindrical source}$$

(A3)

$$r^2\mathcal{D} = r^2 l\,|V_p| = r^2 l\,\frac{B}{r^2} = lB \longrightarrow \text{spherical source}$$

where A and B are constants($V_p = A/r \rightarrow$ cylinder and $V_p = B/r^2 \rightarrow$ sphere). Introducing the expressions above into Eq. (A2) we obtain the dispersion equations in terms of temperature:

$$\frac{DT}{Dt} = \frac{lA}{r}\,\frac{\partial^2 T}{\partial r^2} = \mathcal{D}\frac{\partial^2 T}{dr^2} \longrightarrow \text{cylinder}$$

(A4)

$$\frac{DT}{Dt} = \frac{lB}{r^2}\,\frac{\partial T}{\partial r^2} = \mathcal{D}\frac{\partial^2 T}{\partial r^2} \longrightarrow \text{sphere}$$

c. Conduction, dispersion, convection

Combining the effect of both conduction and dispersion means adding the right-hand side of Eqs. (A1) and (A4). Setting further

$$\frac{DT}{Dt} = \frac{\partial T}{\partial t} + V_T\frac{\partial T}{\partial r},$$

where $V_T\frac{\partial T}{\partial r}$ is the convective term, we end up with Eqs. (4) and (5).

Acknowledgement. My thanks to my colleagues Armbruster, Hötzl and Merkler for their support and cooperation on this and other groundwater problems.

References

Kobus H. (1980)
Ausbreitung von abgekühltem Wasser in Grundwasserleitern. Beitr. Statussem. Wärmepumpen und Gewässerschutz. Umweltbundesamt, 19.- 20.11.1979, Berlin.

Hoopes J. A., Harlemann D. (1965)
Waste water recharge and dispersion in porous media. Hydrodyn. Lab. School Eng., Rep. 75. MIT, Cambridge

Raimondi P., Gardner G. H. G., Petrick C. B. (1959)
Effect of pore structure and molecular diffusion on the mixing of miscible liquid flowing in porous media A.I.CH.E - S.P.E Fundam. Concepts of Miscible Fluid Displacement: Part I, Preprint 43, San Francisco 12/59

Carlslaw H. S., Jaeger J. C. (1959)
Conduction of heat in solids, 2nd edn. Oxford Univ. Press, London

Verruijt A. (1972)
Stationary heat transport etc. In: Fundamentals of transport phenomena in porous media. IAHR, Elsevier, Amsterdam

DATA PROCESSING BY TEMPERATURE MEASUREMENTS

J. Gerlach
Bundesanstalt für Wasserbau, Karlsruhe
Kußmaulstr. 17, D-7500 Karlsruhe

In order to evaluate flow phenomena in dams, temperature measurements are carried out. For this a multitude of different measurement systems are used. For research, over long time periods, great quantities of temperature test data must be registered and processed. The aim of data processing in this connection is basically a graphic preparation or a representation of the measured data.

In order to accomplish this task the temperatures must be registered, processed and stored in order to be represented as:

1. Temperature-timegraphs; or
2. Temperature distributions in longitudinal, transversal or horizontal sections.

The temperature measurements data can be obtained with various temperature sensors, either manually or automatically with measurement systems.

Generally, the equipment for measuring body temperatures and equipment for infrared-radiation systems must be distinguished.

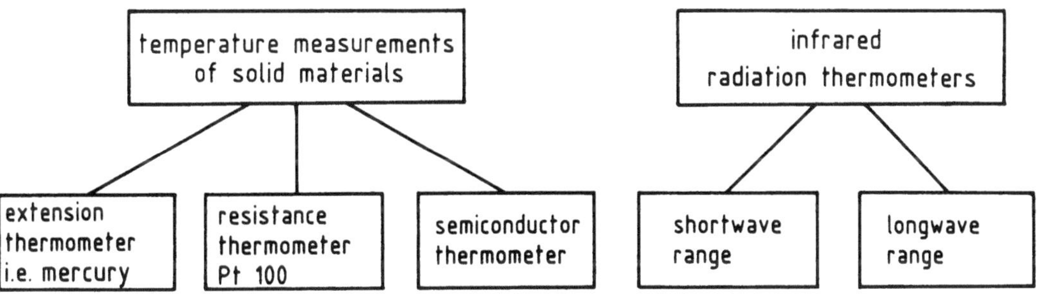

Fig. 1: Measurement systems

Lecture Notes in Earth Sciences, Vol. 27
G.-P. Merkler et al. (Eds.)
Detection of Subsurface Flow Phenomena

Measurement of solid materials temperatures:

The temperature sensors for solid materials have to be installed in or directly on the object of measurement. These objects are here the dry or wet sand of a dam, the air or the water. We used extension thermometers and resistivity or semiconductor thermometers.

Extension thermometers work with a metal or a fluid which extends according to temperature alteration. This extension is analogously indicated on a Celsius scale. The data can be read visually and registrated manually. In order to process the data, the test data have to be given via terminal to the Siemens computer, the host computer of the Bundesanstalt für Wasserbau. There, the data are stored and represented graphically with the available software.

The second method of temperature measurement is based on the change of the electric resistivity of metals or semiconductors due to the alteration of temperature. The measuring instrument provides the temperature sensor with the electric current, which is changed by sensor. In the measuring instrument the electrically measured data is transformed with a specific calibration curve into a temperature value. This value is indicated on a digital display. It is possible to register the temperature value manually, or continuously with an analogous plotter, Figure 2 shows such an instrument. The sensor is connected with the measuring instrument and the display by a long transfer cable.

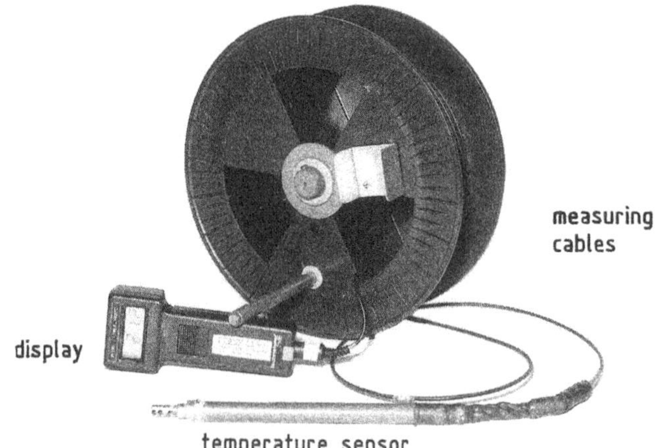

Fig. 2. Temperature-measuring instrument by electric resistivity

Often several sensors are installed at one measuring site. Each sensor is controlled by a manual scanner. A periodic registration of many sensors is only possible by the automatic control of a computer. The requested time intervals are given by aid of computer. The computer-integrated timer activates the scanner at the desired time; sensor and measuring instruments are joined by the scanner. In order to process the analoguous measuring signals they are converted into digital data of 8 or 12 bits. These digital data may be stored in the computer on discs or cassettes. With the help of software these data can be transferred by transmission tube line or by minidisc to the Siemens host computer. This computer has a large hardware capacity to store these data; and here they are processed graphically. Figure 3 shows a measuring unit used in the laboratory of the Bundesanstalt für Wasserbau. It consists of a Hewlett-Packard computer with a monitor and minidisc, as well as the measuring instruments with scanner and measuring cables for the sensors.

Fig. 3. Automatic measuring unit

The schematic plan in Fig. 4 shows how the use of other sensor types can be connected to the measuring unit. In addition to the temperature sensors, self-potential sensors, water-pressure sensors and discharge measurements are registered. As about 400 sensors have to be registered, an expansion unit for data processing is required. The test data are stored on minidiscs. These minidiscs can be read by the Siemens host computer and the data can be processed into graphs.

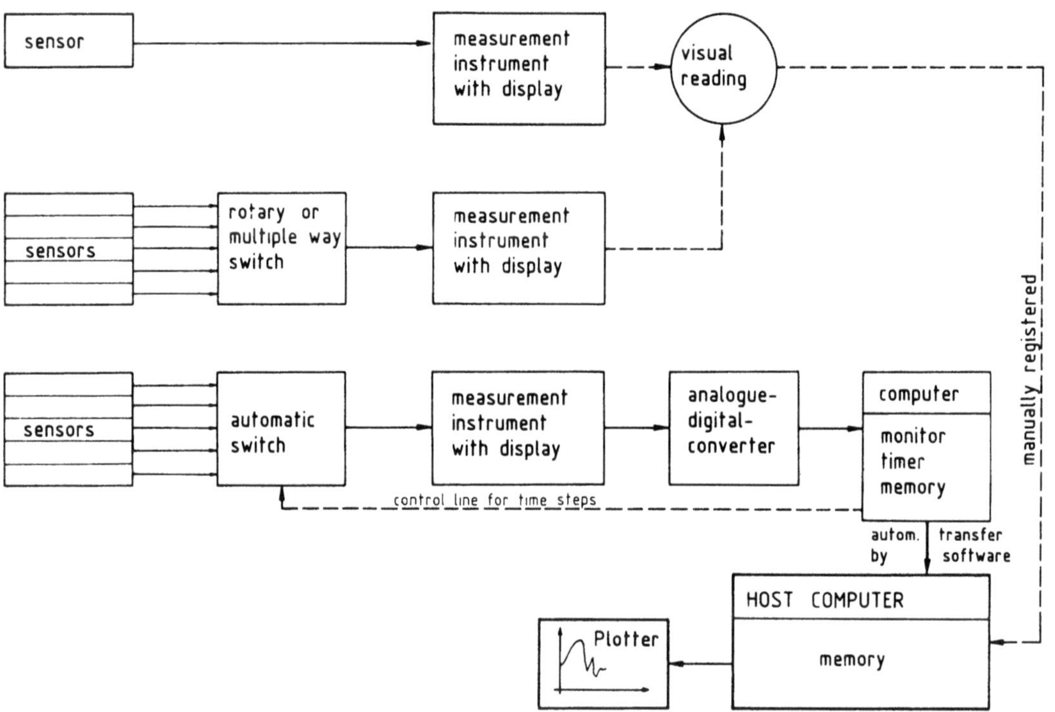

Fig. 4. Measurement of resistivity or semiconductor thermometers

For laboratory and in situ measurements measuring equipment from the Geotherm-Kappelmeyer Co. was used. It is movable and can function by main current as well as by electric battery. It is possible to measure 24 temperature sensors simultaneously. Four sensors are installed on one transmission tube line. In this system the test data are stored on a minicassette which is able to store about 5000 items of test data. Before processing the data have to be transferred to a disc in order to be stored in the Siemens host computer.

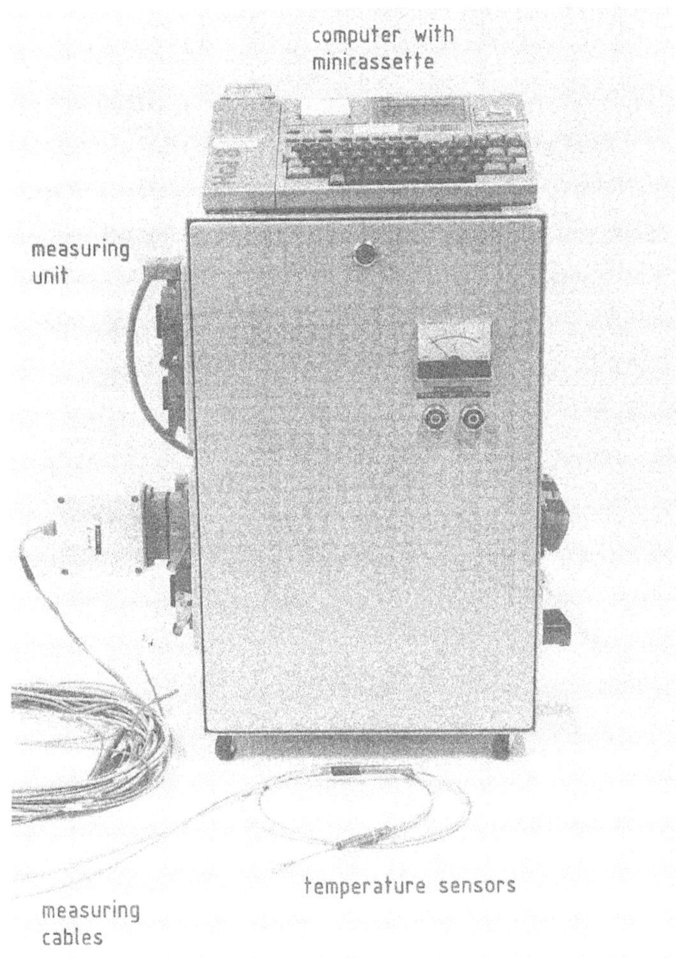

Fig. 5. Laboratory and in situ measurements

For in situ measurements the measuring equipment of "WAS" company, called "Datensammler MDSII", can also be used (Fig. 6). Four separate sensors can be linked to this absolutely isolated system. The test data are registered in the random-access memory of the integrated computer, whose capacity runs up to 56 kb. For adjustement and control of the system a portable computer is used. The stored test data are transmitted to this computer, too, and stored there on minidiscs. Using special software it is also possible to represent test data graphically on a screen. These graphs may be printed on a graphic printer.

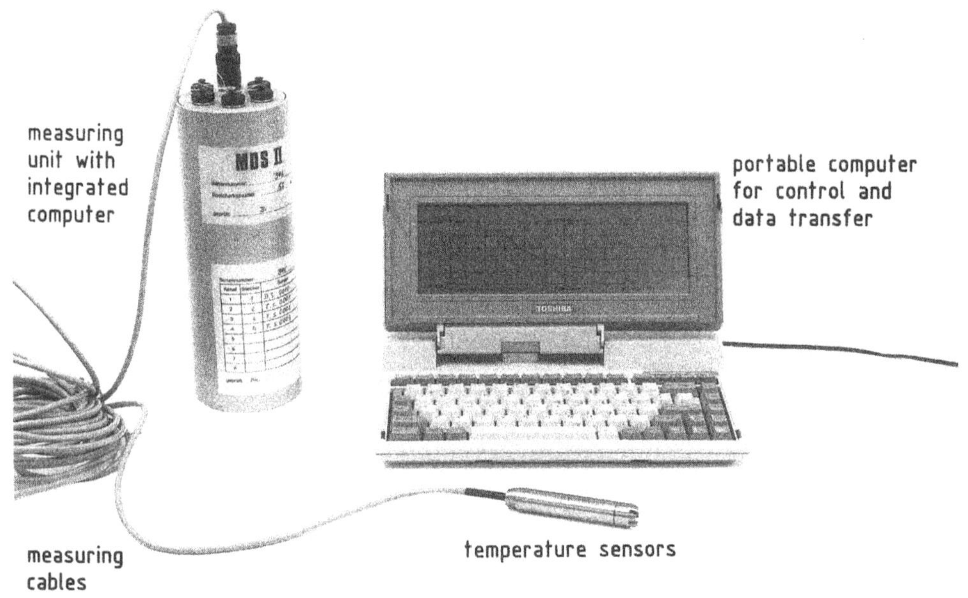

Fig. 6. In situ measurement; "Datensammler MDSII"

Radiation thermometers:

Radiation measurement is based on infrared temperature radiation, which is reflected by the surface of any substance. This temperature radiation is transmitted by a lens combination to a so-called detector, a radiation thermometer, which transforms the radiation energy into an analogous electric signal. Considering the emission coefficient it is possible to determine the surface temperature of the object to be measured by this electric signal.

Fig. 7 shows the infrared thermovision device of the AGA company. This instrument works in the shortwave range of the infrared spectrum. It registers the medium radiation intensity of the considered surface. Together with a known reference temperature or the emission coefficient, it calculates the medium surface temperature. The measurements to be considered are decomposed 25 times per second into 100 horizontal and 70 vertical pixels by two prisms which rotate at high speed. In this manner the surface radiation of each pixel is succesively fed to the nitrogen-cooled detector. It registers the radiation intensity in the shortwave range of the infrared spectrum and transforms it into analogous signals. These signals can be represented as gray-strip photography on a monitor. It is also possible to store them on a videotape and to transfer them from there or directly to personal computer.

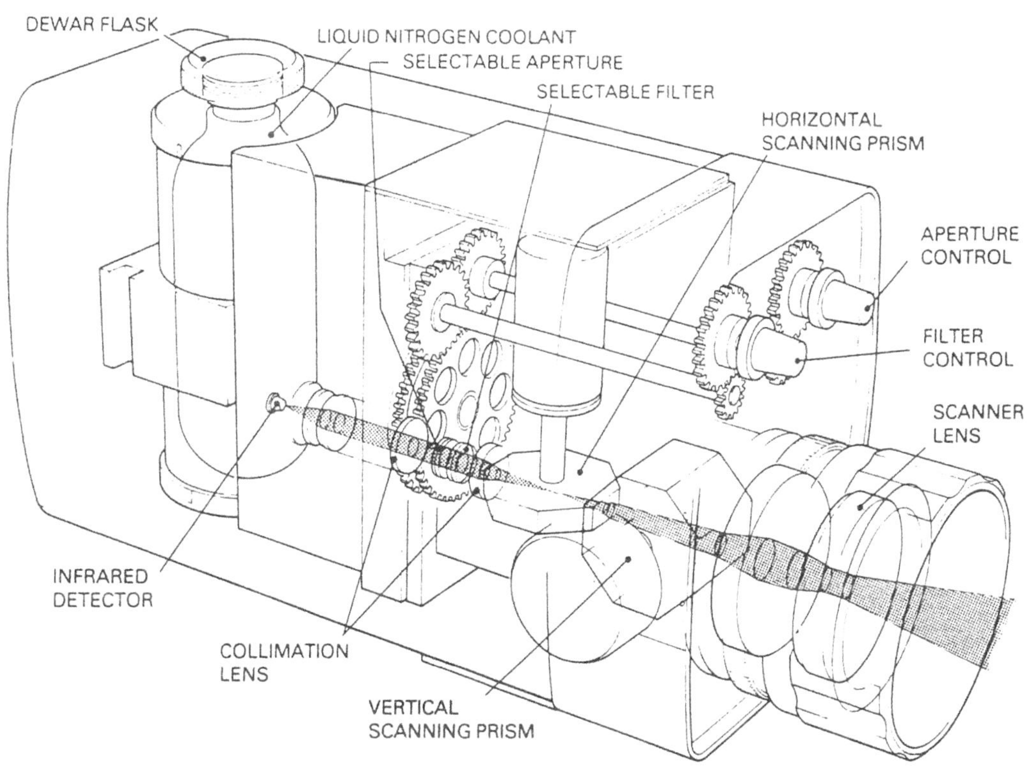

Fig. 7. Infrared measuring device

For computer-aided data processing the produced analogous signal of each pixel has to be registered, digitized, assigned and stored on discs. The processing speed of the analogous digital converter or the computer allows one to store the test data of only two pictures per second on discs, containing 128 times 64 pixels with a decomposition of 8 bits. In relation to the real-time processing some picture information is lost. Several pictures can be exposed in order to attain better radiation information of the pixels. On a disc of 360 kb it is possible to store 42 unit frames, which causes a frequent change of discs. The hard-disc system of recent soft- and hardware allows a superior storage capacity and a real-time processing of pictures. These instruments have been purchased now by the BAW.

Different colors can be assigned to the digital test data of each pixel and represented on the screen or issued by a color printer. With a program for transformation the digital test data of one picture can be transferred to the host computer of the BAW. Further processing of these data at the color work station installed here, is possible in this way.

On the multispectral scanner (Fig. 8) of the DEDALUS company the registration of the surface radiation is carried out in lines by a rotating prism. The scanner is installed in an airplane, so that the registration of the area results from the airplane's own motion. The radiation intensity of the registered pixels is transmitted to several detectors, which analyze various spectral ranges. The analogous signals produced by the detectors are stored on magnetic tape and transformed, after the flight, with an 8-bit analogous digital converter into 672 pixels per line. These pixels are separated into channels and recorded on magnetic tapes.

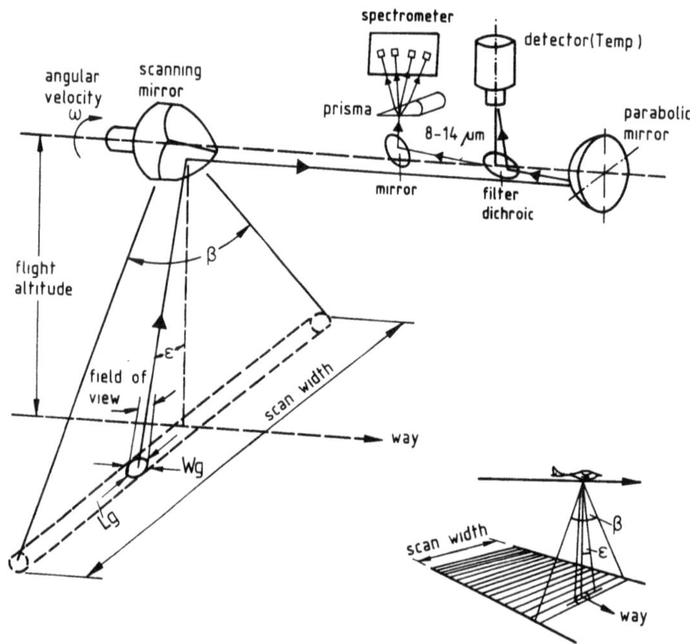

Fig. 8. Schematic plan of the multispectral scanner

Further processing has to be done with special software at the host computer of the BAW. The processing of aerial photographs or picture information of the AGEMA camera is carried out pixel by pixel while assigning certain color graduations to the stored digital data. These color graduations can be represented on screen, as mentioned before, or issued by a color printer. As the emission coefficient is unknown, it is only possible to determine the radiation intensity of the area, but not the exact surface temperature. To achieve a better evaluation the digitized results of measuring various spectral ranges can be superposed or different ranges can be analyzed more exactly.

APPLICATION OF THERMAL IR-TECHNIQUES FOR RECONNAISSANCE OF DAM AND BARRAGE DEFECTS IN AN EARLY STATE, ANALYSIS OF DUMP SITES AND TUNNEL CONDITION SURVEYS

M.Sartori[1]

1 Introduction

Since 1973 SPACETEC applies synoptic measurement techniques especially IR-recordings to examine large objects.

This technique shows defects on dams and barrages in an early state, areas on dump sites saturated with gas or liquids and allows a condition survey of railway or road tunnels.

Up to now security control of such buildings were carrid out by visual methods or by measurements and core probes with relative large steps. In addition to this expensive method of security control, the critical disadvantage is the missing information in the area between the measurements and core probes.

A security control should not be based on such procedures on dams, barrages and tunnels. The presented example on waste dumps leads to the localization of pollution sources and cannot be regarded as a security control.

The application of IR techniques allows an examination covering the whole visible surface by aid of airborne, or stationary measurement platforms. Although the results of this synoptic, surface covering method are less accurate than in-situ measurements, the advantage of discovering anomalies, connected with succeeding in-situ measurements, predominates.

In this way a consequent and economic procedure now exists for the security control on large buildings mentioned in the title.

[1] SPACETEC Datengewinnung GmbH, Freiburg, FRG

Lecture Notes in Earth Sciences, Vol. 27
G.-P. Merkler et al. (Eds.)
Detection of Subsurface Flow Phenomena

2 The Recognition of Dam Defects in an Early State

The supposition for the detection of dam defects in an early state is a surface covering IR-recording of the dam and the surrounding area.

With IR-airborne line scanner recording in the spectral range between 8 and 14 µm surface temperature data are collected. Feasible measurement conditions lead to information about heat flows through the surface of the building. The method is based on the influence of a dam defect on these heat flows.

The detection of a damage with the IR-technique requires:

1. The definition of dam damage consequence in an early state.

2. The definition of damage indicators for these damage consequences.

Consequences of dam damages in an early state are:

1. An anomaly of humidity in the dam body.

2. An anomaly of humidity in the surrounding area.

3. An extraordinary seapage stream in adjacent water body.

The following damage indicators are taken into consideration:

1. Anomaly of surface temperature of the dam or in the surrounding area.

2. Characteristic values of thermal inertia of a near-surface layer above a humid part of the dam body.

3. Anomaly of the vegetation.

4. Anomaly in the surface temperature of an adjacent water body.

The damage indicators (1), (2) and (4) can be aquired by surface temperature recordings. The indicators (1) and (4) can be acqiured with a single IR-recording, but the acquisition of the indicator (2) requires

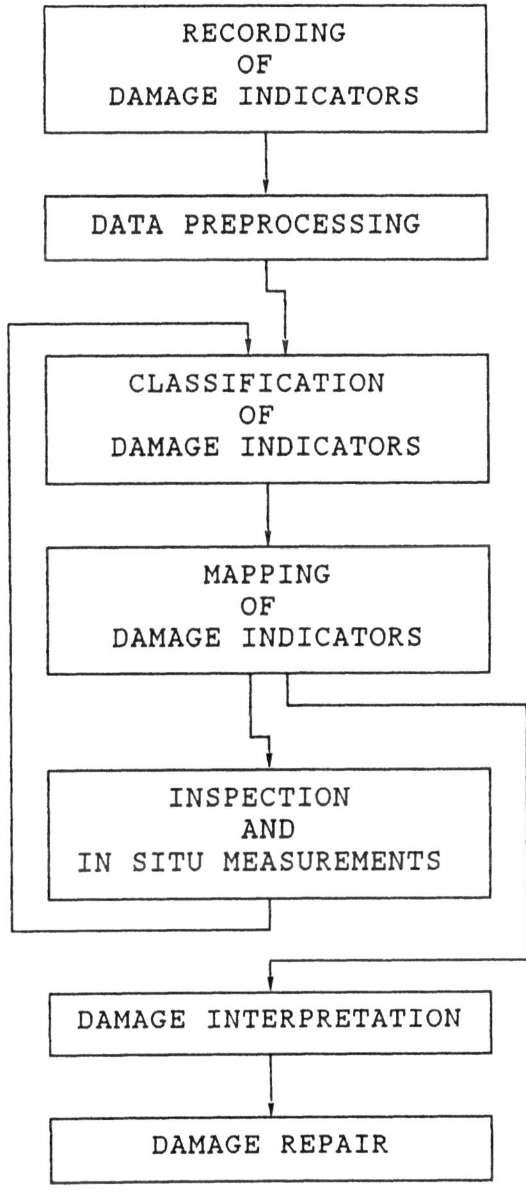

Fig. 1: Procedure of detecting and classification of damage indicators until damage repair.

a minimum of two surface temperature recordings. The vegetation anomaly can be detected through an anomaly in the reflected sun radiation. This damage indicator (3) requires a multispectral radiation recording in the range of the visible light and the near-infrared. The measurement and interpretation of all four indicators leads to an enhancement of the damage classification result (Fig. 1).

The cooperation between the favorable waterway engineering and research institute (BAW) and SPACETEC leads to extensive research and development activities of damage classification on the basis of the four damages indicators and connected measurements along dams on Oberrhein, Main-Donaukanal, Mittelkanal and Elbe Seitenkanal over a total length of more than 400 km.

3 Barrage Examination

Like the examination of dams, the damage indicator "thermal inertia" for the examination of barrages is still the center of attention. Similar to dam examination thermal inertia of parts of the building can be computed on the basis of a minimum of two surface temperature recordings compiled at different daytimes. The temporal development of the temperature field of a near-surface layer of the building depends on the structure and exposition of the barrage. First, this situation must be analyzed in order to obtain the optimal timetable for the IR-recordings. The examinations on the "Edertalsperre" required four surface temperature recordings of the whole barrage surface within a 24 h periods. An image mosaic, with a spatial resolution of 7 cm on the 240 m x 45 m surface, must be compiled for each recordings.

A multitemporal classification of the four surface temperature recordings leads to a thematic map of the thermal inertia and therefore to the localization of wet parts of the building (Fig. 2).

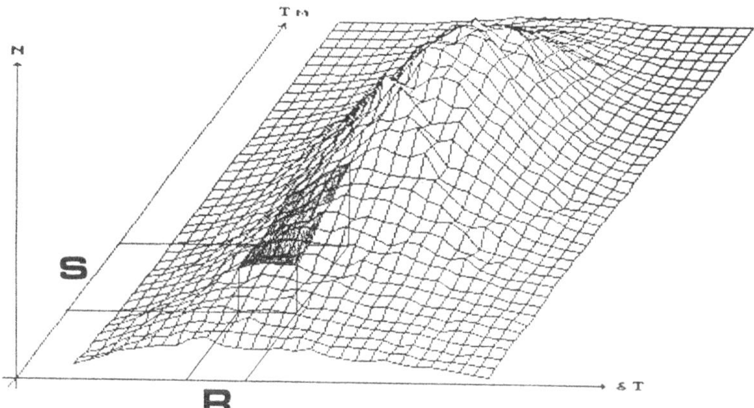

Fig. 2: Selection of wet structures.
The coordinate transformation of two surface temperature images from x, y into temperature coordinates leads the identification of wet parts of the building (structures with temperature coordinates in the range of R and S).

4 Waste Dumps

It is necessary to localize pollution sources in old waste dumps. The main interest thereby is focused on one of the following damp-specific factors.

1. Content of Gas;
2. Humidity and seapage outlet;
3. Inhomogeneity of the waste dump;
4. Chemical/biological processes.

The basis of the application of synoptic measurement processes is the interaction of the dump-specific factors mentioned above with the dump surface. Again, specific indicators (visual, surface temperature-time function, vegetation state and composition) on the dump surface are necessary to identify those interactions.

Outlet of gas on a waste dump affects the thermal inertia of soil and the growth of vegetation. Anomalies in vegetation growth and composition can be pointed out by measuring and spectral classification of the reflective sunlight.

Seapage symptoms and inhomogeneity in the dump composition affect the formal inertia of a dump layer and lead to a different temporal behaviour of surface temperatures.

Chemical and biological processes are connected with energy changes, the buildup and the generation of heat sources which lead to an influence on the surface temperature of the waste dump.

The examination of the waste dump requires a measurement technique which comprehends those indicators covering the whole surface of the dump. Operational surveys require a fast and economic solution for measurement, interpretation and mapping of results.

Parts of the measurement techniques applied for dams and barrages were modified by SPACETEC for the examination of waste dumps. Again, the damage indicator "thermal inertia" of a near-surface soil layer was applied. An individual combination of IR-recordings and in-situ measurements is used for the investigation of dump sites.

In contrast to dam examination with the installation of the line scanner in a fixed-wing airplane, the scanner was mainly installed in a helicopter for the examination of waste dumps.

In the year of 1984, SPACETEC applied a multitemporal classification for the examination of the waste dump "Georgswerder" in Hamburg (FRG). IR-recordings were made within a period of 24 h at three different daytimes. The interpretation of the multitemporal classification of two nighttime IR-recordings leads to the display of the contours of an old barrel depot (Fig. 3). The sucessful indicator was the thermal inertia of a dump site layer with seapage anomalies.

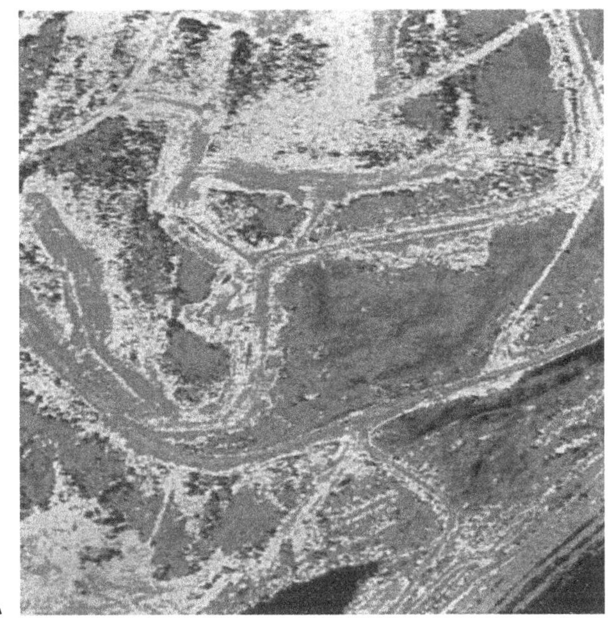

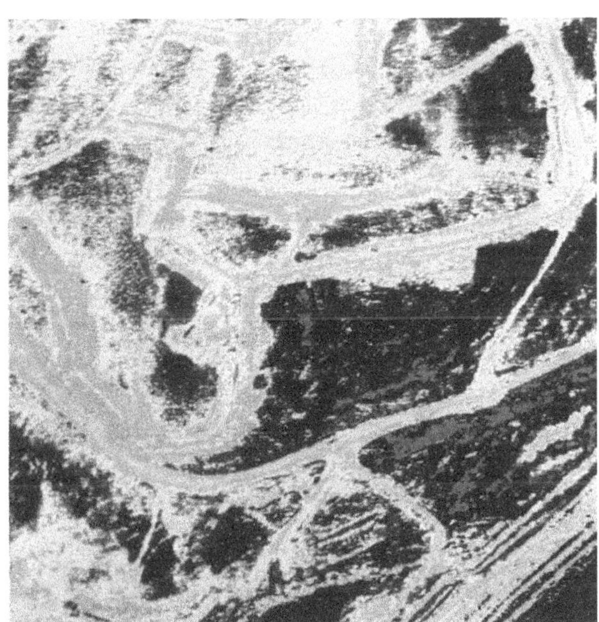

Fig. 3: Examination of dump sites. A 15,000 m^2 area of the Georgswerder dump site in Hamburg, FRG shows after multitemporal classification (image A) of two surface temperature images the contours of an old barrel depot (red pattern in the center of image B).

In contrast to difficult multitemporal classification of different IR-recordings, individual temperature surface images often show numerous interesting thermal structures on the dump surface. Such imagery can be very useful for a first evaluation.

5 Tunnel Condition Surveys

In contrast to dam, barrage, and waste dump applications mentioned before, a diverse application procedure of thermography for tunnel conditions survey was chosen.

Based on the existence of a stationary heat flow through the tunnel lining between the rock and the air in the tunnel interior, dependece of the lining-surface temperature on heat diffusivity of the lining is used for the tunnel condition survey.

This heat flow is caused by temperature changes in the outside air and is affected by cracks and voids in the lining, poor backfill, damp and the contact between lining and rock.

Three essential differences between the procedure in the tunnel and the investigations mentioned before are:

1. In the case of the stationary heat flow only one IR-recording is necessary.

2. The investigation of much larger building volume is possible.

3. An additional, stationary measurement program for preparation is necessary.

The tunnel condition survey using thermography was developed at SPACETEC since 1982 and provides a detailed, fast and comprehensive assessment of the factual state of a tunnel. The survey can be performed during normal breaks in traffic. The measuring instruments are fitted on a survey vehicle and permit a continuous non-contact investigation of the entire tunnel surface. During traffic breaks, the survey car travels through the tunnel at a speed of about 6 km/h, records a temperature image and visual surface data with a spatial resolution of 1 cm^2, of the entire tunnel lining surface. Survey data are recorded on magnetic tape.

As mentioned before, cracks and voids in the tunnel lining, poor backfill, variations in the contact between lining and rock face, and dumps will modify heat flow between the rock face and the air inside the tunnel.

Figure 4 originates from an investigation of the Hasenbergtunnel (German Federal Railways) in "Stuttgart" (FRG). The survey was performed under quasi-stationary heat flow conditions in the direction from the rock to the tunnel interior. The image shows the classification result as an unwinded tunnel lining.

The classification of surface temperatures presents areas (Fig. 5) affected by water behind or in the lining. It becomes possible to follow the distribution of water, penetrating a fractured concrete layer, situated behind the 35-cm-thick inner lining.

As the heat flow in the tunnel is produced by changes of external air temperature, the occurence of the stationary heat flow conditions are awaited through observing the results of the preparartion-measuring program (Fig. 5). Data from different tunnels are transmitted through the telephone circiut to the SPACETEC laboratory in Freiburg.

Under these controlled measurement conditions, the appearance of stationary or quasi-stationary heat flow leads to the start of the survey vehicle.

The preparation-measuring program delivers comprehensive data of the regarded tunnel sections. So it is possible to control different tunnels simultaneously to obtain start decision data for the disposed survey vehicle.

The completed measurements of the survey vehicle are evaluated after preprocessing the data stored on magentic tape. Presently, inventories of tunnels with length up to 8 km can be performed during a nightly traffic break lasting 4 h.

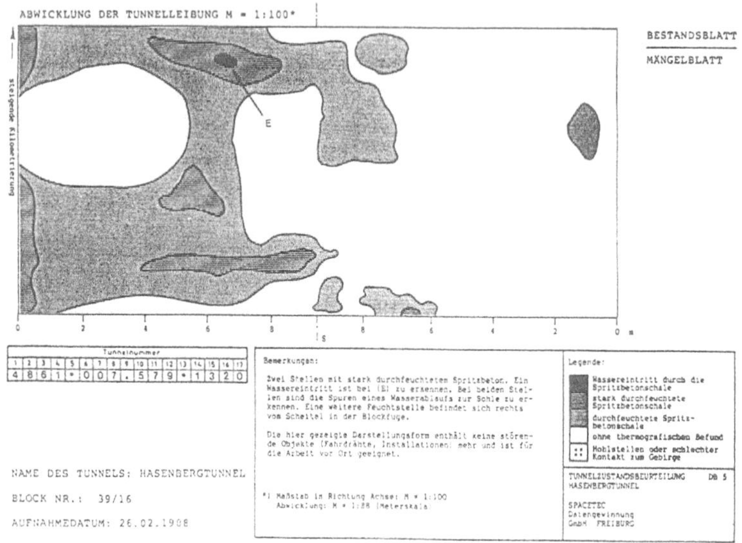

Fig. 4: Tunnel condition survey. The result of the examination of a 634-m-long part of the Hasenbergtunnel (German federal railways, S-Bahn, (Stuttgart)) is displayed in 64 printout sheets. Each sheet shows a 10-m-long, unwinded part of the lining. In the example of the block No. 39/16 gray areas are wet zones in the 20-cm-thick outer concrete lining, situated behind the 35-cm-thick inner lining. The black area E shows a water flow through a fracture of the outer concrete lining (not visible part, situated more than 35 cm beyond the inner lining surface).

The classification (Fig. 4) is carried out with a mathematical-physical model. A large part of the research and development work for the model and survey runs is performed in an experimental tunnel. Many tunnel sections were built and examinated under different heat flow conditions.

The tunnel-model experiments and the interpretation in the mock-up tests on parts of tunnel constructions are the results of joint development by Amberg Engineering Ltd in Zürich (Switzerland) and SPACETEC Datengewinnung GmbH in Freiburg (FRG). The test work mainly contained the solution of heat flow and climate problems. The simulation of stationary heat flow was realized by controlled rock and air temperature. The rock behind the models were equipped with tubes for heating and cooling and the space in front of the lining models is air-conditioned.

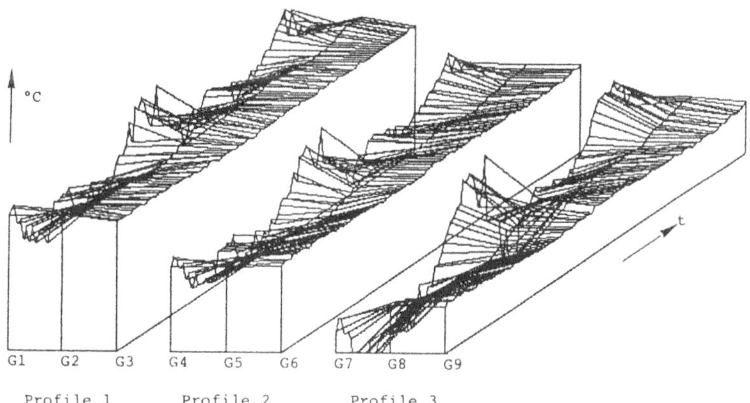

Fig. 5: The preparation program. The temporal development of temperature in the lining between the air in the tunnel (G1, G4, G7) and the rock (G3, G6, G9). The installation of three measurement profiles in the "Hasenbergtunnel" enables the identification of a quasi-stationary heat flow in the tunnel lining.

The development of this technique was decisively influenced by pilot programs of the German and Swiss federal railway societies. Presently, this technique has been developed further to a very sensitive instrument, regarding the following damage indicators.

1. surface temperature;

2. The visual impression of the lining surface;

3. Tunnel profile.

Tunnel profile data with the same high spatial resolution as surface temperature data lead to an important improvement of tunnel condition surveys.

The advantage correlating data of diffusivity, tunnel profile and surface items is presently used in the realization of new equipment for new survey vehicle generation.

DETECTION OF RIVER-INFILTRATED WATER FLOW BY EVALUATION OF HYDROGEOLOGIC, HYDROCHEMICAL AND HYDROTHERMAL DATA USING NUMERICAL GROUNDWATER MODELS

K. Zipfel[1] and U. Horalek[1]

1 View of Situation

The drinking water supply of the city of Mainz results partly from groundwater of the catchment area "Eich" about 30 km south of Mainz in the northern part of the Upper Rhine Valley (Fig. 1).

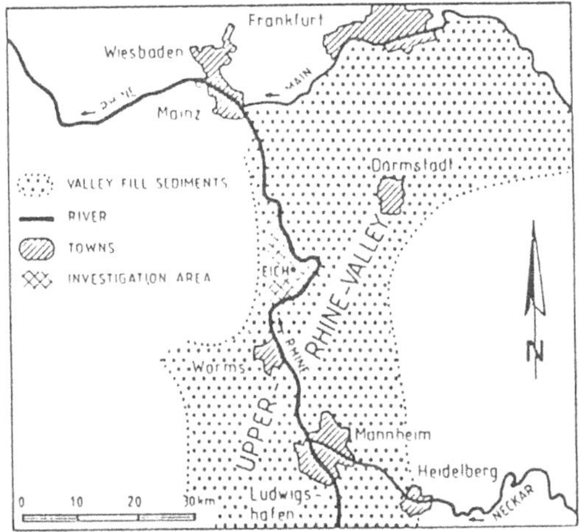

Fig. 1: Northern part of the Upper Rhine Valley

Sandy to fine-gravelly sediments with a thickness of about 80 m along the river Rhine form an efficient storage for the groundwater inflow from the inland and from the Rhine depending on its time-variant depth.

The development of this groundwater reservoir had been planned for a long time. In 1981 the waterworks Eich of the Stadtwerke Mainz AG began work.

[1] Technologieberatung Grundwasser und Umwelt GmbH
Björnsen Beratende Ingenieure, Kurfürstenstr. 87 A,
D-5400 Koblenz

Lecture Notes in Earth Sciences, Vol. 27
G.-P. Merkler et al. (Eds.)
Detection of Subsurface Flow Phenomena

Presently about 10 million m^3/year groundwater are pumped from a well gallery parallel to the Rhine and are transported by a distant water pipeline to Mainz. The nine (about 70 to 80 m deep) wells of this gallery are situated about 300 to 600 m away from the river (Fig. 2).

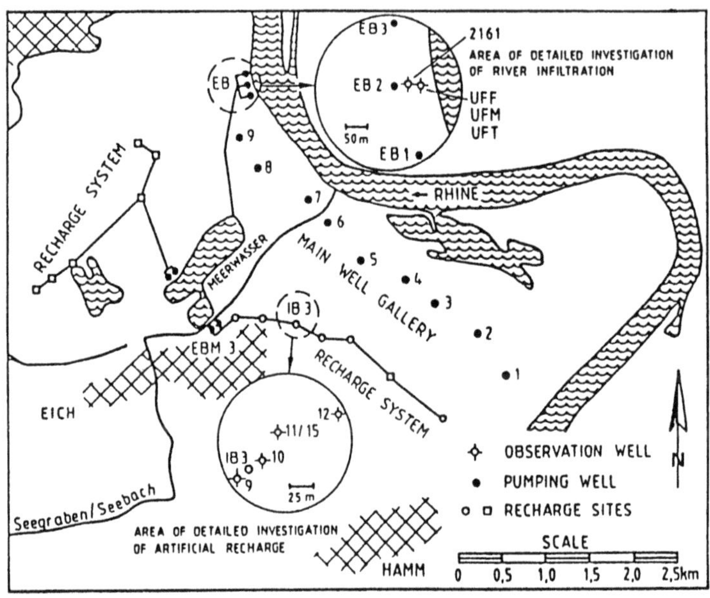

Fig. 2: Extended and local areas of investigation

Because of the importance of the agricultural use of the catchment area and the necessity to conserve some fields with significant ecological characters, measures have been taken from the first to avoid extended effects on the goundwater system in the area by groundwater extraction.

The measures are composed of artificial recharge systems (ditches and injection wells) inland of the well field. To obtain water for infiltration special wells were drilled near the Rhine to the north of the main well gallery (EB1, 2, 3 in Fig. 2). An existing dredging lake "Meerwasser" is part of the artificial recharge system, too.

The well-planned development of these groundwater resources offered a favourable opportunity to obtain profitable knowledge and experience for other plans and measures from systematic observations and investigations in the first stage of work of that pilot project. Therefore, these investigations have been sponsered in a long-term research program by the German Environmental Agency since 1981.

Due to the complex problem, the research project consists of three integrated investigation phases. Hydrochemical, hydrobiological and hydrogeologic-hydraulic conditions and processes are integrated in an interdisciplinary analysis.

In addition to the investigations of this extended groundwater system, different local areas have been selected for extensive investigation of special problems. Figure 2 gives an overview of the total area with two local areas of investigation. This report concerns the detailed investigations of the river infiltration in the northern local area.

2. Use of Models

Under the mentioned conditions it was evident that an interpretation of the long-term groundwater situation and the quality of the pumped water would only be possible by using predective groundwater models.

Therefore, a numerical groundwater model was used already during the design stage of the pumping and recharge well system. The results of this model were the basis of the executed measures. Undoubtedly, the permanent management and the conservation of the quality of drinking water require far-reaching knowledge and information e.g. on the time-variant groundwater flow towards the wells and the probable quality differences of the pumped water.

Accordingly, some other models related to the basic model system were developed and tested. Their characteristics were a more detailed discretization, and additional boundary conditions were used to calculate groundwater flow and mass transport.

For the local areas special models with three-dimensional simulation of groundwater flow and mass transport were developed and tested due to the results of observation.

Because of the detailed observation data the influence of smaller inhomogeneities and anisotropies in the underground could be taken into account. The result of a mass transport calculation is given in Fig. 3. The inflow of river infiltration is shown over depth, whereby the

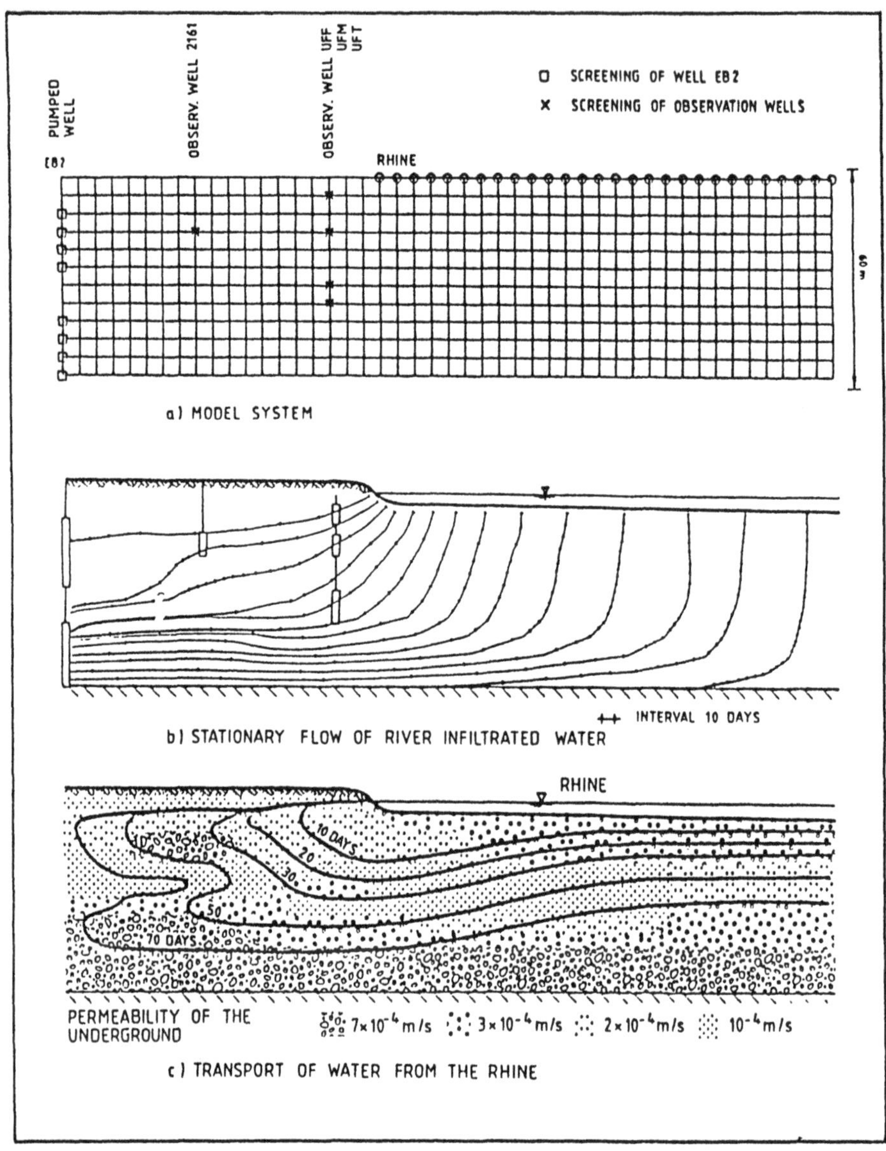

Fig. 3: Model investigations of vertical groundwater flow and transport near the Rhine

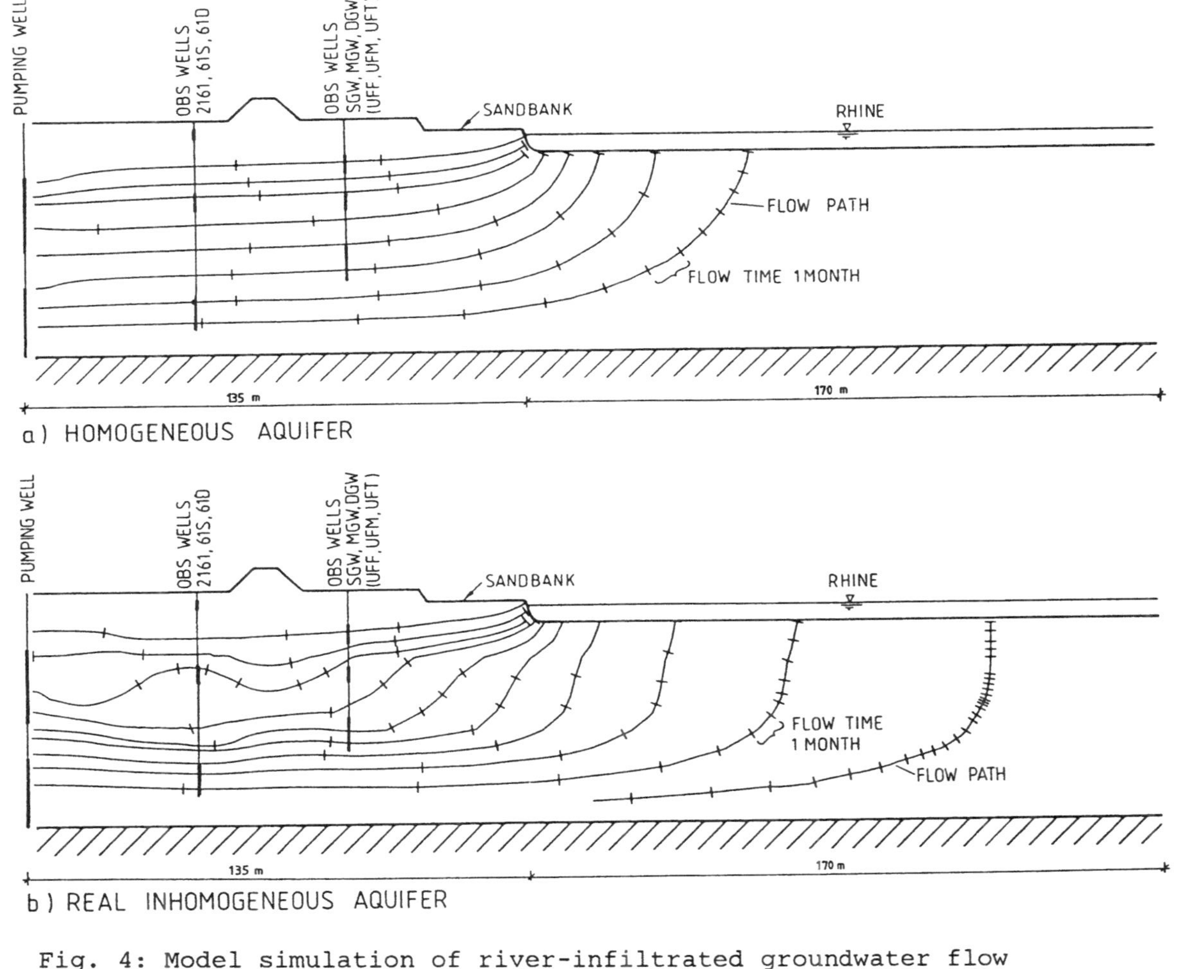

Fig. 4: Model simulation of river-infiltrated groundwater flow

criteria of inhomogeneity, obtained by detailed observation, are considered.

With this knowledge additional observation wells with different screening over depth were installed at important inland locations. Better knowledge of the underground structure and the hydraulic boundary conditions resulted in additional consequences to the possible model variations. This resulted in better simulation of the influence of the riverbank and inhomogeneities of the underground.

Figure 4 demonstrates different possibilities of the river-infiltrated groundwater flow to the nearby pumping well under homogeneous and probable inhomogeneous hydrogeologic conditions.

Nevertheless, these results of model simulation made obvious the possible variability of the river-infiltrated flow. The basis for the development of reliable and simple model systems simulating all variations of similar flow situations requires additional information about the flow characteristics. Using only geologic and hydrologic data a calibration of such groundwater-flow and mass-transport models seemed impossible.

3. Use of Hydrochemical and Hydrothermal Data

Because of the known complexity of the conditions of a multivarious influenced system, the investigations were carried out as integrated research from the beginning. Three parts of the investigation were combined: a hydrogeologic/hydraulic part, a hydrochemical part and a hydrobiological part.

To obtain usable data, special observation instruments were installed in the northern area of river infiltration under consideration. In a cross-section of three pumped wells and the Rhine several observation wells were built. To distinguish the flow characteristics over depth at the 60-m-deep pumping wells, separate observation systems with differing depths were necessary. Especially near the riverbank the different flows over depth had to be recognized. Therefore, an adequate system of three observation wells was installed about 20 m from the upper border of the riverbank (see Fig. 5). The wells had depths of 10 m (SWG = shallow groundwater, observation well UFF), 20 m (MGW = medium groundwater, observation well UFM) and 40 m (DGW = deep ground-water, observation well UFT) and each were screened at the end over a limited length of about 5 to 10 m.

Unfortunately, the riverbank is not inclined proportionally. Therefore, the real border of the river has a distance between 20 and 70 m from the observation wells with changing water levels of the Rhine.

To register the conditions of groundwater flow more distant from the riverbank, too, another system of observation wells with different depths is situated at half the distance to the wells (observation well system 2161).

Since 1983 the water from the observation wells has been analyzed continuously at least each month. So numerous hydrochemical and biochemical data are available which have been evaluated separately and used to recognize the natural flow system and its possible simulation.

In this report only part of the data and results are considered. To obtain a general view of the flow characteristics of river infiltration possible tracer or representatives of Rhine water quality were sought. In this respect the chloride contents in the water seemed to be an optimal indicator because of the significant concentration between 50 and 300 mg/l in the Rhine. On the other hand, the parameter "Adsorbable Organic Halogen (AOX)" could be used as a criterion for the typical organic content of Rhine water.

Both have been shown to be characteristic quality parameters, indicating the influence of river infiltration. Their time-dependent variation, however, with numerous effects till now did not allow sufficient and detailed conclusions on the different flow paths and transport periods of infiltrated water in the underground.

Figure 6 demonstrates this aspect with the time series of chloride contents in Rhine water and in the groundwater from observation wells with different distances to the riverbank. Due to the extensive change in the chloride content in the Rhine water, only a rough approximation of the dependencies between distance from the river and depth with the amount and duration of the flow of infiltrated water seems possible

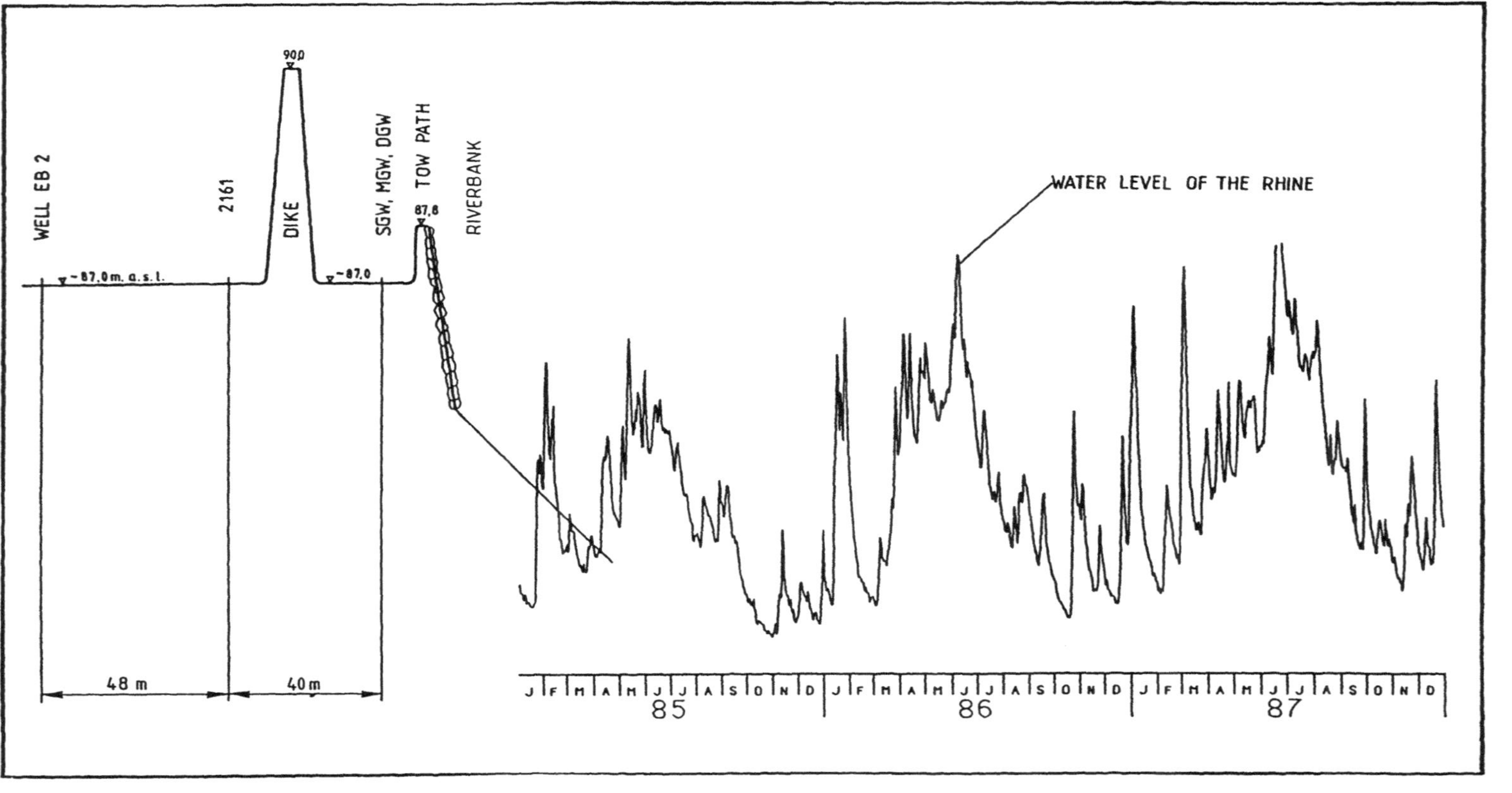

Fig. 5: Cross-section trough the investigated area near the Rhine

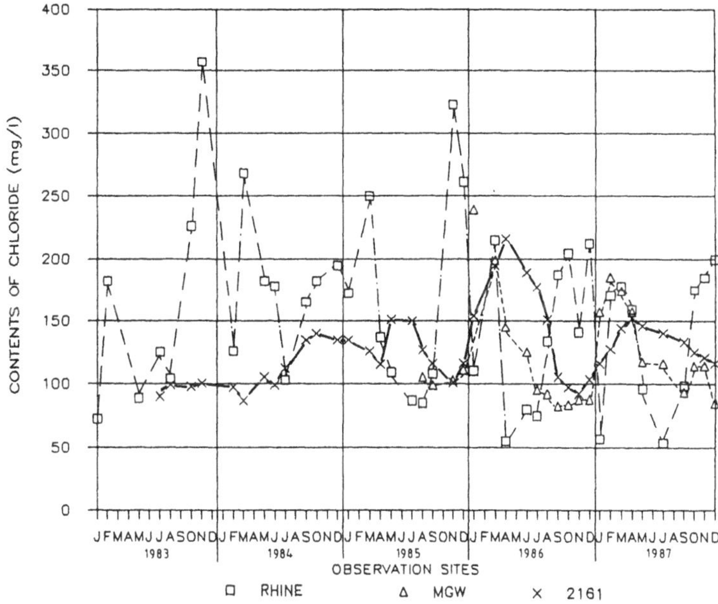

Fig. 6: River infiltration area. Contents of chloride in the water

Pursuing the time variation of Rhine water temperature, remarkable characteristics occuring each year could be recognized. Figure 7 with the measured time series of Rhine water and groundwater at different depths (observation site near the Rhine) shows the periodic changing of temperature over the last 3 years but also the differences in height and the displacement of maxima and minima between the measuring points.

Already from these results reasonable conclusions on the different flow characteristics over depth near the riverbank are possible. In connection with the hydrogeologic and hydrologic conditions this knowledge brought essential improvements of the model simulation regarding inhomogeneous underground conditions. The large flow times of the infiltrated water are recognizable from the displacement of the temperature fluctuations. The calculated flow situation in comparison with the results explains the influence of the underground structure and the riverbank conditions.

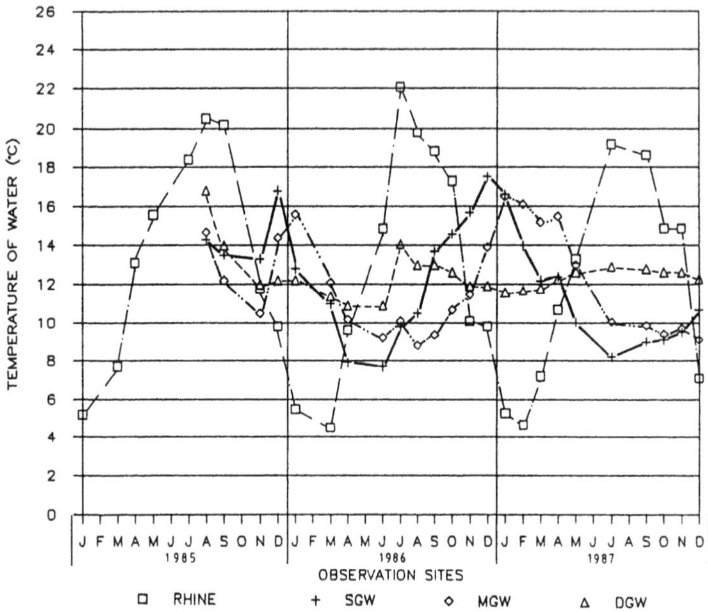

Fig. 7: River infiltration area. Time-variant temperature of water

The measured temperature profiles over depth at the observation sites with diferent distances from the Rhine at different times give additional information.

In Fig. 8 the temperature profiles have been plotted, which were measured in four different months of 1987 at the observation site near the river over a depth of 40 m. First, one can recognize the significant temperature changes over depth. Comparing the data with the temperature of the Rhine at the same time, it is evident that the water in the underground must have been flowed at different times from the Rhine. Also, at lower depths significant differences in temperature from Rhine water are present. Especially the low temperature in summer demonstrates the slow movement of water in the underground. At greater depths the temperature fluctuations are very small. This demonstrates the results of the flow simulation with the model, where a mixing of water with different flow time from the Rhine in this section was recognizable.

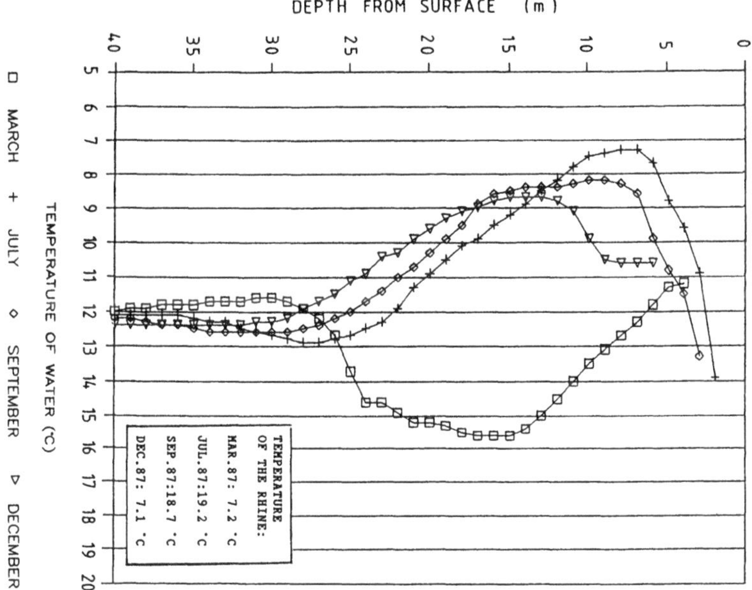

Fig. 8: Profiles of temperature 1987. GW site: near the river

The Profiles at the observation site distant from the river (Fig. 9) show the same conditions but with another temperature displacement.

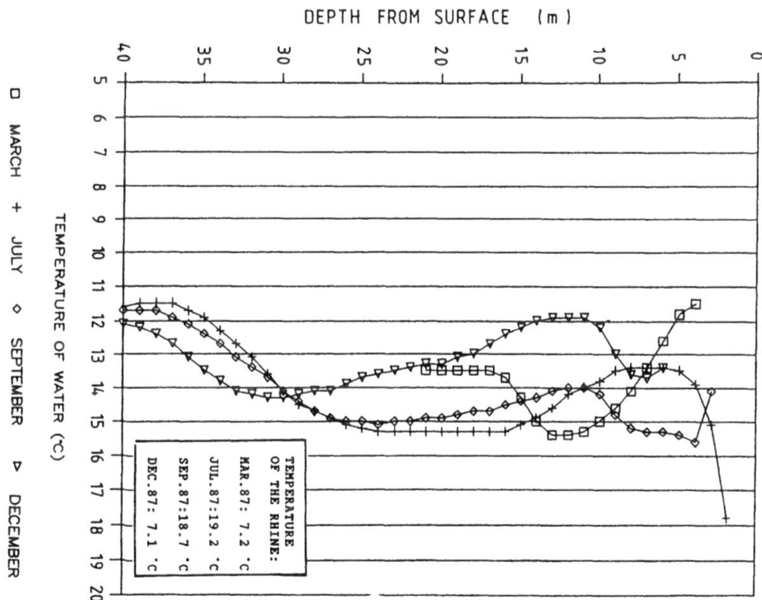

Fig. 9: Profiles of temperature 1987. GW site: distant from the river

Here, it is particulary significant that a longer time for the flow of water from the Rhine over this distance is needed and that the influence of surface water can be only recognized at a depth of about 3 to 5 m.

4. Conclusions

The observations and evaluations up to this time let one assume that the water temperature can be used as a representative parameter of bank-infiltrated flow near the river. In connection with hydrogeologic and hydrologic conditions and supported by hydrochemical information, the development of model systems to describe groundwater flow and mass transport near rivers seems possible. Due to the observations up to this time, also usable results are obtained on the time-variant quality of pumped water in wells near surface waters.

The essential results of this investigation program consist of the development and testing of simple models as frequently as possible to predict the flow pattern to such pumping wells near rivers and the time-variant quality of the pumped water.

On the other hand, essential problems are also the development and testing of useful installations and methods of observation of groundwater flow and water quality in similar situations. This concerns a suitable determination of relevant locations for the installation of only a few expensive observation wells of different depths to measure groundwater flow and quality over depth. Suitable proposals to the manner and size of necessary observations must also be made.

Even though there are many differences between various cases of groundwater use near rivers, many conditions are also similar and characteristic. Therefore, the results of these current investigations can be transferred to other similar installations and projects. In this way suitable plans and measures can be started very early in order to avoid disadvantages effects and uneconomic constructions and management. Thus, an optimal development of usable water resources is possible without unexpected consequences and irreversible effects on the environment.

References

Gatz, K.W. (1985): Besonderheiten der künstlichen Grundwasseranreicherung in der Anlage Eich. -DVGW-Schriftenreihe Wasser 45: 41-49

Gatz, K.W.; Horalek, U.; Lorenz, G.; Mühlhausen, D.; Obst, U.; Zipfel, K. (1987): Erfassung der raumzeitlichen Veränderung der Strömungs- und Qualitätsverhältnisse in einem ufernahen Grundwasserbereich des Oberrheingrabens. - GWF-wasser/abwasser 128: 104-111

Müller-Trimbusch, P. (1983): Wasserwirtschaftliche Ersatzmaßnahmen im Einflußbereich eines neuen Wasserwerkes. -GWF-wasser/abwasser 124: 369-371

Obst, U.; Mühlhausen, D.; Lorenz, G.; Gatz, K.W. (1985): Chemische und biochemische Untersuchungen zur natürlichen und künstlichen Infiltration im neu erschlossenen Wassergewinnungsgebiet Eich/Rhein. - GWF-wasser/abwasser 126: 443-477

Zipfel, K. (1983): Groundwater resources development avoiding conflicts of usage: example of modern planning strategies. -In: Groundwater in Water Resources Planning. Proc Int Symp convened by UNESCO in cooperation with the National Commitee of the Federal Republic of Germany for the International Hydrological Programme: 661-672

Zipfel, K. (1985): Anwendung von Simulationsmodellen zur Ausbreitung von Schadstoffen in einem Uferfiltratswasserwerk mit künstlicher Grundwasseranreicherung. -DVGW-Schriftenreihe Wasser 45: 277-295

Zipfel, K.; Horalek, U. (1989): Modelling time variant groundwater quality in a catchment area influenced by agriculture and river infiltration. -In: Groundwater contamination: use of models in descision making. International groundwater modeling center, Kluwer Academic Publishers: 533-540

Application of Geoelectrical Measurements in Engineering and Round Table Discussion

Geophysical Methodology for Subsurface Fluid Flow Detection, Mapping and Monitoring: An Overview and Selected Case History

Dwain K. Butler[1]

1 Introduction

In recent years, geotechnical applications of and research in geophysical methodology for subsurface fluid flow detection, mapping and monitoring have increased significantly in the United States. Self potential (SP) and other geoelectrical techniques are the primary tools of the developing methodology. Prior to 1970, the primary usage of the SP method in the US was for geothermal and mining exploration applications. Publications by Russians A. A. Ogilvy and V. A. Bogoslovsky in English-language journals, during the period 1968-1972, presented case histories of detection of anomalous seepage from reservoirs, which stimulated interest in the US. Soon, successful application of the technique to mapping anomalous seepage from water retention structures in the US, led to interest in the broader field of geotechnical applications of the technique. A special session on geotechnical applications of the SP method at the 1984 Annual Meeting of the Society of Exploration Geophysicists further increased interest in the technique and speeded the technology transfer. In 1986, a Workshop on the SP Method was held at the US Army Engineer Waterways Experiment Station, Vicksburg, Mississippi; attendees from the Corps of Engineers, Tennessee Valley Authority, and the Bureau of Reclamation discussed current practice, lessons learned, future plans, and outlined areas for cooperative effort.

In addition to practical applications of the SP method the Corps of Engineers is sponsoring research to investigate time and environmental factor effects on metallic and nonpolarizing electrodes, to quantify the relation between flow quantity, geoelectric structure and SP measurements, and to develop SP-source modeling for geotechnical applications. The practical applications have included seepage assessment studies at numerous dam sites and levees and investigations at hazardous waste disposal sites. Much of the work has been accomplished using metallic electrodes, due to the simplicity of heir installation and subsequent utilization by field office personnel. Data processing techniques have been developed which attempt to compensate for time-varying electrode polarization effects. Also, some initial attempts to understand the differences between SP anomaly magnitudes, relative to background levels, measured with metallic and nonpolarizing electrodes have been made.

2 Outline of Methodology

Geotechnical problems or conditions which can be addressed by the geophysical seepage assessment methodology includes:

----

1. U.S. Army Engineer Waterways Experiment Station, P.O. Box 631, Vicksburg, Mississippi, USA

Lecture Notes in Earth Sciences, Vol. 27
G.-P. Merkler et al. (Eds.)
Detection of Subsurface Flow Phenomena

1. Anomalous seepage involving dams, levees, locks;
2. Flow path/fracture/joint mapping in karst areas;
3. Fracture/flow path mapping from hydrofracturing and pumping tests;
4. Leakage from containment lagoons/settlement ponds;
5. Contaminant plumes from hazardous waste sites/landfills;
6. Landslide process monitoring.

While the methodology described here is general in nature, the case histories presented in this chapter will address aspects of areas (1)-(4) above.

Geophysical programs for anomalous seepage assessment consists of the following concepts: a strategy addressing detection, mapping and/or monitoring; a scope defined by program constraints and designed to map flow paths in plan or in three dimensions. A monitoring strategy implies that geophysical anomalies are detected as a function of time and can be correlated with reservoir level. A program which attempts to map seepage paths in three dimensions is necessarily more complex than one to map paths in plan. Individual geophysical methods that are utilized either in a primary or supporting (secondary) role in the methodology are indicated below:

SP Method--primary role;
Electrical resistivity profiling--primary role;
Electromagnetic (EM) profiling--primary role;
Pole-dipole resistivity surveying--primary role;
Ground penetrating radar--primary/supporting role;
Seismic refraction and reflection--supporting role;
Electrical resistivity sounding--supporting role.

Methods are selected from this list to meet site and project specific conditions. The program is designed to detect anomalies which are due to the fluid flow, the path followed by the fluid, or both. The SP method, for example, detects anomalies produced by the fluid flow. Electrical resistivity profiling, can be used to detect the path (such as a fracture zone or solution cavity) regardless of whether flow is occurring, although the nature of the anomaly will depend on whether or not the path is filled with fluid. In any event, an integrated, multiple method program is preferable to relying on a single geophysical method.

The geophysical program must be integrated with the overall geotechnical investigation program. Planning for the overall geotechnical program is best accomplished with early input from the geophysical program to guide placement of exploratory boreholes and piezometer installations. Results of exploratory borings, dye testing, and piezometer data are then used to finalize and/or refine the geophysical program results. Also, the results of the geotechnical "ground truth" may even be used to guide planning for further geophysical investigations. The great power and value of applied geophysics lies in its use to extrapolate from "point" ground truth data, integrate multiple, point ground truth data, visualize trends which are often missed with ground truth data, and detect localized anomalous conditions which escape detection by all but extremely close-spaced (hence prohibitively expensive) drilling programs.

3 Case History: Clearwater Dam, Missouri (USA)

Fig. 1 is a plan map of a portion of the left abutment of

Clearwater Dam, Missouri, showing an area of anomalous seepage. While the anomalous seepage was and is not considered to pose a threat to the dam, the seepage area was described as "unsightly both physically and psychologically." A geophysical survey and monitoring program was planned as part of a comprehensive seepage assessment program to map and monitor anomalous seepage paths and to assess foundation conditions at the dam (Butler 1985; Koester, Butler, Cooper and Llopis 1984). This case history will present only selected results which illustrate (1) mapping subsurface fluid flow paths from a pump test at the site, (2) use of complementary geophysical methods along a survey line on the left abutment ridge, and (3) portrayal of the results of the overall geophysical program in the form of an anomalous seepage path map.

Clearwater Dam was built in the 1940's for flood control purposes and is located on the Black River near Piedmont, Missouri. The earthfill dam is 154 ft (47 m) high with a crest length of 4225 ft (1288 m) and has a storage capacity of 391,000 acre-ft (approximately 0.48 km^3). Rock below the floodplain and abutments of the dam is a dolomitic limestone which is cherty, intensely fractured, and highly weathered, particularly in the abutments. Top of the limestone is pinnacled, and air-, water-, and clay-filled cavities exist below the rock surface. Top of rock is typically about 50 ft (15 m) below the surface of the left abutment ridge indicated in Fig. 1.

During a pumping test of borehole LK-10, located on the crest of the dam, seepage was induced at an upstream location as shown in Fig. 2. Also shown in Fig. 2 are two SP survey lines designed to demonstrate the capability of the SP method to map subsurface fluid flow at the site. SP measurements were taken before and during the pumping test. Results of the SP measurements for the downstream array are presented in Fig. 3. Pumping was initiated at 1745 hours on 17 March, with a pumping rate of 250 gal/min (945 l/min), and sustained through 0830 hours on 20 March. Reservoir level was at elevation 494 ft msl (mean sea level) during the pumping test. There was no significant mean temperature change and no rainfall during the test. Since the SP data were acquired with metallic electrodes, the magnitudes of the SP values are biased by the electrode polarization potential. Assuming that the polarization potential is constant for all electrodes in the array, subtracting one data set from another data set will eliminate the electrode polarization potential bias. The SP anomaly profiles in Fig. 4 are relative to the SP profile data acquired at 0900 hours on 17 March (prior to initiation of the pumping tests). Using the criteria that flow paths will be indicated by negative potential areas surrounded by relative positive potentials, three possible flow path anomalies can be identified (electrodes 3-4, electrodes 6-8, and electrodes 10-11). The upstream SP data, similarly analyzed, indicated two anomalous areas.

Using the simple technique of just assuming straight line flow paths from the pump test borehole under the SP anomalous areas, the flow path map in Fig. 5 results. The flow path map indicates a logical flow path for explaining the observed surface seep induced by the pump test. Also, the flow paths suggest an orthogonal to subothogonal flow system, which may be indicative of an orthogonal, solution-widened fracture system in the limestone.

Along the crest of the left abutment ridge, seismic refraction surveys, pole-dipole resistivity surveys, and SP surveys were conducted. Using a modification of the graphical interpretation

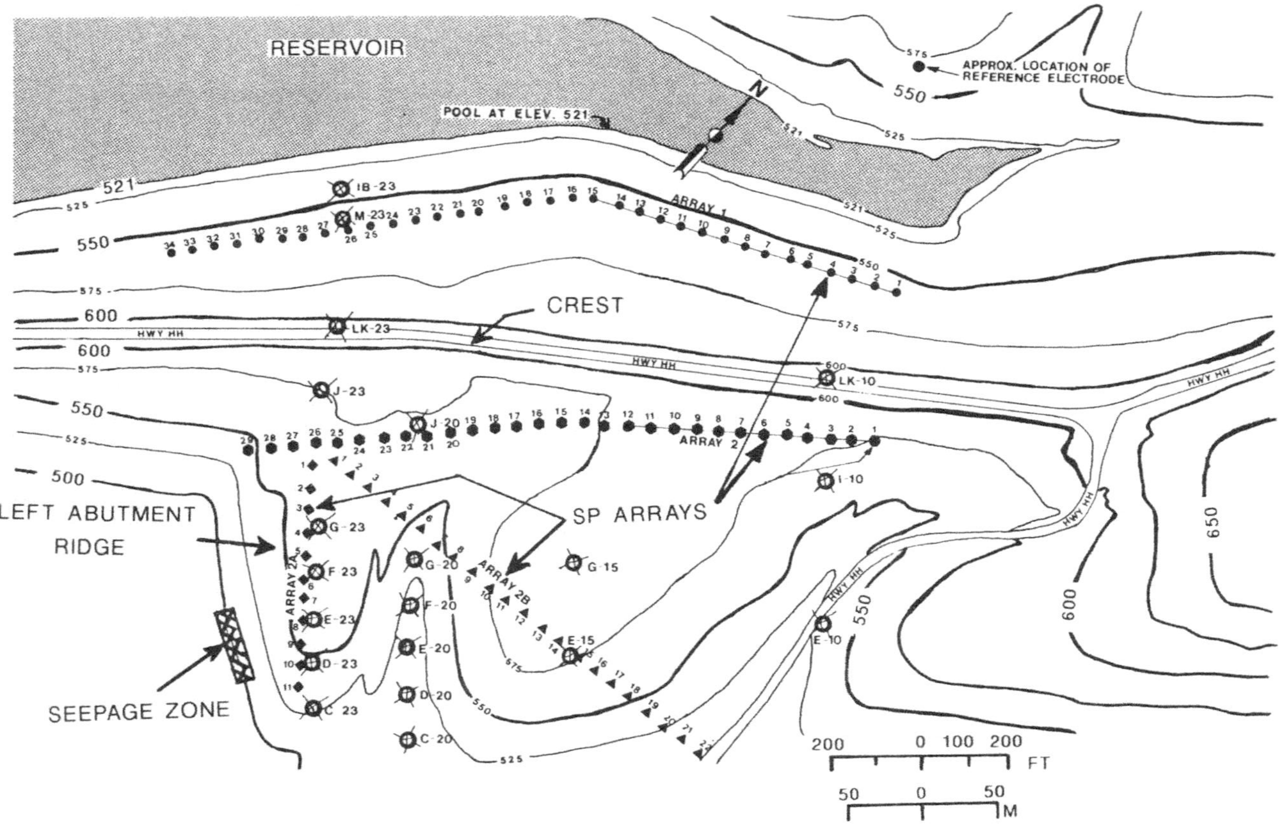

Fig. 1. Plan view of left abutment area of Clearwater Dam, Missouri, showing seepage zone and self potential survey arrays.

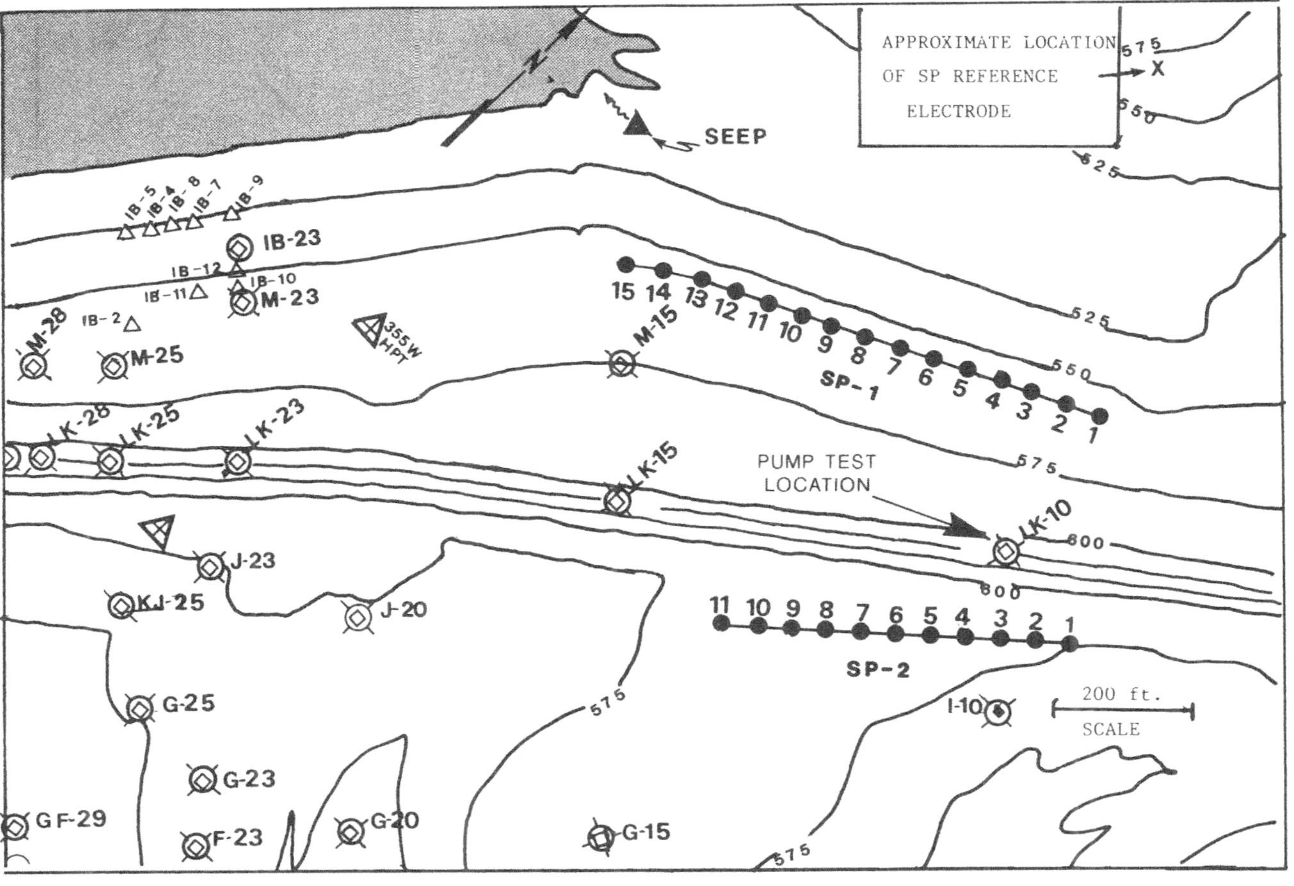

CLEARWATER DAM, MISSOURI -- PUMP TEST GEOMETRY

Fig. 2. Plan view of pump test location, SP arrays, and induced seep location.

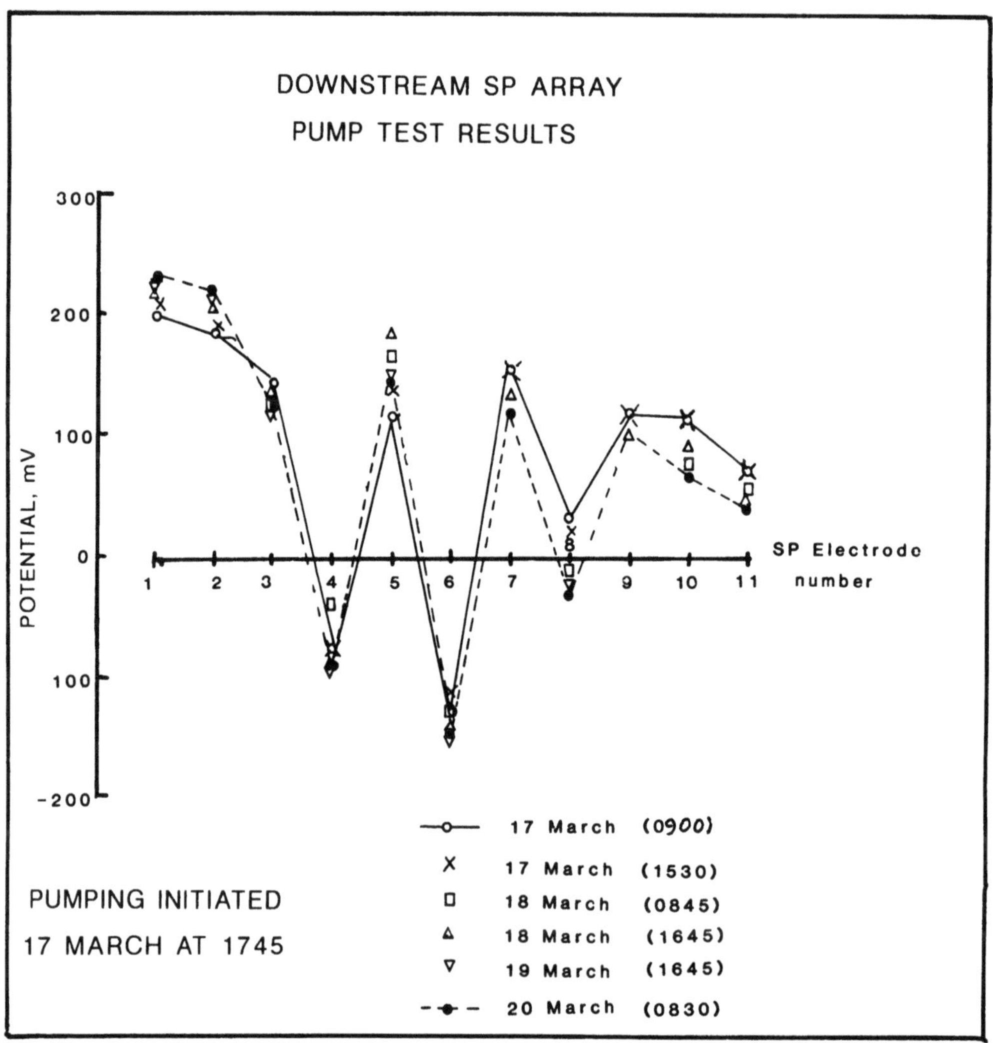

Fig. 3. Downstream array pump test SP data.

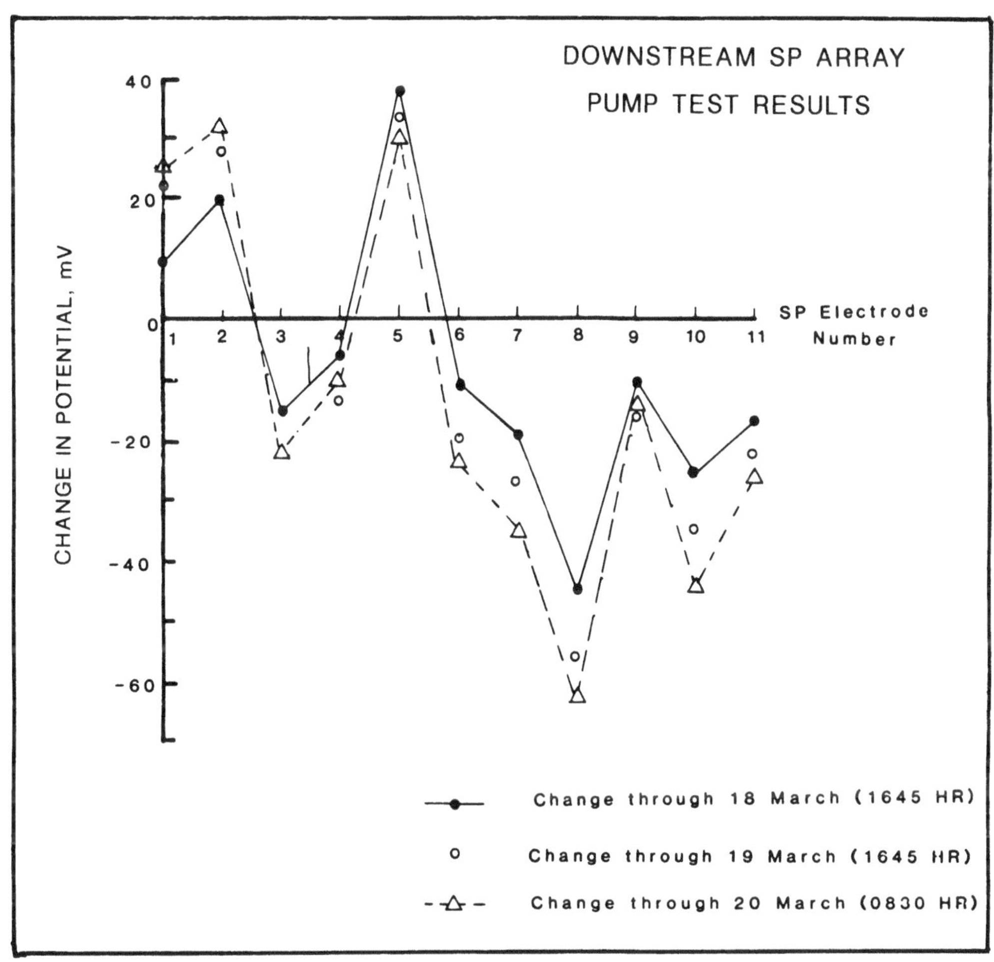

Fig. 4. SP anomaly profiles for downstream array relative to pretest profile data.

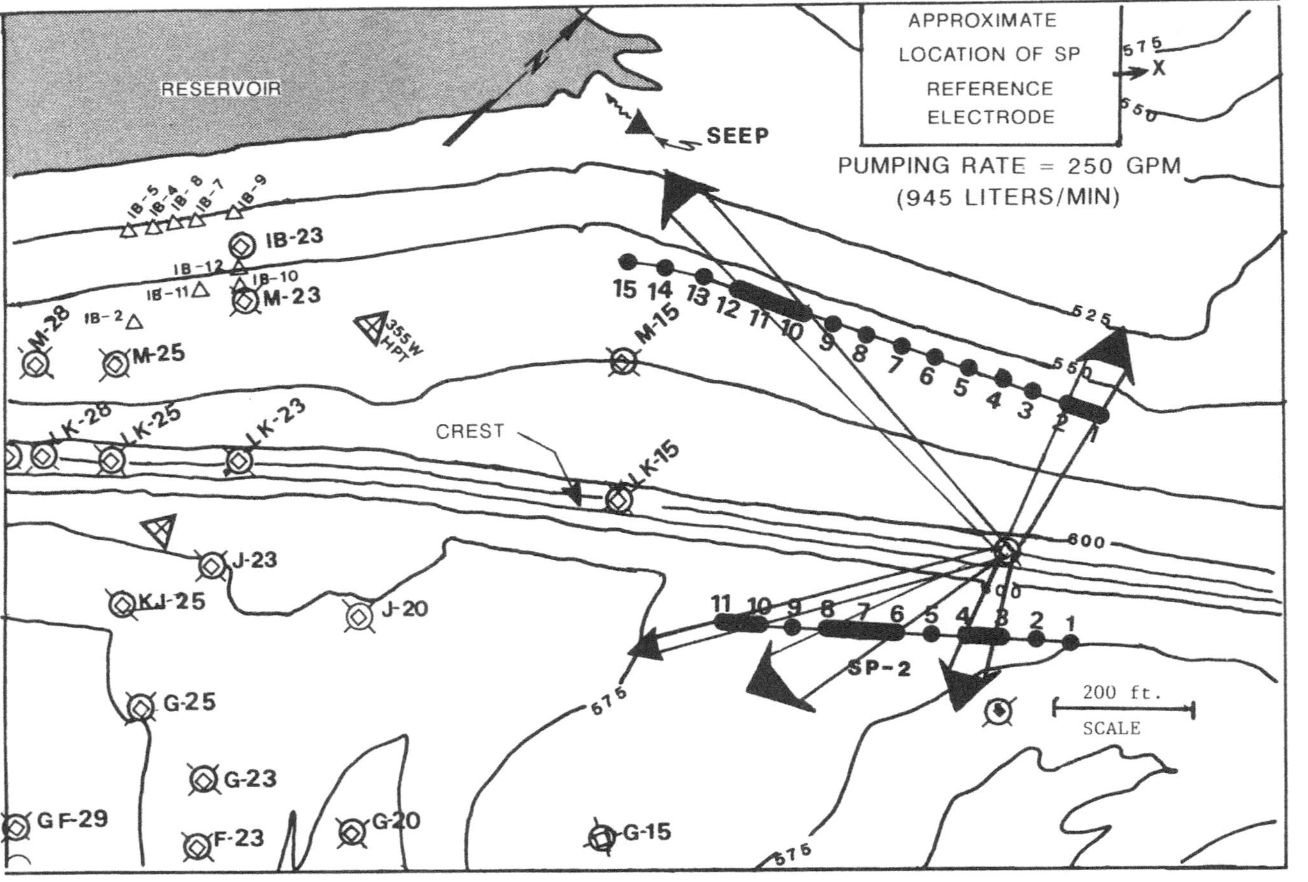

PUMP TEST RESULTS

INTERPRETED FLOW PATHS

FIG. 5. Interpreted flow paths during pump test based on induced SP anomalies.

technique of Bristow (1966), the resistivity anomaly cross-section shown in Fig. 6 was deduced (see Butler et al. 1982). The representation is termed an anomaly cross-section, since the "normal" resistivity variation with depth is used as a trend "line" and only anomalies relative to the trend line are identified. The anomalies are indicated as high or low, and the numbers in the anomalous areas indicate the number of intersecting "hemispherical" shells which define the anomaly. The top of rock and the water table were determined from piezometer data, drilling data and seismic refraction surveys. A significant cluster of anomalies is located between profile positions 50-75 ft. This cluster occurs at the water table, with high anomalies above and low anomalies below the water table. A plausible interpretation for this cluster of anomalies is a partially water- and/or clay-filled (lows) and partially air-filled (highs) cavity system. Reservoir level during the resistivity survey was low at 494 ft msl. If this interpretation is correct, the air-filled portion of the cavity system at this location represents a significant seepage capacity. The location is indicated as a "recommended drilling site." The high and low resistivity anomalies at and above the top of rock are interpreted as limestone pinnacles or "floaters" (boulders) and residual clays (weathering products).

An SP survey was conducted along the left abutment ridge also, indicated as Array 2A in Fig. 1. Surveys were conducted during low reservoir pool levels (502-509 ft msl) and during high pool levels (>520 ft msl). Flow at the downstream seepage area increases significantly when the reservoir pool level exceeds 510 ft msl. The three SP data profiles shown in Fig. 7 obtained during low pool levels illustrate the typical repeatability of SP data at the site during low or no flow conditions. The data vary about a mean value of approximately 25 mV, reflecting variations due to soil-type variations or other types of geologic variability along the profile line. Results for the SP surveys conducted during high pool level conditions are also shown in Fig. 7. The high pool level data exhibit considerably greater magnitudes and variability than the low pool level data. Four negative anomaly areas relative to the low pool level data are evident in the high pool level data. The variability in the high pool level data directly reflects changes in the reservoir pool level.

Similarly, the SP data from all the arrays shown in Fig. 1 were examined to identify anomalies in high pool level data relative to low pool level data. Results of this analysis, along with the results of electrical resistivity surveys, seismic refraction surveys, and piezometer data were used to deduce a possible flow path map as presented in Fig. 8. Flow path maps are extremely valuable for planning further exploratory boring and piezometer installations and for remedial measure planning if deemed necessary. For the Clearwater Dam case, the comprehensive seepage assessment served to define the seepage paths and hence the placement of piezometers to monitor flow conditions. Given the extremely complex spatial and temporal variations of piezometer data commonly associated with seepage through carbonate rock with extensive solution features, it is doubtful that boring and piezometer data alone could be interpreted to yield a flow path map comparable to Fig. 8. The only immediate remedial measure deemed necessary, since the anomalous flow path was well below the base of the dam in the foundation rock, and since the seepage quantities emerging were not excessive and contained no "fines," was to install a gravel drain along the toe of the left abutment ridge to intercept the anomalous seepage and convey it to a concrete channel. Placement of an upstream clay blanket is planned for the future.

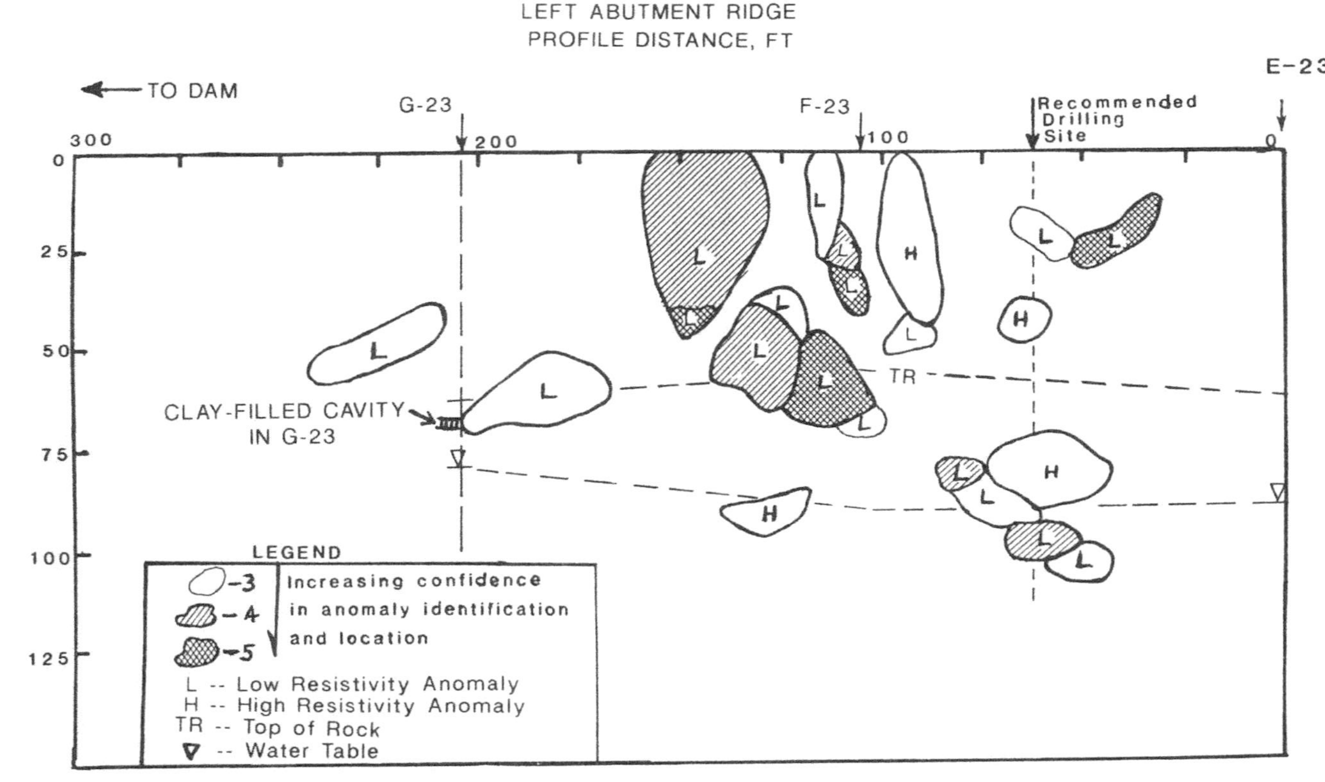

FIG. 6. Resistivity anomaly cross-section for the left abutment ridge.

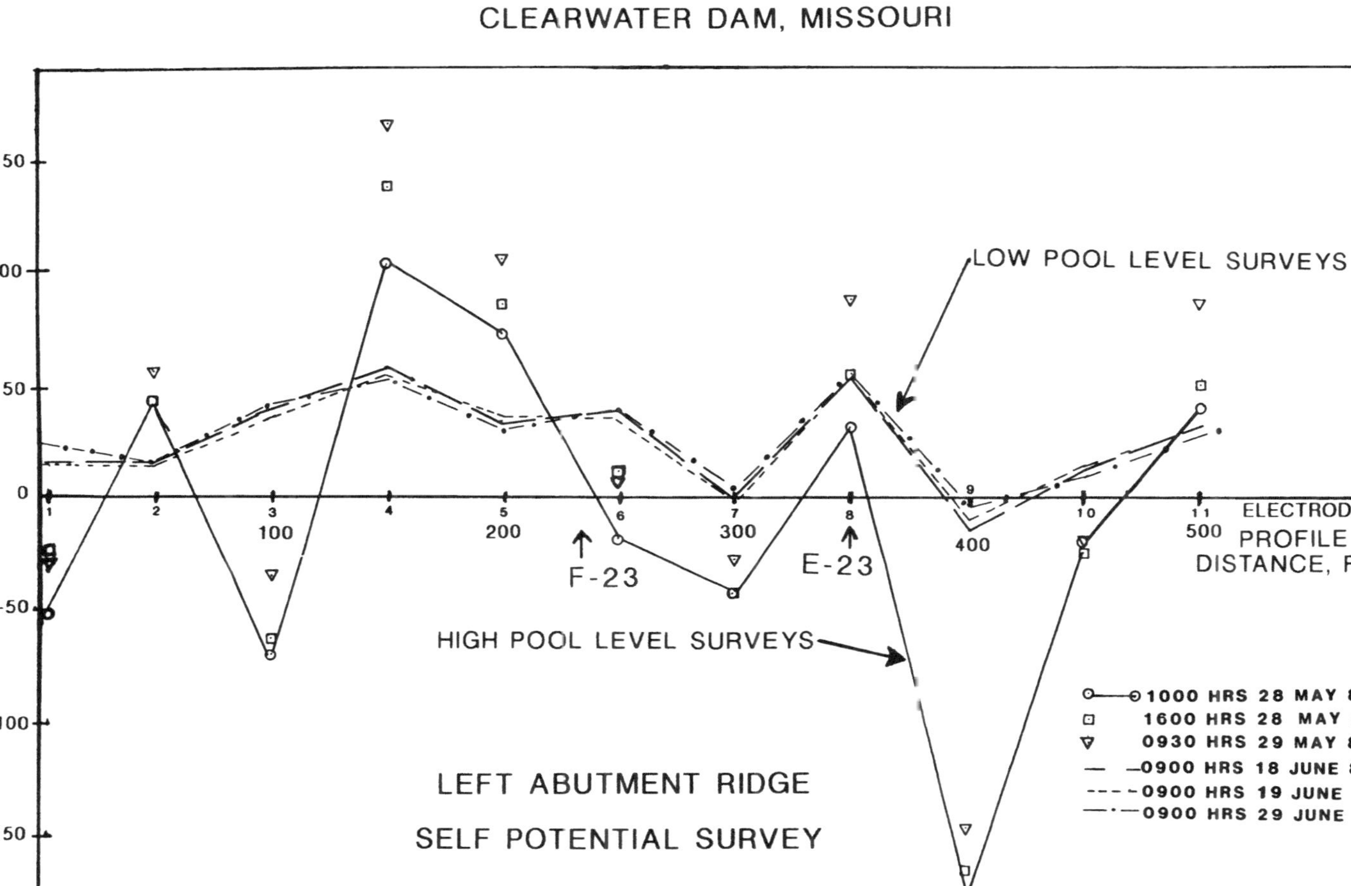

FIG. 7. Low and high pool level SP survey profiles for the left abutment ridge (Array 2A).

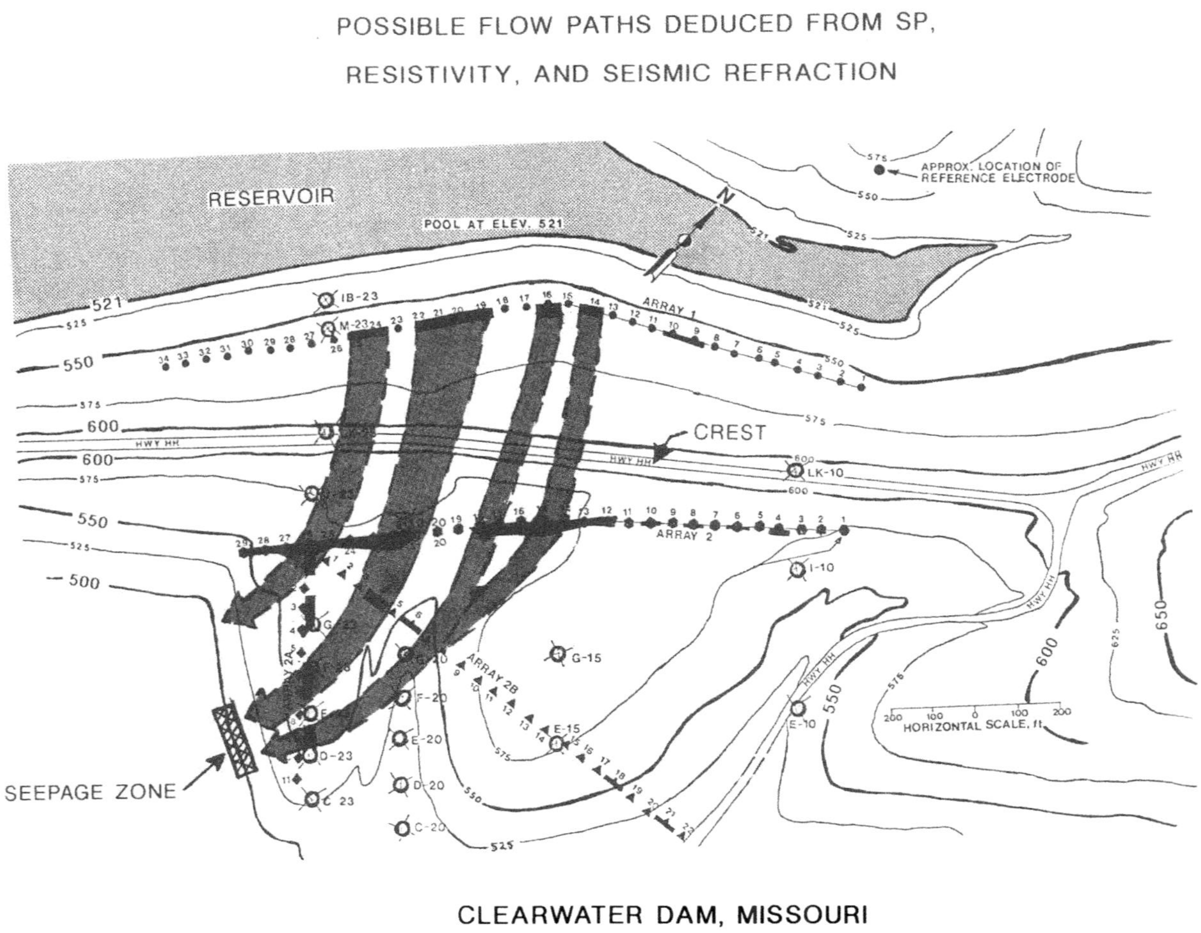

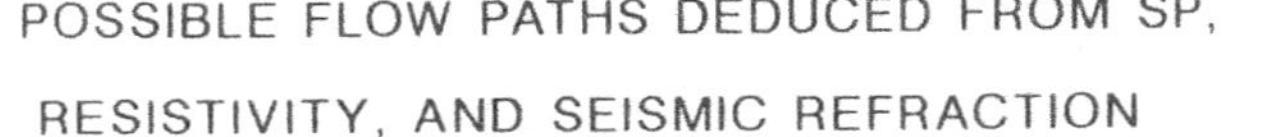

FIG. 8. Flow path map deduced from the geophysical program.

4 Conclusions

A geophysical program, such as that described in the case history, can contribute significantly to an anomalous seepage assessment effort. Results of geophysical programs have been utilized effectively to explain often confusing and apparently conflicting piezometer data and to produce a logical and verifiable flow path map of anomalous seepage. The flow path maps and other subsurface information deduced from the geophysical program results are used effectively to plan further exploratory borings and piezometer installations and remedial measures. Additional references to and discussions of the methodology described in this chapter are given by Butler (1984), Cooper and Koester (1984), Butler et al. (1984), Cooper et al. (1982), Erchul (1988), Llopis et al. (1988), and Llopis and Butler (1988).

Acknowledgment. The work presented in this paper was conducted under the Dam Safety Assurance Program and the Repair, Evaluation, Maintenance and Rehabilitation Research Program of the United States Army Corps of Engineers. Permission was granted by the Chief of Engineers to publish this information.

References

Bristow CM (1966). A new graphical resistivity technique for detection of air-filled cavities. Stud Speleol 7:204-227

Butler DK (1984). Geophysical methods for seepage detection, mapping and monitoring. Proc 54th Annu Int Meet Soc Exploration Geophysicists, Atlanta, Ga

Butler DK (1985). Geophysical methods applied to detect and map seepage paths at Clearwater Dam. REMR Bull, U.S. Army Corps Eng 2, 2

Butler DK, Gangi AF, Wahl RE, Yule DE, Barnes DE (1982). Analytical and data processing techniques for interpretation of geophysical survey data with special application to cavity detection. Misc Pap GL-82-16, US Army Eng Waterw Exp Stn, Vicksburg, Miss

Butler DK, Wahl RE, Sharp, MK (1984). Geophysical seepage detection studies, Mill Creek Dam, Walla Walla, Washington. Misc Pap GL-84-16, US Army Eng Waterw Exp Stn, Vicksburg, Miss

Cooper SS, Koester JP (1982). Geophysical investigations at Gathright Dam. Misc Pap GL-82-2, US Army Eng Waterw Exp Stn, Vicksburg, Miss

Cooper SS, Koester JP (1984). Detection and delineation of subsurface seepage using the spontaneous-potential method. Proc 54th Annu Int Meet Soc Exploration Geophysicists, Atlanta, Ga

Erchul RA (1988). Geotechnical applications of the self potential method, Rep 1: The use of self potential in the detection of subsurface flow patterns in and around sinkholes. Tech Rep REMR-GT-6, US Army Eng Waterw Exp Stn, Vicksburg, Miss

Koester JP, Butler DK, Cooper SS, Llopis JL (1984). Geophysical investigations in support of Clearwater Dam comprehensive seepage analysis. Misc Pap GL-84-3, US Army Eng Waterw Exp Stn, Vicksburg, Miss

Llopis JL, Butler DK (1988). Geophysical investigation in support of Beaver Dam comprehensive seepage investigation. Tech Rep GL-88-6, US Army Eng Waterw Exp Stn, Vicksburg, Miss

Llopis JL, Deaver CM, Butler DK, Hartung SC (1988). Comprehensive seepage assessment: Beaver Dam, Arkansas. Proc 2nd Int Conf Case Histories in Geotechnical Engineering , St. Louis, Mo

ELECTRICAL SURVEYS USING THE METHOD OF THE NATURAL ELECTRICAL FIELD; NEW INVESTIGATIONS

A.A. Ogilvy , E.J. Ostrovskij, and E.N. Ruderman[1]

1 Introduction

The relationship between physical and chemical fields existing in the earth, becomes especially clear regarding the potential minima of the natural electrical field (self-potential or streaming potential) as a function of different effects in:

1. Zones with a high metalliferous content;

2. Transition zones from the permanent frost to thawed areas;

3. Localization of metamorphic rocks with a high content of sulphitic and graphitic minerals resp. ores;

4. Filtration zones of the groundwater in soils with a definite content of sand and clay or other ingredients with a solid phase associated with a high water exchange.

Negative anomalies of the natural potentials (SP) are related to deposits of graphite and apatite-ilmenite and titanomagnetite, deposits of copper and sulphurous ores, paragenesis of lead, zinc and silver, pipes of cimberlithic composition and to sulphitic zones enriched by hydrocarbons.

Positive anomalies of SP mark zones with hydrothermal activity. Generally they cause an increase of temperature and of the circulation of the subsurface fluid. Both may be the reason for developing the potential fields on the earth's surface. The negative anomalies of SP may also be related to convective movements, introduced by terrestrial heat sources. This fact is indicated by a descendent movement of cool water in a convection system (Ishido et al., 1983).

---

[1] Faculty of Geology, Moscow State University

Lecture Notes in Earth Sciences, Vol 27
G.-P. Merkler et al. (Eds.)
Detection of Subsurface Flow Phenomena

Electrokinetic processes connected with the movement of subsurface water may be used to forecast earthquakes, supposing a mechanism of developing SP variations (streaming potentials), which have originated by flowing water through extension zones (Corvin et al., 1979; Corvin et al., 1981).

Such an increased application of the SP method, beginning with the mapping of geochemical contrasts, characterization of several types of deposits, up to a forecast of cimberliths, hydrocarbonic deposits, geothermal provinces and the prediction of earthquakes, supports the opinion that SP measurements have maintained their leading position with regards to the complex of geophysical and geochemical methods.

During the last years publications have also increased in foreign countries. This development with respect to theoretical and applied aspects demonstrates a new interest for an inexpensive and reliable instrument. The possibility of realizing remote sensing of geochemical processes (redox events, replacing the oxides by sulphides, exchange of cations in clay minerals) favoured the usefulness of the SP method in modern prospecting surveys. The possibility to correlate SP anomalies with magnetic and radioactive anomalies has also increased the interest in the physical nature of SP.

After Bolviken (1981) there is a redox potential field in the earth. In some characteristics it is similar to the gravimetric and magnetic field. Any process, which develops oxides, influences the magnetic properties of the rocks. According to Haggerty (1977), especially the oxides are the main carrier of the magnetic properties of vulcanic rocks (especially of the basalts). It is well known that gossans in ore deposits are those parts, in which secondary development of uranium mineralization occurs. Besides, the so-called hematitization resulting from intensive oxidation by the action of the ionization radiation of radioactive minerals in the surrounding sphere is also known. If iron silicate of double valency becomes reduced by the redox reaction, the uranium becomes oxidized from a six-valency state into a four-valency one. This product accumulates as hematite in the surrounding rocks. Thus, these rocks acquire a red colour. SP anomalies near deposits of sulphides have been well known for a long time. However, the interest in using this electrochemical phenomenon for prospecting surveys has decreased due to the introduction of methods using an artifically electrical field. Electrochemical processes, associated with natural galvanic elements, until now only were investigated to solve the problem of the hydrothermal development of sulphitic orebodies.

In approximately all the former theories the redox mechanism of the sulphitic minerals over- and underlying the groundwater level was considered as the main reason for developing SP. The deposits of pyrite, pyrrothite, chalcopyrite, magnetite, covellite and graphite produce SP anomalies with amplitudes of some hundreds of millivolts. If there is only a small electrical conductivity of the sulphitic particles, the SP method is the most effective one for determination of the mineralization.

Two models were proposed to explain the developmet of SP potentials over sulphitic deposits. The first model is based on the redox cell. It was proposed by M. Sato and H. Mooney. Later on, it was further developed by Bolviken et al. (1981).

With regards to this model the orebody was considered as an indolent dipole electrode without having a share in the geochemical reactions. The electrical current develops as a result of the redox events simultaneously occuring, induced by the oxidized agents near the boundary surface and by the oxidation of the reduced agents beneath the surface. Thus, in an orebody there are two reactions; each only comprising half of the galvanic elements. The reaction of the anode type occurs in the orebody and the reaction of the cathode type occurs on its surface. The orebody itself acts as a conductor of electrons from the anode to the cathode of the galvanic element. With regard to the second model the orebody acts as a dipole electrode, too, but also takes part in the electrochemical reaction.

The reactions occur in both parts of the galvanic elements, in the lower and upper sphere of the orebody simultaneously. The character of these reactions is likewise a reaction of anodes as well as of cathodes. The reaction of the anode beneath the surface is only a galvanic corrosion and must be reduced to aeration of that part of the orebody, which is situated near the groundwater level. The reaction of cathode means a reduction of the radical of oxygen in the upper sphere of the electrode. This is because the oxygen of the atmosphere, which is dissolved in the water, acts as a motive power of the element. The process of corrosion favours the movement of the ions to the anode through the groundwater. Sivenas et al. (1982) demonstrated, that the electrochemical potential is a combination of the above mentioned reactions. Regionally, the cell of oxygen concentration dominates (ZSK).

2 Physical Fundamentals of the Self-Potential Method

2.1 Cell of Oxygen Concentration; Source of the SP Field

Different models of the development of self-potentials depend on the assumption, that the orebody forms a dipole electrode. In this model the cathode reactions occur in the upper, more oxidizing sphere and the anode reactions occur mainly in the lower, more reducing part of that electrode (Fig. 1). If oxygen is present at the cathode, the electrochemical reactions occur simultaneously and form a rather marked self-potential. From the theoretical point of view, there exist different reactions. These reactions may occur at both poles and may produce nearly the same potentials of the orebody, the goundwater and the physicochemical condition of the system. These factors explain in particular the existence of different models of the electrochemical elements. They are called the cell of the oxygen concentration and the sulphitic galvanic cell.

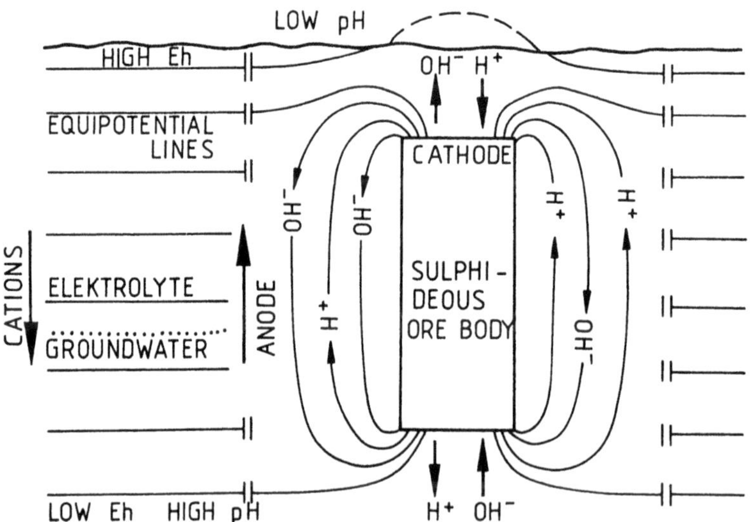

Fig. 1: Model of the equipotential lines in the homogeneous part of the lithosphere and in the neighbourhood of an orebody. Eh-redox potential

If there is any orebody or another electrical conductor in the sphere of the different oxygen concentrations, the electrons move out of the zone, which is then characterized by a deficit (anode) to the more oxidized zone (cathode). These electrons are formed and become absorbed as a result of the reactions occuring at the ends of the conductor.

The electrons move from the anode to the cathode, compensating these electrons, they are produced by the electrolyte at the anode or they become absorbed by the elctrolyte at the cathode. In the vicinity of the anode the reduced materials are oxidized. Thus, electrons develop. They "swim" to the cathode, whilst the material in the neighbourhood of the cathode becomes reduced by the absorption of the electrons at the anode. These two reactions occur simultaneously. In the vicinity of the orebody, i.e. in the sphere of the electrolyte, the cations may move upwards, from the anode to the cathode and the anions may wander downwards from the cathode to the anode. In this case, a galvanic cell develops. The potential difference produces a natural current and must be reduced to variations of the Eh (redox-potential) of the solution, directly in the vicinity of the contact of the opposite ends (anode-cathode) of the orebody. The ion H^+ and the ion OH^- are the most important carriers of charges in the upper part of the lithosphere. The condition, concerning the reaction and the electrochemical properties of both these ions, influences the entire electrochemical process. According to Sivenas et al. (1982), the system, cathode conductor anode, represents a cell of oxygen concentration. If oxygen reacts upon the cathode, a substitution of the electrons occurs in the elctrode and a by-product of this reaction is energy set free at the elctrode. The sequence of the reaction charges as a function of the pH parameter of the solution.

Let us assume an acid as well as a basic medium although both these suppositions are not excluded mutually in situ.
The acid electrolyte:

anode- $2H_2O \rightarrow O_2 + 4H^+ + 4e^-$

cathode- $O_2 + H^+ + 4e^- \rightarrow 2H_2O$

At the anode, two water molecules divide into one oxygen molecule, four protons and four electrons. Four electrons move along the conductor to the cathode. Here, they take part in the cathode reaction. Oxygen diffuses into the solution. At the cathode, the oxygen of the surroundings (secondary rocks) generates two water molecules. Besides the oxygen, the reaction is accompanied by four electrons of the anode and four ions of oxygen. The oxygen ions, originating at the anode, enter the solution and diffuse the electrolyte. With regards to the whole reaction, two water molecules are originated. They are accompanied by streaming electrons, producing a potential difference of almost 1.229 V. The oxygen ion acts as a charge carrier in the electrolyte. A higher concentration of H^+ develops at the anode, i.e. at that point where they are generated. A lower concentration develops at the cathode, i.e. at that point where the H^+ ions react upon the oxygen and water molecules become generated. A relatively high density of diffusion exists in the spheres of the anode and the cathode of the solution. However, the diffusion ends in the mean sphere and there does not follows any variation of the concentration. The total current of the cell is given by:

$$I = I_D^{H+} - I_L^{H+}$$

where I is the current of the cell, I_D^{H+} the diffusion current, I_L^{H+} the conducting current.

As long as a pure current of electrons (I>0) exists, the diffusion current dominates the conducting current and has the same direction as the total current of the cell. The diffusion current supports the reaction. It guarantees the continuous influx of oxygen ions to the cathode sphere of the solution. The conduction current is not as powerful as the diffusion current. The signs of these currents are reversed. Influenced by the conduction current, the oxygen cation wanders to the anode (negative electrode of the natural element).

Basic or neutral electrolyte:

Anode- $4OH^- \rightarrow O_2 + 2H_2O + 4e^-$
Cathode- $O_2 + 2H_2O + 4e^- \rightarrow 4OH^-$

The hydroxidic ion coming from the cathode deteriorates at the anode. Thus, oxygen develops and diffuses into the solution. It consists of two water molecules and four electrons. They wander along the conductor to the cathode and take part in the cathode reaction. At the cathode, the oxygen from outside reacts together with two water molecules of the solution and four anode electrons, generating four hydroxidic ions. The potential difference of the cell amounts to 1.229 V. The hydroxidic ion acts as the charge carrier in the solution. In the solution, its concentration is approximately constant, but it rapidly diminishes at the anode and increases at the cathode. The total current in the cell is given by the following equation:

$$I = I_D^{OH^-} - I_L^{OH^-} .$$

The diffusion current is more powerful than the conducting current. It has the same direction as the total current of the reaction and guarantees a continuous influx of the hydroxidic groups to the anode sphere of the solution. In the spheres of the anodes and the cathodes of the solution, the diffusion current is powerful and dominates over the conducting current. In the remaining part of the solution, the diffusion stops and there is no variation of concentration. The conduction current is of subordinated importance only. Its direction is contrary to the diffusion current and it influences the distribution of the concentration unessentially. The hydroxidic ion wanders together with the conduction current to the cathode (positive electrode of the natural galvanic element). If the gradient of the concentration of H^+ or OH^- is absent, the natural galvanic element stops existing theoretically and the potential of the cell E becomes zero. Departing from the model, the influx of oxygen ions and gaseous oxygen is limited in nature. But from the geological point of view the influx of oxygen, dissolved in the groundwater thoroughly, is possible. Only a few of the generated H^+ and OH^- ions find their way to the cathode and to the electrolyte; the others become adsorbed by the natural electrolyte.

In summary the relative distribution of the active ions in the electrolyte as well as the intensity of their development and the cooperation of currents, coexisting in the system, subordinate itself the cell of oxygen concentration.

2.2 EMK of the Cell of Oxygen Concentration

The EMK is defined by the Nernst equation:

$$E = E^0 - \frac{RT}{nF} \ln (a_{red}/a_{ox}),$$

where E is the redox potential in V at temperature (T), pressure (P) and concentration (C); E^0 redox potential in V under normal conditions: t = 25°C; P = 1 atm; a = 1; R = gas constant = 1.987 cal/K mol; T = absolute temperature (K); F = Faraday number (6487 cal); n = number of electrons exchanged by the redox process; a_{red}/a_{ox} activities causing the development of the EMK in the spheres of oxidation and reduction with regard to the generalized reaction of the half-cell. Accepting normal conditions the Nernst equation acquires a simplified form:

$$E = E^0 - \frac{0.059}{n} \log k; \qquad k = a_{red}/a_{ox}.$$

Acid medium:

cathode: $O_2 + 4H^+ + 4e^- \longrightarrow 2H_2O$

$$E_k = E_k^0 + \frac{RT}{4F}k\text{-} \ln (a_{O_2})_k (a_{H^+})_k^4$$

$$E_k = E_k^0 + \frac{RT}{4F}k\text{-} \ln (a_{O_2})_k + \frac{RT}{4F}k\text{-} \ln (a_{H^+})_k^4$$

$$E_k = E_k^0 + \frac{2{,}3\ RT}{4F}k\text{-} \log (a_{O_2})_k - \frac{2{,}3\ RT}{F}k\text{-} (pH)_k$$

anode: $2H_2O \longrightarrow 4H^+ + 4e^- + O_2$

$$E_a = E_a^0 - \frac{2{,}3\ RT}{4F}a\text{-} \log (a_{O_2})_a (a_{H^+})_a^4$$

$$E_a = E_a^0 + \frac{2{,}3\ RT}{4F}a\text{-} \log (a_{O_2})_a + \frac{2{,}3\ RT}{F}a\text{-} (pH)_a$$

The following potential is produced:

$$E = E_k - E_a$$

E_a and E_k are the electrical potentials of the cathode and the anode:

$$E = (E_a^o) + (E_a^o) + \frac{2,3\ R}{4F} \log \frac{(a_{O_2})_k^{T_k}}{(a_{O_2})_a^{T_a}} + \frac{2,3\ R}{F} \log \frac{(a_{H^+})_k^{T_k}}{(a_{H^+})_a^{T_a}}$$

Neutral or basic medium:

Cathode: $O_2 + 2H_2O + 4e^- \longrightarrow 4OH^-$

$$E_k = E_k^o - \frac{2,3\ RT_k}{4F} \cdot \log \frac{(a_{OH^-})_k^4}{(a_{O_2})_k}$$

$$E_k = E_k^o - \frac{2,3\ RT_k}{4F} \cdot \log (a_{OH^-})_k + \frac{2,3\ RT_k}{4F} \cdot \log (a_{O_2})_k$$

anode: $4OH^- \longrightarrow O_2 + 2H_2O + 4e^-$

$$E_a = E_a^o - \frac{2,3\ RT_a}{4F} \cdot \log \left\{ (a_{O_2})_a / (a_{OH^-})_a^4 \right\}$$

$$E_a = E_a^o + \frac{2,3\ RT_a}{F} \cdot \log (a_{OH^-})_a - \frac{2,3\ RT_a}{4F} \cdot \log (a_{O_2})_a$$

$$E = E_k^o - E_a^o + \frac{2,3\ R}{4F} \log \left\{ \frac{(a_{O_2})_k^{T_k}}{(a_{O_2})_a^{T_a}} \right\} - \frac{2,3\ R}{F} \log \left\{ \frac{(a_{OH^-})_k^{T_k}}{(a_{OH^-})_a^{T_a}} \right\}$$

2.3 Sulphurous Galvanic Elements in Situ

The accelerated corrosion of two sulphurous elements is called galvanic oxidation, if both elements (or one of them) are chemically active. The elements must bc connected by an electron conductor. The groundwater has to conduct the electrical current or guarantee its streaming by ion diffusion. Only with regard to these conditions can the galvanic potential of the sulphurous elements become developed by galvanic oxidation. The elements should from either a uniform sulphurous body or should exist as two different sulphides or rather two various alloys of sulphides. The orebody is the electron conductor and provides for the transport of the charges upwards to the cathode. The lower part of the orebody reacts chemically upon the surroundings and becomes the anode of the galvanic element. The nature and intensity of the reaction are determined by the sulphide, the composition of the solution and the physicochemical conditions of the system. The sulphide is stable as long as the absorption of the free electrons continues. This process takes place at the cathode and is induced by the sulphide. The reduction of the oxygen or its exemption at the cathode is accompanied by a spontaneous decomposition (corrosion) of the anode. Nevertheless, there are also some other reactions at the cathode (for example, a decomposition of the sulphide at the cathode) with respect to absorption, if free electrons are developed by decomposition of the anode. In zones of the electrolyte (secondary rocks, groundwater) the cations wander upwards from the anode to the cathode and the anions move downwards from the cathode to the anode. This process may be modeled by an electrical circuit, consisting of four components. These components regulate the electrical current (Fig. 2). The electrical current directs the "productivity" of the process. The above mentioned four components may be considered as single resistivities. R_1 is the resistivity of the electron current inside the sulphide. R_1 changes as a function of the composition of the orebody. R_2 is the kinetic parameter. It controls the processes of corrosion at the anode (boundary between sulphide and electrolyte). It decreases in dependence on increasing improvement of the reaction. The most powerful electrical potentials develop in the course of such processes, characterized by a minimum sum of these four resistivities. The sulphides must be the least stable elements of the geological medium. The resistivities R_1 and R_3 will be minimized, if there is a very small distance only between the ranges of the anode and the cathode. But in situ, the distance may be several kilometers. Thus, the natural currents must travel long ways.

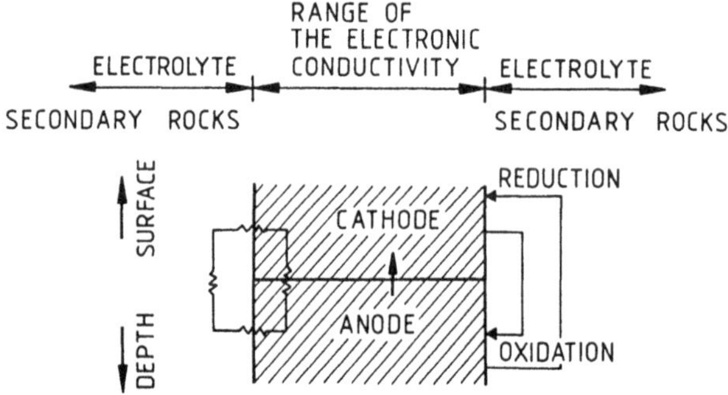

Fig. 2: Model of the galvanic oxidation of the sulphides.

The sulphides with various electron potentials form galvanic pairs. They enrich the solution in the secondary rocks by several ions. Svesnikov et al. (1967) explain that minerals with a powerful electrochemical potential form the cathode and those, characterized by a smaller potential, form the anode. Sometimes the galvanic corrosion produces a powerful potential (up to 350 mV) in south-east Missouri). Normally, the oxidation of the anode depends on the contact surface of each element with the electrolyte. If, in comparison with the sulphurous cathode, the surface of the sulphurous anode is not too large, the anode becomes intensively corroded. Such a sulphurous orebody also becomes oxidized galvanically, if its potential of oxidation differs from the potential of the surroundings. Besides, there may be a natural oxidation of the sulphurous electrode by the solution of the surroundings, showing a higher Eh than the electrode potential. If, however, the Eh of the solution becomes smaller than the electrode potential, the orebody may be oxidized by the solution. Blain et al. (1977) postulated similar mechanisms of electrochemical variations of nickel-sulphides in a hypergenetic medium: all the anode reactions are oxidation reactions of the orebody (oxidizing sulphides release useless electrons). The cathode reaction absorbs the released electrons and is only the reduction of the atmospheric oxygen. Thornber (1975) explains similarly the metamorphism of a massive orebody of the Kambalda type. His model contains an EMK. This EMK is developed by the reduction of the dissolved oxygen, which is in the groundwater of the upper part of the orebody near the groundwater level. The corrosion cell is generated by two types of anode processes:

1. Reactions on the surface of the upper part of the sulphurous body. Here, the sulphides become oxidized to sulphates;

2. Reactions beneath the surface. Here, the primary sulphides deliver iron into the solution and become more sulphurous.

The electrons are conducted through the orebody to the surface and the groundwater supplies the electrolyte for the cell. So, two kinds of sulphurous oxidations exist: either the end of the anode of the mono-sulphurous electrode becomes decomposed, influenced by the electrolyte, and the oxygen becomes reduced at the end of the cathode or, if there are two sulphides in one orebody, the sulphide with the lower reduction potential becomes an anode and decomposes. The other sulphide becomes a cathode. Here, the reduction of the oxygen occurs. The high-valency, nearly metallic character of the chemical compound presumes a similarity between the decomposition in the dissolved electrolyte and the corrosion of the metals. There is a complete analogy, regarding the sulphide as an alloy (melt) of a metal and sulphur.

The common reaction, concerning a sulphide, consisting of two-valency metals, may be expressed as follows:

anode- $MeS \rightarrow Me^2 + 2e^- + S^0$
or $(x/4)MeS + xH_2O \rightarrow x/4MeSO_4 + 2xH^+ + 2xe^-$ (for very powerful potentials)

cathode - $MeS + 2e^- \rightarrow Me^0 + S^2$

The products of the reactions show the following variations:

$MeS + 2xH^+ + 2xe^- \rightarrow MeS_{1-x} + xH_2S$
or $MeS + H_2O + 2e^- \rightarrow Me^0 + HS^- + OH^-$ (for very low potentials)

The feeding of the cations to the cathode and the anions to the anode happens by the groundwater and follows independently of the galvanic reaction; it runs off on both ends of the body. The system is just like a large battery with the same configuration of those orebodies, which form the cell of oxygen concentration. The metal ions Me^0 enter the solution as a result of the anode reaction. By this event, a concentration of metal cations is developed in the solution over the anode. This occurs together with the current of positive exchange carriers to the surface. This concentration produces halos with lower values of the redox potential. Sulphur ions S^0 are concentrated at the cathode, thus developing sulphides, enriched by sulphur. If the reduction occurs at the cathode, then the same conditions as in the cell of oxygen concentration are found. However, if the electrode is decomposed, metal ions M^0 will settle down at the cathode, as a consequence of the absence of oxygen and other oxidizing agents. Indeed, some tons of copper have been detected while exploiting compact sulphurous ore in Sweden (Sivenas et al., 1982). The above mentioned copper ore, situated between the fragments of the ore and the mineral grains inside the moraine, was similar to cement over the covered outcrop. There have been reports of findings of pure copper in clay above the outcrop of an ore vein, of a pyrite deposit in Norway. Both deposits are distinguished by intensive SP anomalies. Any oxidation must be excluded, since the pure metals were detected over the sulphurous bodies, situated in a glacial province (perhaps in a province characterized by permafrost).

The electron potential of the sulphurous material at the anode is equal to the potential of the pure metal, including the chemical potential of the metal inside of the sulphide:

$$Me^{n+} + ne^- \overset{MeS}{\rightarrow} Me^0 \ (MeS);$$

$$E^0 = E_1^0 - \frac{2,3\ RT}{nF} \log a_S^{2-} - \frac{2,3\ RT}{nF} \log a_{Me}^{n-}.$$

The potential of the electrodes at the sulphurous cathode is equal to the potential of pure sulphur, including the chemical potential of the sulphur inside the sulphide.

$$S^0_{(MeS)} + ne^- \overset{MeS}{\rightarrow} S^{2-};$$

$$E = E_3^0 - \frac{2,3\ RT}{nF} \log a_S^{2-} + \frac{2,3\ RT}{nF} \log a_S^0 \ (MeS).$$

2.4 EMK of the Sulphurous Galvanic Element

With regard to the common equations of electrochemistry, the EMK of the sulphurous galvanic element may be obtained as follows:

$$E = E^0 - \frac{RT}{F} \ln \frac{\pi_i a^{ni} \text{ (products of reaction)}}{\pi_i a^{ni} \text{ (reagents)}} .$$

With regard to the galvanic oxidation:

cathode - $1/2O_2 + 2H^+ + 2e^- \rightarrow H_2O$

anode - $MeS \rightarrow Me^{2+} + S + 2e^-$

element - $MeS + 1/2O_2 + 2H^+ \rightarrow Me^{2+} + S + H_2O$

$$E = E^0 - \frac{2,3\ RT}{2F} \log \frac{(a_{Me}^{2+})_a}{(a_{O_2})_k^{1/2} (a_H T)_k^2}$$

$$E = E^0 - \frac{2,3\ RT}{2F} \log \frac{(a_{Me}^{2+})_a}{(a_{O_2})_k^{1/2}} - \frac{2,3\ RT}{F} (pH)_k$$

The EMK of the sulphurous galvanic element is determined by E^0. This value depends on the specific participating reactions. Additionally, the concentration of the metal ions in the vicinity of the anode, the activity of the oxygen and the pH value at the cathode influence the EMK of the element.

3 Ore Potentials

Two mechanisms, which do not contradict one another and may exist in one and the same orebody, are the cell of the oxygen concentration (ZSK) and the sulphurous galvanic element (SGE). The cell of oxygen concentration is generated by the concentration gradient of the oxygen in the electrolyte. A milieu of oxidation is a prerequisite for the existence of an SGE. Moreover, therefore, a cell of oxidation concentration is also necessary.

Fixed, specific electrochemical conditions are necessary for the existence of a sulphurous galvanic element. However, it is not possible to realize these conditions in each medium. Therefore, the potential of the sulphurous galvanic element is more seldom found than the potential of the cell of the oxygen concentration. The EMK, generated by the sulphurous galvanic element, may be more powerful than the EMK, generated by the cell of the oxygen concentration. The potential, measured over the orebody, is equal to the sum of the potentials of the sulphurous galvanic elements and the cell of oxygen concentration. For cooperation of the two mechanisms, it is not necessary that its cathodes are situated at the same side; its anodes, however, must be situated side by side. Sivenas et al. (1982) call the potential of the natural galvanic element (geobattery), which is generated by the orebody, the "ore potential". It consists of the potentials of ZSK and SGE:

$$EP = SGE + ZSK.$$

Therefore, the sulphurous bodies are called polyelectrodes. Inside of these polyelectrodes, electrochemical reactions of different types may occur. The power W of the electrochemical cell is given by the following equation:

$$W = nFV,$$

where $V = E_{cell}$, n = number of electrons, F = Faraday constant.

4. Sulphurous Galvanic Elements in Situ

Each conducting sulphurous body acts as a galvanic element and produces an electrical current. Electrical and magnetic parameters of the sulphides are used in prospecting geophysics. Pyrite, markasite and pyrrhotite have low specific electrical resistivities. Pyrrhotite shows magnetic properties. Besides, all the sulphides are of good polarizability. However, there is no experimental evidence, concerning the variations of the electrical properties of the rocks, which are directly influenced by the existence of sulphur. Olhoeft (1981) stated, that the chemical properties of sulphur and its chemical compounds influence the electrical conductivity of the rocks. It is well known that the volatility of sulphur and oxygen influences the magnetic properties. The sulphur itself reacts very well (with regards to corrosion) and is a good electrical conductor.

In the surroundings of the sulphurous orebody, two types of electrochemical fields may exist. They are caused either by the orebody itself or by the existence of a vertical redox potential and a decreasing pH value in the groundwater. Influenced by the electrochemical field of the earth, the anions move upwards and the cations downwards. The reversed movement of these particles is effectuated by alternating relations with the field of the electrodes. In this case, the path of the ions is equal to the vectorial sum of two electrochemical powers. In part of the electrode, the primary field dominates. In the vicinity of the orebody, however, the secondary field is predominant. The primary and secondary power have a reversed direction and the same values at a defined distance to the electrode. The resulting power is equal to zero and the ion concentration, produced by the electrical field, will be minimized. In the vicinity of the orebody the products of the cathode reaction diffuse (sometimes becoming oxidized) and are influenced by the environment. The products of the anode reactions may also diffuse and may become reduced. The streaming groundwater supports these processes and guarantees the passage of the electrical current.

The above mentioned processes occur in a homogeneous isotropic medium only. This condition is very seldom valid in situ. With regards to many inhomogeneities, the boundary surface between the sedimentary layers and the primary rocks is most interesting. On this surface, the direction of the current lines changes very strongly due to the relatively small specific electrical resistivity of the sediments, in comparison with the primary rocks. The current density is more powerful on the boundary surface of the orebody and less above it. Zones characterized by a higher current density are marked by a higher ion concentration of the electrochemical cell. Govett (1976, 1984) detected the diagnostic characteristics of covered sulphurous deposits in Manitoba while investigating deposits consisting of compact sulphides:

1. Deeply situated deposits are characterized by an anomaly "H^+", including two maxima on the boundaries of the central minimum ("hare's ear").

2. Deposits at a mean depth are characterized by three minima;

3. There are anomalies of the electrical conductivity, reciprocally to the anomalies "H^+";

4. There are further anomalies with respect to the distribution of the metals. With regard to deeply situated deposits, their shape is nearly the same as that characterizing the anomaly "H^+"; for deposits at a mean depth, however, the shape is like the anomaly of the electrical conductivity.

Such a form of anomalies corresponds completely to the common conception of electrochemical processes. This proves the real existence of electrochemical cells in situ and the possibility of their discovery. In the region of Missouri, all the sulphides of stratified deposits, which are impregnated by lead, sphalerite, markasite, pyrite and chalcopyrite, are good electron conductors. ZnS is the only exception. The current is not hindered and flows in both the vertical and horizontal direction. This is important with regards to the uniformity of the potentials measured on the ores. The ores are situated at a depth of 300-330 m, however, due to their extension (500-650 m), signals may exactly measured with an amplitude up to -55 mV on the earth's surface. By measuring in situ a law was established for the first time, evidently concerning the anomaly "H^+" with the shape of a hare's ear (Govett 1976). There are characteristic minima (up to -40 mV) at a distance of 200 m from the orebody. It is possible that such laws have already been established by former measurements, but their bases were reduced to the extraordinary high potentials of the background. It is possible that such secondary extreme values are valid for more deeply situated deposits only. Figure 3 demonstrates the projection of the

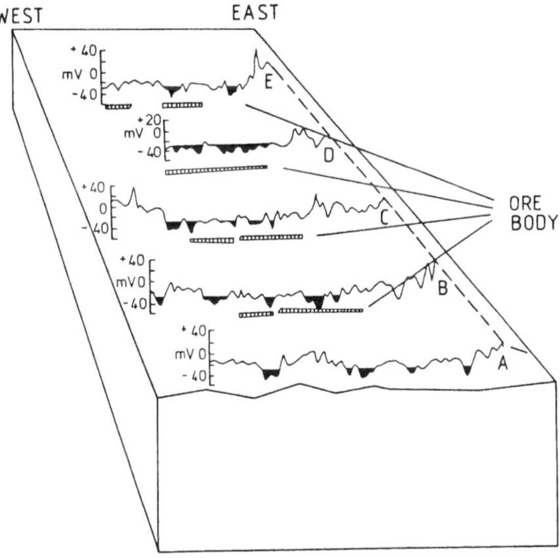

Fig. 3: SP curves, measured on the earth's surface in the vicinity of the ore pit (Elliot et al.)

ion current lines covering the same region as the detected secondary anomalies. The cause of the secondary minima is the high ion concentration in these zones. In spite of this fact, the secondary extreme values are a function of the laws induced by the sulphurous bodies. Borehole measurements were done, using a high pressure calomel electrode and a non-polarizable electrode of $Cu/CuSO_4$ on the earth's surface, for confirmation of the detected anomalies.

The results of the SP measurements agreed well with the above mentioned theory of the natural galvanic elements. The negative potential (-360 mV) characterizes the sulphurous zone, which was influenced by the electrochemical activity of the boundary surface between the ore and the dolomite and was caused by the electron current through the ore. The secondary minimum (10 mV) is a function of the ion current through the fluid, which fills the pore volume of the sedimentary secondary rocks. The measurements done in boreholes of the pyrite deposits in Norway show an analogous picture (Bolviken, 1981). The anomaly with the steep flanks was called the "potential of the electron current" and the flat anomaly "potential of the ion current". We assume that the positive current streams upwards in the secondary rocks resp.

downwards in the orebody. This observation corresponds completely to the conception of a polarity of the primary electrochemical field of the earth. The results of the measurements in situ confirm the existence of two mechanisms which generate the electrical potential by the sulphurous body: the cell of the oxygen concentration (ZSK) and the galvanic element (SGE) (Sivenas et al., 1982). Both cells may exist in a single orebody.

The cathodes may or may not be identical, but the anodes must be situated in different zones of the orebody in any case. Ore zones may be identified by using the sum of the potentials (ore potential), generated by ZSK and SGE. But it is not possible to detect the existence of any galvanic oxidation only, i.e., the SGE alone. Results of subsurface observations demonstrated the existence of sulphurous galvanic elements (SGE) on the flanks of small orebodies in Viburnum. Perhaps this fact is related to the ores, which are enriched by marcasite. Up to now, the exact reason for the development of the SGE is unknown, because there are too many factors, which influence the EMK. The main factors are the mineral specifics of the ore and the geological situation. Perhaps marcasite generates an active SGE on the boundary surface adjoining the crystals of lead. This SGE is related to local potential distributions. Consequently, positive anomalies (up to 120 mV) develop in the zones without ores and negative anomalies (up to -205 mV) develop above the orebody at the flanks of the deposits.

5 Electrokinetic Model for the Development of SP

Electrical fields are generated by the filtration of a fluid through a porous medium. The reasons are the development of an electrical double layer (DES) on the boundary between the solid and the fluid phase, as well as the absorption of the ionized fluid by the surface of the solid material. The potential difference (filtration potential) is a function of the pressure, which generates the movement of the fluid. These SP anomalies are interpreted on the basis of the electrokinetic model; they are related to the infiltration of groundwater in soils of sand and clay or other sedimentary rocks, as well as with areas of hydrothermal activity and zones of ensuing earthquakes. The filtration potentials generate a natural electrical field in situ. Their anomalies indicate zones with a good water exchange. Filtration potentials, developed by filtration of the groundwater in extension zones, may be the causes of SP variations, which were observed before an earthquake in California (Corvin et al., 1981). The development of electrical and magnetic fields may be explained by diffusion of the groundwater in the epicentre of the earthquake. The authors (Ishido et al., 1981, 1983) offered a simple theory to explain this phenomenon. It based on the capillary model of a porous medium. The authors described the fundamental relation between the pressure gradient of the solution and the induced electrical field as a function of a electrokinetic potential, the viscosity, the dielectric constant of the solution and the water permeability of the rocks.

The same electrokinetic phenomenon is the fundamental explanation for the development of SP above sources of warm water. SP anomalies were calculated with regards to their generation by hydrothermal convection. Thus, variations of filtration potential on the boundary sheet were demonstrated. The boundary sheet separated zones with differences in temperature of 100 and 200 C (Ishido et al., 1981).

The electropotentials in ore geophysics were considered only recently due to the extraordinary high values of the SP amplitudes over some orebodies, which were described in the literature. Nayak (1981) obtained high amplitude values of negative SP anomalies (-750 up to -1940 mV), while prospecting for sulphurous ores in quartzites of north east India. A borehole was drilled to investigate the geological reasons. The groundwater circulated in the quartzites without restraint and developed electrokinetic potentials. These potentials protect the elctrochemical ones; they then associate with the sulphurous orebody again. Extremely high amplitudes of natural potentials were measured in Peru (1800 mV). Formerly, the highest SP anomalies (-780 mV) were known for graphites (Sivenas et al., 1982). Anomalously high SP anomalies may be observed, if there are electrokinetic potentials, additionally related to the dip and strike of the conducting orebodies, provided they are situated in a zone of intensively moved sweet groundwater. The amplitude of the negative potentials increases on the top of the hills. The minima correspond to the highest elevations of the relief. Therefore, the correlation with topographic maps is the most effective method to separate the electrokinetic and the electrochemical anomalies. Commonly, the electrochemical potentials may be situated across the strike of the hills and ravines.

6 Electrical Double Layer

The flow of a fluid through a permeable medium generates a potential difference at the ends of the flow lines. It develops at the contact of two phases with a different chemical composition and is accompanied by a separation of the charges. On one side of the boundary there is a concentration of positive charges; on the other side, there is a concentration of negative charges. Such a system is called "electrical double layer". Supposing that side, which is charged negatively is the boundary of the solid phase, then the other side, bordering upon the electrolyte, is charged positively. Altogether, the boundary will be neutral, if the amplitude of the charge density in the solid phase is as high as the amplitude of the charge density in the electrolyte. The signs must be reversed, however. With regards to the model of the electrical double layer, developed by Stern (1945), the charges of the electrolyte are fixed at the solid phase. They partially diffuse into the solution and generate the so-called diffusion layer of Gouy. This part of the electrical double layer corresponds to those ions, which are influenced by arranging electrical powers and by non-arranging thermal forces. The fixed part of the electrical double layer is called the "layer of Stern". It consists of two parts, the internal and the external "covering of Helmholtz". The inner part of it consists of ions, fixed by absorption of the solid material. The external covering of Helmholtz consists of hydrate ions, concentrated mainly at the electrolyte. Figure 4 shows the distribution of the electrical potential inside the solution as a function of the distance from the boundary of the solid phase. It is known that non-soluble oxides in a watering solution generate an electrical surface charge σ_0, caused by the dissociation of the hydroxide groups:

$$M(OH)_n \rightleftharpoons /M(OH)_{n-1}O/^- + H^+,$$

where H^+ and OH^- are the ions determining the potential.

The surface charge depends on the proton concentration (pH value of the solution). The ions, determining the potential, are completely in the liquid phase. These ions react chemically with the solid material and generate the charge density σ_0 at the boundary of the solid phase. With regards to definite conditions, the layer of Stern may contain more neutralizing charges than necessary to compensate σ_0 . This results in a diffuse layer of Gouy, containing a charge with the same sign as σ_0.The variation of the potential is shown in Fig. 4. It is possibly a function of the absorption of the neutralizing ions of the covering of Helmholtz.

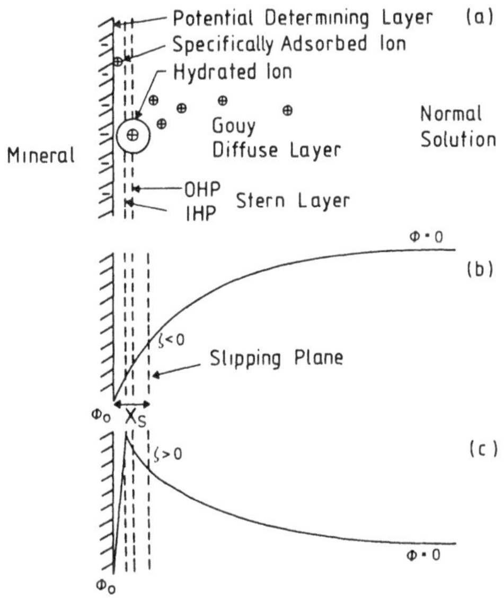

Fig. 4: a) The Stern model of the electrical double layer;
b) The potential variation according to the Stern model (in the Stern layer the potential varies linearly);
(c) The potential variation when the Stern layer contains more (positive) charge than is required to balance the (negative) charge on the solid
OHP outer Helmholtz plane
IHP inner Helmholtz plane (after Ishido et al., 1981)

7 ζ-Potential

Moving along the solid phase, the electrolyte may shift the external, more diffuse part of the electrical double layer with regards to the internal, more solid part. That sheet, where displacement occurs, is called "shifting sheet". The potential generated in this shifting sheet is called the "ζ -potential". There are many different phenomena, which are characterized by the movement of two sheets, contacting the shifting sheet. Examples for such electrokinetic phenomena are the electroosmosis, the electrophoresis and the filtration potential. With regards to classical thermodynamics, the potential is the same fundamental constant in all these kinetic phenomena. To determine the potential, it is common to use analytical expressions for the potential gradient in a porous medium, representing a bundle of capillaries. Usually, such a model is used to describe the permeability or the electrical conductivity of the porous medium. However, the real structure of the medium is not so simple, as in the model of the capillaries. Thus, this model may be regarded as a first approximation. As a scheme, it is presented in Fig. 5. For calculation, we assume:

L = length of a tortuous pore channel, A_f = the free cross-sectional area available to flow, η = porosity, t = tortuousity, $S(m^{-1})$ = inner sheet, m(m) = hydraulic radius:

$$\eta = A_f L_f / AL$$
$$t = L_f / L$$
$$S = S_f / AL$$
$$m = A_f L_f / S_f = \eta S^{-1}$$

S_f = whole sheet of the inner part of the porous area (m^2).

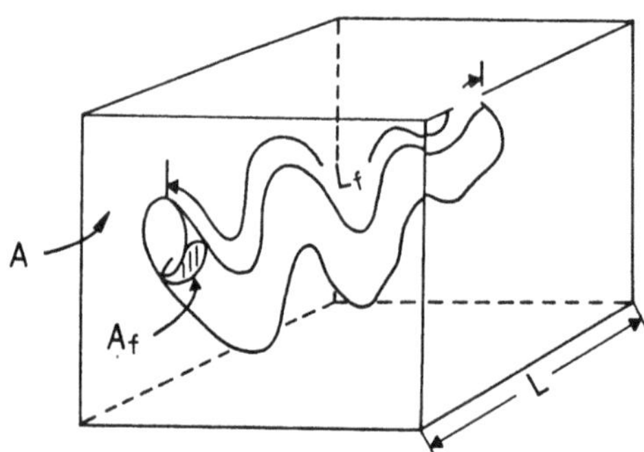

Fig. 5: Capillary model.
A Cross sectional area of the model. I overall length of the model; A_f free cross-sectional area available to flow; L_f length of the tortuous pore channel

The specific electrical conductivity of the sample is:

$$L_{ee} = \eta t^{-2} \sigma_f + t^{-2} S \sigma_F,$$

where σ_f and σ_S = specific conductivities of the fluid inside the capillaries ($\Omega^{-1}m^{-1}$) resp. the specific conductivity of the surface (Ω^{-1}).Assuming the conductivity of the matrix is less than the conductivity of the pore fluid, the Archie equation becomes valid for many rocks, if conductivity of the surface is very small:

$$L_{ee}/\ \sigma_f \cong \eta^2$$

Assuming $t \sim \eta^{-1/2}$, the coefficient $L_{e\nu} = L_{\nu e} = -\eta t^2 \varepsilon\zeta/\mu$, where ε = dielectric constant of the fluid (F/m), μ = viscosity of the fluid (Pa, s), ζ = potential at the slipping plane of the electrical double layer (EDS).

We presume a laminar streaming of the fluid and the hydraulic radius should be larger than the thickness of the electrical double layer. The equation for $L_{\nu\nu}$ corresponds to the equation of Darcy: $L_{\nu\nu} = k/\mu$, where k = permeability (m^2).

The most common expressions for the current I and the streaming fluid with the density J, connecting the electrical potential gradient $\nabla\phi$ and the pore pressure ∇P , are:

$$I = -\ L_{ee}\ \nabla\phi - L_{e\nu}\nabla P;$$

$$J = -\ L_{\nu e}\ \nabla\phi - L_{\nu\nu}\nabla P;$$

where L = empirical coefficients.

$L_{ee}\nabla\phi$ represents the law of Ohm and $L_{\nu\nu}\nabla P$ the law of Darcy. The terms with $L_{e\nu}$ and $L_{\nu}e$ correspond to the electrokinetic effect $L_{e\nu} = L_{\nu e}$ with regards to the reciprocal relation of Onsagar (Sivenas et al., 1982). By substitution, we obtain:

$$I = -(\eta t^{-2}\sigma_f + t^{-2}S\sigma_S)\ \nabla\phi + \eta t^{-2}(\varepsilon\zeta/\mu)\ \nabla P;$$

$$J = \eta T^{-2}(\varepsilon\zeta/\mu)\ \nabla\phi - (k/\mu)\ \nabla P.$$

These are fundamental equations for the description of electrokinetic effects in a porous medium (Ishido et al., 1981).

We may use the streaming potential (filtration potential) for the determination of the ζ-potential, if the solid particles of the two electrodes are immovable and the electrolyte is moved by the influence of a constant potential difference. Assuming stationary conditions and Δp = const. the electrical current I disappears and we obtain the following relation:

$$(\mu t^{-2}\sigma_f + t^{-2}S\sigma_S)\ \nabla\phi/d = \eta t^{-2}\ (\varepsilon\zeta/\mu)\ \Delta P/d,$$

where d = distance between the electrodes. After having measured the filtration potential, we may calculate the ζ-potential as follows:

$$\zeta = \frac{\eta t^{-2}\sigma_f + t^{-2}S\sigma_S}{\mu t^{-2}} \cdot \frac{\mu}{e} \cdot \frac{\Delta\phi}{\Delta P}\ .$$

At least, there is a linear function between the filtration potential and the presssure, involving the movement of the fluid, up to the value of 13.3 Pa. ζ depends on the pH value of the solution. That means, those ions, which determine the potential of the system mineral-water, are the same H^+ and OH^- ions of the system "simple oxides-water". The main differences with regard to the ζ-potential of the various kinds of minerals are the different values of the isoelectrical points (IEP) (pH value with respect to the zero charge) along the curves, demonstrating: ζ-potential = f(pH value). The IEP values of various kinds of rocks and minerals are presented Table 1 (Ishido et al., 1981). The high IEP value of enstatite and dunite, with regard to the IEP of quartz, may be a function of high IEP values of MgO, the main components of these minerals (Fig. 6). With regards to the

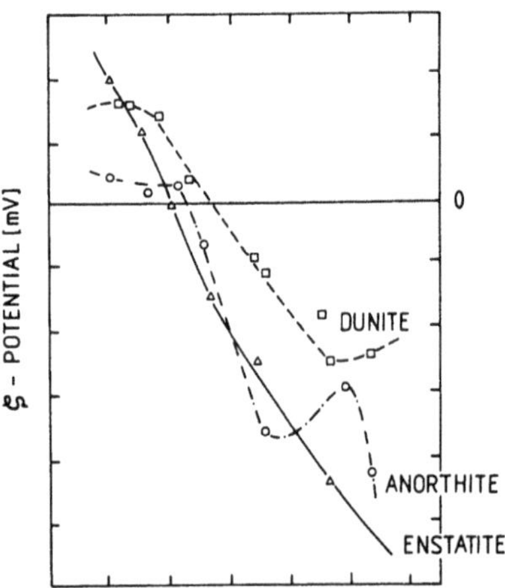

Fig. 6: Variation of the ζ-potential of anorthite, enstatite and dunite as a function of pH in aqueous solutions of 10^{-3} N KNO_3. Temperature is set 45°C (Ishido et al., 1981).

table, we may conclude that the IEP values of most rocks and minerals decrease to low pH values as a function on an increasing temperature (the IEP values of quartz and orthoclase decrease from 2.6 to 2 resp. from 2.5 to 1 as a function of an increasing temperature from 20° up to 45 °C).

Table 1: IEP's (IEP = isoelectrical point) of minerals and rocks (after Ishido et al., 1981).

Minerals or Rocks	IEP		
	R. T.*	45°C	70°C
Quartz	2.6	2.0	
Orthoclase	2.5	1.0	
Albite		1.5	
Anorthite		4.5	
Enstatite		4.1	
Granite		1.5	
Granodiorite		1.5	
Andesite		1.8	
Gabbro		2.5	
Dunite		5.3	
SiO_2 (vitreous silica)	2.5		
MgO	12.4		
$Fe(OH)_2$	12		
Al_2O_3	9.2	8.8	8.6
Fe_3O_4	6.5	6.1	5.7
TiO_2	6.0	5.7	5.5
Kaolinite	≃1		
Montmorillonite	≃2		

*Room temperature.

The value of the ζ-potential is reversed proportional to the concentration of the electrolyte. The values of the ζ-potential change, because of non-calculable variations of the electrolyte concentration as a result of penetrating CO_2 from the atmosphere, if electrolytes are no longer added to the solution. For typical crystalline rocks, the ζ -potentials are usually negative in solutions with a pH value of >2. The absolute value increases with decreasing electrolyte concentration and increasing temperature.

8 Filtration Potential

The current density in a homogeneous medium is described by:

$$I = -\eta t^{-2} \sigma \nabla (\phi - \frac{\varepsilon\zeta}{\sigma\mu} P),$$

where $\sigma = \sigma_f + m^{-1} \sigma_S$; σ_f = conductivity of the pore fluid; $_S$ = conductivity of the surface; m = hydraulic radius. The relation

$\varepsilon\zeta/\sigma\mu$ is called the coefficient of the filtration potential. It presents the contribution of the pressure gradient, induced by the electrokinetic connection, to the electrical field. Besides, the coefficient of the filtration potential is equal to the filtration potential per pressure unit $\Delta\phi$ / ΔP, determined experimentally. The coefficient of the filtration potential must be known in order to quantitatively estimate the electrokinetic effect, which is induced by water diffusion into the earth's interior.

The coefficient of the filtration potential does not depend upon ζ, μ, ε, σ_f, σ_f only, but also on the geometric factor m (hydraulic radius). The effective conductivity σ of the pore fluid increases, if m decreases. $\Delta\phi/\Delta P = \frac{\varepsilon\zeta}{\sigma\mu}$ decreases for low values of m, if σ increases (Fig. 7).

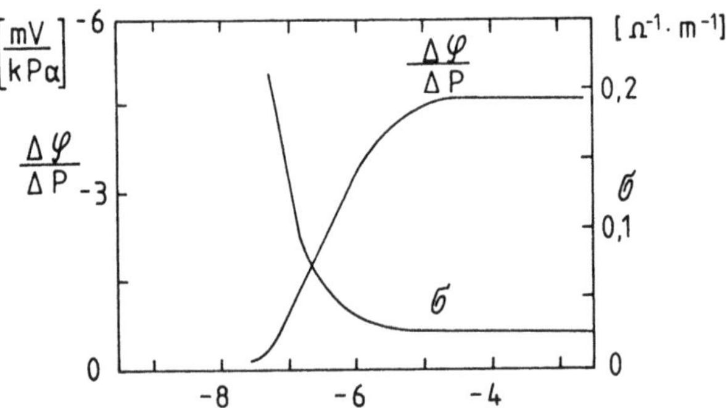

Fig. 7: Variation of the effective pore fluid conductivity ($\sigma = \sigma_f + m^{-1}\sigma_s$) and the streaming potential coefficient ($\Delta\phi/\Delta P = \varepsilon\zeta / \sigma\mu$) as a function of hydraulic radius m in quartz-water (pH=7; 10^3 N KNO_3; 45°) system (Ishido et al., 1981).

The value $\Delta\phi$ begins to decrease at $m = 10^{-5}$. $\Delta\phi$ increases as a function of diminution of the electrolyte concentration, since the influence of the surface conductivity σ_s/σ_f becomes higher at low concentrations. Indeed, this dependence is valid for laminar streamings only, within the pores and joints. The flow will be turbulent in channels with a large hydraulic radius, for example in a system of rock joints. In such a system there is a relatively low pressure gradient and $\Delta\phi$ does not increase linearly up to a definite ΔP-value. $\Delta\phi$ decreases with increasing ΔP.

Bogoslowskij et al. (1972) demonstrated experimentally, positive filtration potentials at low differences of pressure. The curves $V_f = f(\Delta P)$ show negative values with increasing pressure. Their course is parallel and marked by a constant coefficient of -22 in the negative region; the absolute value of V_f diminishes at ΔP = const, if the joints become larger (Fig. 8).

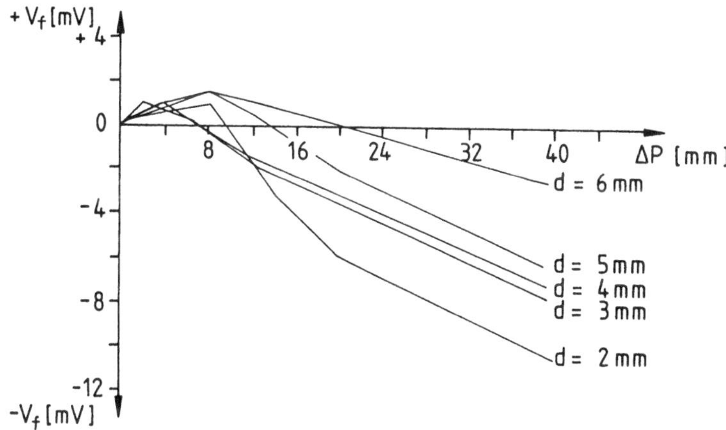

Fig. 8: $V_f = f(\Delta P)$ dependence for different openings of fissures (d); P = pressure drop.

The dependences between V/ ΔP, the water permeability and the NaCl solution permeability of sand are shown in Fig. 9.

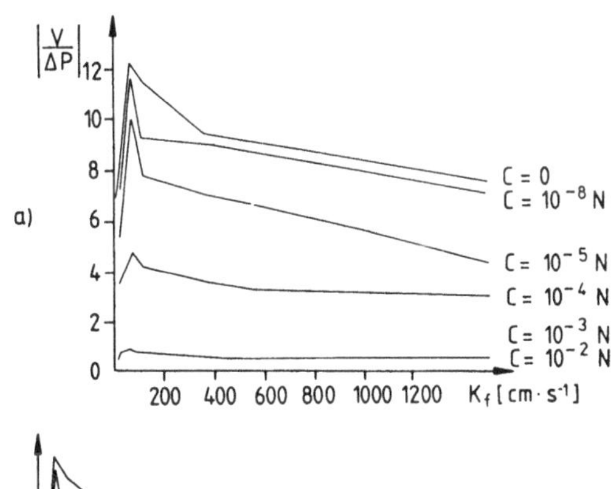

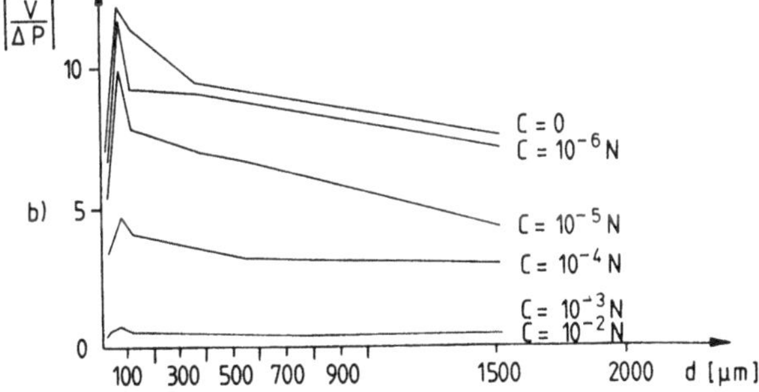

Fig. 9: $V/\Delta P$ as a function of the permeability K_f (Fig. a), the grain diameter d of quartz-sand (Fig. b), water dest. (C = 0) and the solution of NaCl ($C = 10^{-n}N$).

The relation V/Δ P increases (initial permeability of 20 Darcy) and reaches its maximum at a value of 60-70 Darcy. A further increase of the permeability leads to a diminution of V/Δ P up to an asypmtotic value. It is interesting to analyze the relationship between the grain size of the loose rocks and the quantity V/Δ P (Fig. 9). The highest amplitude of this relation belongs to a grain size of 250-315 μm. The porosity influences the value V/Δ P reversed proportionally, with regard to the influence of the permeability and the grain size (Bogoslovsky et al., 1972).

The absolute values of the filtration potentials strongly diminish with increasing clay content. A clay content of up to 1% diminishes the absolute value of the filtration potential up to 1.5-2 times; a clay content of 6 up to 10 % diminishes this value 5-6 times, however. The graphics of $V_f = f(\Delta P)$ point to negative values and remain further on linear, if the clay content and P increase (Fig. 10).

According to Bogoslovsky et al. (1972) such a behaviour of the absolute filtration potential values depends on the decrease of the medium permeability and the existence of adsorption potentials. Acceptance of the latter is confirmed by the fact that an increasing clay content leads to an increasing electrical potential, even if no filtration occurs. The presence of medium-grained sandstone in the joints leads to an increase of the filtration potential. It reaches the highest value at 40 % joint fillings.

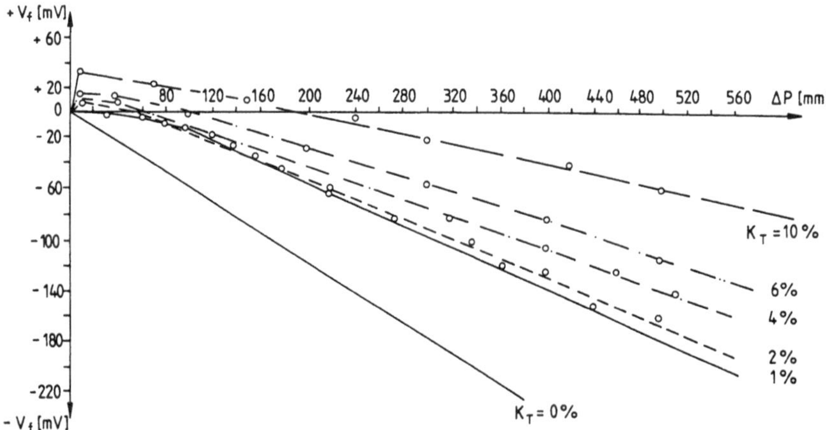

Fig. 10: $V_f = f(\Delta P)$ upon the clay content K_T in the filler.

The laws governing the development of the filtration potential in jointed rocks are not well known, at the present time. There are perhaps lower filtration potentials, influenced by a non-hindered flowing through a joint in comparison with the movement of the same quantity of water in a porous medium. Often the joints are filled by a redeposited material. Anomalies of the filtration potential must be expected, if this material consists of medium-grained sandstone. On the other hand, clay sediments cause a diminution of its permeability and consequently a diminution of the filtration potentials. Although we can not expect high amplitudes of the filtration potential, much water streams through fine-grained material. The existing of clay-containing material causes an inversion of the filtration potential field. Clay-containing layers are distinguished by high positive values of the potential, with regards to the natural field, since diffusion and absorption have been observed prospecting and engineering geophysics. In the following we observe positive filtration potentials, which cross the upper or lower parts of the clay-filled joints; there exists here also a flow-through of water.

A streaming of NaCl solution through quartz sand (grain diameter of 125-3000 μm) is marked by a diminution of the absolute values of the filtration potentails. Theoretically, this fact may be explained as follows. Presuming an anology between the electrical double layer and a plate condenser, the potential difference between its plates is:

$$\zeta = \frac{4\pi eL}{D}$$

where L = distance of the plates (cm); e = charge on the single plates (C/cm); D = dielectric constant of the medium between the plates.

Streaming water through the capillary with a small diameter develops a double layer at the boundary of the fluid and solid phase. From this double layer part of the water and the positive charge carriers are carried away. The filtration potential, developing in the course of this process, depends on:

$$V = \frac{P\zeta D}{4k\mu},$$

where μ = viscosity of the fluid; k = specific electrical conductivity.

The expression for the ζ-potential in a cylindrical tube is as follows:

$$\zeta = - 1202.5 \frac{V}{P} \cdot \frac{d^2 l}{R(1-\varepsilon)^2 s^4} ,$$

where R = specific resistivity of the layer (Ωm); ε = porosity of the sand; S = thickness of the layer; d = diameter of the sand grains (cm).

Presuming (1) ζ is not a function of the grain diameter d and (2) the resistivity is constant, the filtration potential becomes reversed proportional to d^2 for the capillary tube of the length l. The relation $V/\Delta P$ is proportional to the ζ-potential and reversed proportional to the grain diameter, because k and μ are not a function of d. The permeability also depends on d. It diminishes with decreasing grain diameter:

$$k = \frac{cd^2\gamma}{\mu} ,$$

where c = constant without dimension; γ = weight of a fluid unit; μ = dynamic viscosity.

The following empirical relation is valid:

$$V/P = 0.119\ k^{-0.077}.$$

That is, there is a negligible enlargement only, if the permeability diminishes.

Figure 11 shows the dependence $\Delta\phi/\Delta P = f(T)$ of quartz in a water solution (pH = 6.1×10^{-3} N KNO_3). The amplitude of $\Delta\phi/\Delta P$ increases with the temperature. Even for T = const (44, 56, 74°C) a time variation occurs, i.e. the temperature balance of the charge distribution near the boundary does not appear immediately (Ishido et al., 1981).

The filtration potentials are directly proportional to the potential difference between the non-movable part of the electrical double layer and the solution (ζ-potential). Therefore, these potentials decrease with increasing concentration of the electrolyte solution. The most powerful filtration potentials are caused by streaming of sweet

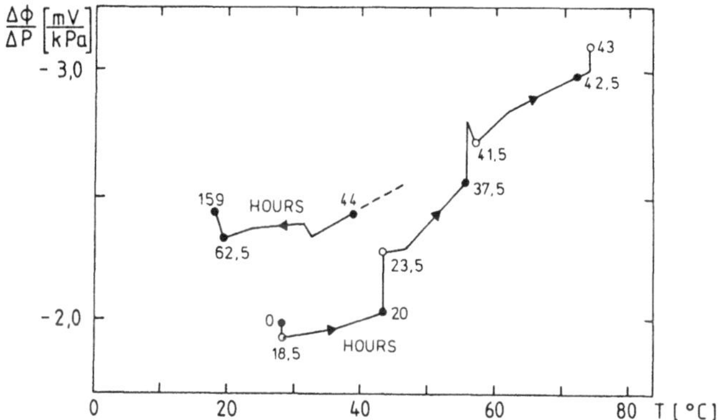

Fig. 11: Variations of the filtration potentials per unit driving pressure as a function of temperature in aqueous solution (pH 6,1; 10^{-3} N KNO_3). The initial value at a temperature is shown by solid circles, and the final value is shown by open circles. Intermediate values at a temperature are not shown. The numbers attached to the data points indicate the elapsed time from the beginning of the experiment to each measurement (Ishido et al., 1981).

the filtration potential, if the mineralization is higher than 5 g/l. The movable sheet of the electrical double layer has positive charges. The positive ions flow with the water streaming. Therefore, a transport of charges is marked by positive anomalies of the filtration potential; the filtration zones are marked by negative anomalies. The filtration potentials increase in the direction of the water streaming. Its intensity is proportional to the hydraulic gradient. The zones of infiltration are marked not only by lower filtration potentials, but by an increasing velocity of the streaming and by positive anomalies of the temperature, too. The latter are related to the influx of water, coming from the upper, warmer horizons (Fig. 12).

The shifting between the solid phase and the fluid pore fillings, as well as the fillings of the joints, is accompanied by an electrokinetic phenomenon, which is induced by the -potential on the boundary of the two phases. The potential difference ΔE, the so-called filtration potential, is proportional to the pressure difference ΔP of the in-situ situation, according to $E = C\Delta P$. The coefficient C depends on many parameters, as for example, on the specific resistivity, the dielectric constant and the viscosity of the fluid, the ζ-potential, the grain diameter, the degree of the joint's opening, the form and curvature of the capillaries. The existence of a pressure gradient is not a sufficient condition for the appearance of the electrical potential on the earth's surface. Fitterman (1979) explains the necessity

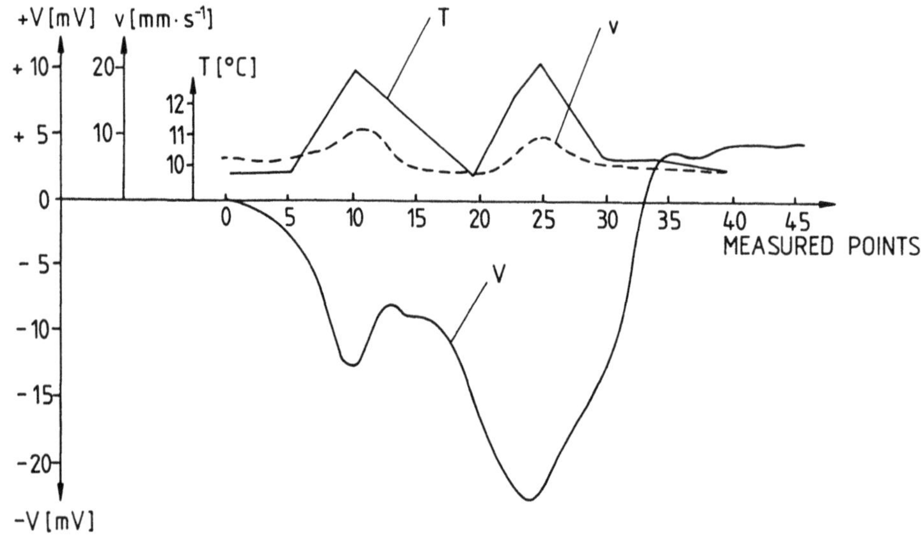

Fig. 12: Curves of the filtration potential V_f, the streaming velocity v and the temperature T.

of having a pressure gradient parallel to the boundary, which separates the spheres with different coefficients of the filtration potential. It is possible to describe the electrical field with regards to these conditions only. The mentioned field is equivalent to that field, which was generated by the distribution of the current dipoles on the surface along the boundary. Fitterman (1979, 1983, 1984) published a mathematical expression pertaining to the distribution of the potential of the surface, induced by flat and spherical sources of pressure and assuming vertical and horizontal boundaries between spheres with different filtration coefficients. The qualitative interpretation may be based on the comparison with anomalies, generated by most simple sources (Fig. 13) (Schiavone, 1984). We obtain asymmetrical anomalies, which cross the vertical boundaries of spheres with different filtration potentials and symmetrical anomalies, which cross horizontal boundaries. The component of the vertical current, caused by the vertical pressure gradient along the vertical boundary, and the radial streaming, running parallel to the horizontal boundary, represent the situation of pumping water from a borehole. Circular anomalies may be observed, which cross vertical cylindrical sources. Such a model approximates the movement of the fluid upwards or downwards along the wall of the borehole. In this case, a positive anomaly is produced by an upwards directed movement of the fluid and vice versa. Supposing a horizontal movement of the groundwater, we observe an increasing potential along the direction of the movement.

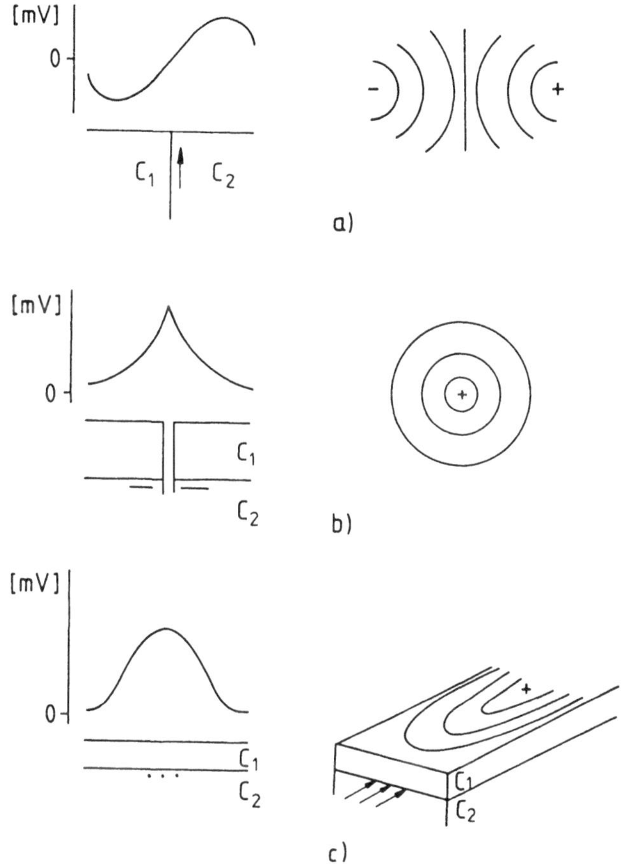

Fig. 13: Anomalies of the filtration potential, induced by the streaming along the boundaries between areas of different filtration coefficients C_i.
a vertical boundary; b pumping of the borehole; c horizontal boundary (Rao et al., 1982).

9 Coefficients of the Electrokinetic and Thermokinetic Binding

The following relation exists between the flux of the electrical current, the streaming of the particles and the heat: the components of the heat flow and the movement of the fluid, parallel to the boundary and separating the spheres of different coefficients of the thermoelectrical and electrokinetical bindings, may generate an electrical current. This current induces a potential field on the earth's surface, equivalent to that field, which is excited by the distribution of the fictive dipoles along the boundary. The coefficients of the bindings are determined as the relation of the induced potential to that potential, excited by the streaming. It is expressed in mV/at, with respect to the electrokinetic coefficients (filtration potential), and in mV/°C resp. in mV/K with regard to the thermoelectrical coefficients.

The flux of the charged particles, induced by the balance of the inductive bindings on both sides of the boundaries, generates an anomaly of the natural field. This effect is investigated by thermodynamics. Formulas, describing the conditions of the electrolytes with inductive bindings, may be presented as follows:

$$\begin{array}{l} \text{flux of cations} \\ \text{flux of anions} \\ \text{flow of the solution} \\ \text{heat flow} \end{array} \begin{vmatrix} J_p \\ J_n \\ J_a \\ J_q \end{vmatrix} = \begin{vmatrix} L_{11}L_{12}L_{13}L_{14} \\ L_{21}L_{22}L_{23}L_{24} \\ L_{31}L_{32}L_{33}L_{34} \\ L_{41}L_{42}L_{43}L_{44} \end{vmatrix} \cdot \begin{vmatrix} -\nabla\mu_p - FZ_p\nabla\phi \\ -\mu_n - FZ_n\nabla\phi \\ -\nabla P \\ -\nabla T/T \end{vmatrix}$$

where ϕ = electrical potential; μ = chemical potential; P = pressure; T = temperature; $L_{ij} = L_{ji}$; F = Faraday constant; Z = valency of the ions.

The left column of the above equations emcompasses the vectors of the fluxes and the right one represents the generalized forces. The matrix L_{ij} consists of binding coefficients with regard to the relation between conductivity and permeability of the medium. Using this relation Nourbehecht (1963) established quantitatively the possibility of exciting anomalies of the natural field by such simple geological processes as the streaming of groundwater in a capillary system and the leakages of fluids through clayey walls (membranes). He also tried to calculate SP anomalies on the basis of the redox potential, associated with the ion pair of the two- and three-valency iron. The concentrations and the mobility of Fe^{2+} and Fe^{3+} were his assumptions. Though the electrokinetic effects were considered as a pure theoretical possibility at first, later on, they could be demonstrated by experiments in situ, with respect to rock mechanics, as well as in geothermal zones, in zones of hot springs and by observation of a highly situated water reservoir in tuffaceous sandstone (Bogoslovsky et al., 1972) in the region of active volcanos (Hawaii Islands) (Wynn, 1984) and in a geothermal zone of the Aleuten (Corvin et al., 1979). In faulted zones the bindings change on the contacts of the rocks, which are characterized by a different lithology or intensity of metamorphism. The movement of the geothermal fluids, filling the pores and joints, changes the composition of the rocks. These changes induce variations of the coefficients of the thermoelectrical bindings. The values depend on the intensity of the metamorphism.

Horizontal inhomogeneities of the binding coefficients may be caused by horizontal variations of the temperature in the vicinity of the heat anomaly. For example, in a hydrothermal province (depth 1500 m) in southern California, the temperature decreases between 190° and 150°C in a horizontal distance of 1.8 km from the centre of this region. This effect is accompanied by a variation of the electrokinetic binding coefficients. Laboratory measurements on rock probes from a geothermal province in Mexico demonstrated that increasing temperatures between 24 ° and 61 °C induce a linear increase of the binding coefficient. The amplitudes of this coefficient are 1.5 mV/at, if this dependence is valid for the geothermal province in southern California.

The value of the binding coefficient is influenced by a mineralization of the pore fluid, too; it also influences the electrical resistivity. It is known that the coefficients of the electrokinetic bindings increase linearly as a function of increasing resistivity of the pore fluid. As shown by Nourbehecht (1963), the coefficients of the thermoelectrical bindings of sediments are 0.086 up to 1.12 mV/°C , using water as fluid. Foreign investigators used a salt hydroxide solution with a mineralization of 1000 mg/l and obtained thermoelectrical coefficients between 0.4 and 1.2 mV/ °C for sandstone. Experimental results show that a point source (T = 49 °C) within a large block of sandstone generates a potential of 20 mV on the surface. Coefficients of thermoelectrical bindings, measured in sedimentary rocks of a geothermal province in Mexico, amount to between 0.01 and 0.18 mV/ ° C (the pore fluid inside the geothermal collector has a mineralization of 25 g/l). The liquefaction of the pore fluid increased the coefficient of the thermoelelectrical bindings between 0.037 and 0.144 mV/° C. This demonstrates a diminution of that parameter by increasing the conductivity of the pore fluid (Cull, 1985).

To evaluate the anomalies of the natural field, which are induced by electrokinetic alternating effects, information is necessary on the ζ-potential and the streaming potential at high temperatures. For determination of the ζ-potential, again, it is necessary to know the chemical composition of the water in the hydrothermal source. By systematic investigations with regard to the chemical composition of the geothermal solutions from typical boreholes in different geothermal provinces, the conclusion follows that the solutions contain only NaCl; its concentration changes between 0.005 and 0.5 mol/l (Ishido et al., 1981). Supposing a NaCl solution with a concentration of 0.02 mol/l as a geothermal solution, the ζ-potential becomes -64 mV at 100 °C and -100 mV at 200 °C. With regard to hydrothermal solutions it is characteristic that the ζ-potential and the streaming potential tend towards zero, if the temperature decreases from 200 °C down to 100 °C and there are $2x10^{-6}$ mol/l Al^{3+} ions in the water of the collector. The temperature decreases from 90 °C down to -20 °C, depending on the concentration between 10^{-4} and $2x10^{-6}$ mol/l; the streaming potential and the -potential change their signs to "plus" at the same temperature. Using these results Ishido (1983) calculated the field distribution of the filtration potentials, which were induced by hydrothermal convection inside the earth's crust. His assumption was that the coefficients of the streaming potentials change by 35 mV/bar, corresponding to a temperature variation of 100 °C along the convective water flow. We can observe an amplitude between 10 and 100 mV on the earth's surface, if a streaming velocity dominates between 10^{-8} and 10^{-7} m/s.

10 New Methods of Quantitative Interpretation: Method of Natural Potentials

10.1 Methods Based on the Well-Known Solutions of the Direct Task for Regularly Formed Bodies

For quantitative interpretation of the anomalies of natural potentials we approximate the source, which forms the anomaly, by any body, which consists of a simple geometric shape, and determine its parameters (depth of the centre of the body; angle between polarization axis and the horizon; shift of the projection of the centre of the body to the horizontal axis at the zero value of the potential). We can do so, using either graphical-analytical methods with the help of characteristic points of the anomaly curves or the method of "trial and error", comparing visually the anomalies with master curves or minimizing the differences between the observed and the theoretical anomalies of the interpretation parameters by a computer-aided iteration process. The calculation of the theoretical anomalies of the natural field, based mainly upon the known solution, has been published by Petrovskij (1928) for a vertically polarized sphere. Later, analytical solutions for layered media and inclined layers were published. Commonly, the polarization was oriented in the dip of the orebody. Another method is used for vertical geologic structures, based on the proposals of Sato et al. (1960), for the electrochemical mechanism. The above mentioned solutions of the direct task are used mainly in ore geophysics. In prospecting for geothermal sources the well-known solutions are limited with regard to vertical contacts, horizontal layers, vertical cylindrical intrusive bodies (Fitterman, 1979, 1983) or two-dimensionally formed bodies (Sill, 1983). The practical problems, concerning prospecting for thermal sources, cannot be solved using such simple models, however. The method of the analytical downward continuation of the field is a common problem in ore geophysics as well as in prospecting for geothermal sources. This method gives positive results also with regards to bodies of irregular forms, if the electrical conductivity of the secondary rocks is extremely small. The relation, concerning the conductivity of the secondary rocks and the objects to be investigated differs between 0.05 and 0.0005. Roy (1959) formulated the main specialities of this method in ore geophysics as follows:

1. The orebodies are characterized commonly by such a form and concentration that the depth to its upper part is equal to or the same as the extreme point of the analytical continuation, presuming the horizontal dimensions of the orebody are small in comparison to its depth.

2. For determining the depth of the interesting body we can use not only the beginning of the oszillation but also the increase of the velocity. The extreme point of the analytical continuation, with respect to the magnetic field, the electrical and the electromagnetic field, corresponds to the concentration of the induced polarization, the charge concentration, the existence of a definite quantity of apparent sources, the depth of the concentration of secondary currents etc.

3. The method of analytical continuation does not depend on the geometry of the problem for the types of orebodies, found mostly in situ.

Roy (1959) published a formula for the approximation of a two-dimensional configuration and for the unit depth of the analytical downward continuation of the field:

$$V(0,1) = 2V\,(0,0) - \frac{V\,(1,0) + V\,(-1,0)}{2}$$

The continuation to greater depths is calculated in units of the grid model by iteration as follows:

$$V\,(0,n) = 4V\,(0,n-1) - /V\,(1,n-1) + V\,(-1,n-1) + V\,(0,n-2/$$
$$n>>2$$

Nomograms, published by Bhattacharya et al. (1981), are used for a quick interpretation of the anomalies of the natural field, represented by an inclined polarized sphere and a horizontal cylinder (Bhattacharya et al., 1981). The polarized sphere or the cylinder with the radius "a" are situated in a homogeneous half-space. The axis of the cylinder is parallel to the y-axis; the depth to the centre is h (Fig. 14). The beginning of the system of the rectangular coordinates is situated on the surface over the centre of the body. The angle between the axis of polarization and the x-axis is α. The straight line, connecting the observation point P and the centre of the body C, forms together with the polarization axis the angle Q (the distance of P to the zero point is x); P_o is the point with potential zero; PC = r; $OP_o = x_o$. The potential in P on the surface has the following form:

$$V_P = M\,\frac{x\cos\alpha - h\sin\alpha}{(x^2 + h^2)^{3/2}} \qquad \text{(for the sphere);}$$

$$V_P = M\,\frac{x\cos\alpha - h\sin\alpha}{x^2 + h^2} \qquad \text{(for the cylinder),}$$

where M = electrical moment of the dipole.

Further on it is possible to prove the following relations:

for the sphere:

$$\frac{|V_{min}|}{V_{max}} = \left[\frac{3\cos\alpha\,(tg^2\alpha + 8/9)^{1/2} + \sin\alpha}{3\cos\alpha\,(tg\alpha^2 + 8/9)^{1/2} - \sin\alpha}\right] \quad x$$

$$x \left[\frac{9/16\{tg\alpha + (tg^2\alpha + 8/9)^{1/2}\}^2 + 1}{9/16\{tg\alpha - (tg^2\alpha + 8/9)^{1/2}\}^2 - 1} \right] = F(\alpha)$$

for a cylinder: $\frac{|V_{min}|}{V_{max}} = \frac{1 + \sin\alpha}{1 - \sin\alpha} = F_1(\alpha)$

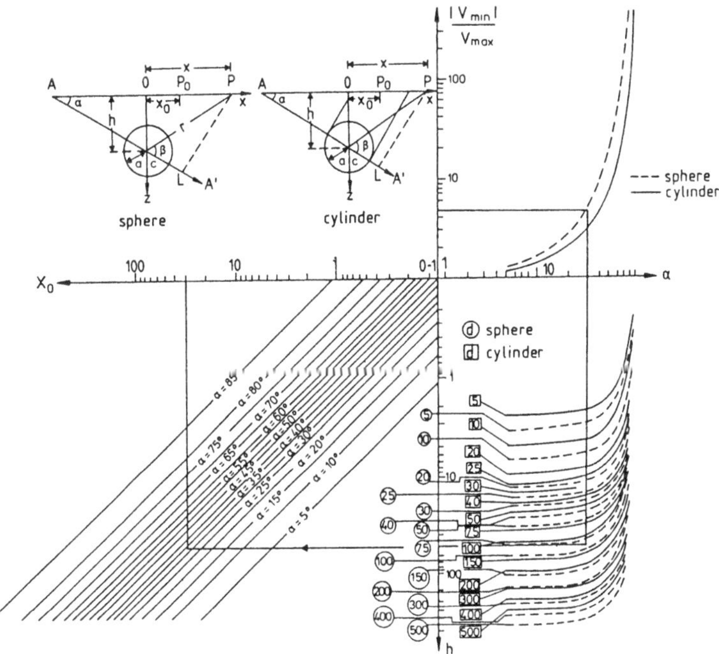

Fig. 14: Nomogram of SP-anomalies for spherical and cylindrical bodies (after Bhattacharya et al., 1981). a = radius; h = depth of the centre c; AA'= axis of polarization; α = angle of the axis of polarization with the s-axis; P_o = the point at which the potential is zero (the axis of the (infinite) cylinder is parallel to the y-axis).

The depth of the object can be determined, if the maximum and the minumum of the potential is valid for x_1 and x_2 ($d = x_1 - x_2$):

$$h = \frac{2d}{3} \frac{1}{(tg^2\alpha + 8/9)^{1/2}} = f(\alpha, d) \qquad \text{(for the sphere);}$$

$$h = \frac{d}{2} \cos\alpha = f_1 (\alpha, d) \qquad \text{(for the cylinder).}$$

Accordingly we have to substitute $V_P = 0$ into the formula for the sphere and replace x by x_o. Now, $h = x_o \; ctg\alpha = \phi (x_o, \alpha)$ (for the sphere as well as for the cylinder).

We obtain two curves, using the formula for the cylinder (Fig. 14, above right). The curves are drawn in a logarithmic scale. We can see the variations of V_{min}/V_{max}, depending on a and d. The curves for a sphere in a double logarithmic scale, with different values of d are also shown (Fig. 14, below right). Curves, valid for the range between 5 and 85°, are presented, regarding the formula of the cylinder (left side, below). In all cases, the coordinate axis has the same scales. The quantitative interpretation now consists of the following steps:

1. First the relation V_{min}/V_{max}, using the measured curve, must be determined.

2. From this point, a horizontal straight line must be drawn until it divides the curve into a sphere or a cylinder.

3. From the intersection point a straight line is drawn vertically downwards until cutting the curve with the attached d, determined from the anomaly.

4. The angle α between the polarization axis and the horizontal line is obtained by cutting the vertical straight line with the axis.

5. A horizontal straight line is drawn, beginning at the above mentioned intersection point, until the line must be drawn with the attached α. Furthermore, a vertical straight line from the intersection point to the crossing with the x-axis. Now, the distance between the origin of the coordinate's system and the point with the zero potential is determined.

In this way the parameters d, h and x_o of the body can be determined. An essential disadvantage of such an interpretation is that we use only part of the information contained in some points of the anomaly curve. Therefore, the reliability of the determined parameters depends essentially on the noise, distorting the measured results. Nevertheless, the accuracy of the method seems to be good enough, used as an operative method. Only Rao et al. (1982) used the frequency analysis for the interpretation of SP anomalies. They obtained the Fourier transform of the anomalies, generated by the inclined plates of finite depth. Assuming the upper and lower boundary of a horizontal layer at the depth h resp. H, we can calculate the SP anomaly, induced by this plate and observed along a profile in the strike direction as follows (Murty et al., 1985):

$$V\ (x) = \frac{I\rho}{2\pi}\ \ln \frac{r_1^2}{r_2^2}\ ,$$

where I = current per length unit; ρ = specific resistivity of the secondary rocks; r_1 and r_2 = distances between the ends of the plates and the observation point. Using x, h and H instead of r_1 and r_2, we obtain the following equation:

$$V(x) = \frac{I\rho}{2\pi}\ \ln \frac{x^2 + h^2}{(x-a)^2 + H^2}\ ,$$

where a = (H-h)/tg Θ (Θ = angle of inclination of the plate).

Using this formula, we calculate the SP anomalies, induced by the plate with the following parameters:
h = 2 m; H = 5 m; Θ = 60°; $I\rho/2\pi$ = 100 mV. If there are V_{max} and V_{min} extreme amplitudes of V(x), we may easily prove that the following equation is valid:

$$V_{max} + V_{min} = \frac{I\rho}{\pi}\ \ln \frac{h}{H}\ .$$

The Fourier transform of F (x) has the following form:

$$F\ (\ \omega) = \int_{-\infty}^{\infty} F\ (x)\ \exp\ (-j\omega x)\ dx.$$

Using this expression the Fourier transform of F (ω) regarding V (x) may be represented as follows:

$$F\ (\omega) = \int\!\!\int_{-\infty}^{\infty} \left(\frac{I\rho}{2\pi}\ \ln \frac{x^2 + h^2}{(x-a)^2 + H^2}\right) \exp\ (-j\omega x) dx,$$

because

$$\int_{-\infty}^{\infty} 1/2\ \ln\ (x^2 + h^2)\ \exp\ (-j\omega x) dx = -\frac{\pi}{\omega} \exp\ (-\omega h)$$

$$\int_{-\infty}^{\infty} F(x-a)\ \exp(-j\omega x) dx = \exp(-j\omega a) \int_{-\infty}^{\infty} F(x)\ \exp(-j\omega x) dx.$$

Using this equation, we obtain F (ω) = R (ω) + jx (ω); R (ω) signifies the real part:

$$F\ (\omega) = \frac{I\rho}{\omega} \left[\exp\ (-\omega H)\ \cos a\omega\ -\ \exp\ (-\omega h) \right].$$

$X(\omega)$ signifies the imaginary part of $F(\omega)$:

$$X(\omega) = -\frac{I\rho}{\omega} \exp(-\omega H) \sin a\omega.$$

The amplitude of $A(\omega)$ and the phase are (ω) given as follows:

$$A(\omega) = [R^2(\omega) + X^2(\omega)]^{1/2};$$

$$\rho(\omega) = \operatorname{arctg}[X(\omega)/R(\omega)].$$

With regard to the above mentioned expressions, we obtain:

$$A(\omega) = \frac{I\rho}{\omega}\left[\exp(-2\omega h) + \exp(-2\omega H) - 2\exp(-\omega H+h)\cos a\omega\right]^{1/2}$$

$$\rho(\omega) = \operatorname{arctg}\left[\frac{\exp(-\omega H)\sin a\omega}{\exp(-\omega h) - \exp(-\omega H)\cos a\omega}\right].$$

We have to analyze the amplitudes and the phases in order to determine the parameters of the plate. We calculate lim $A(\omega)$ regarding $\omega \to 0$.

$$\lim_{\omega\to 0} A(\omega) = \frac{I\rho\,(H-h)}{\sin\Theta} = C_1.$$

Using the modified amplitude $A_1(\omega)$, we get:

$$A_1(\omega) = I\rho\left[\exp(-2\omega h) + \exp(-2\omega H) - 2\exp(-\omega H+h)\cos a\omega\right]^{1/2};$$

whereby $\exp(-\omega H)$ decreases more quickly than $\exp(-\omega h)$, because the depth $H>h$. The influence of the terms containing $\exp(-\omega H)$ may be neglected, if the frequencies are high enough. In this case the following equation is valid:

$$A(\omega) = \frac{I\rho}{\omega}\exp(-\omega h); \qquad A_1(\omega) = I\rho\exp(-\omega h).$$

Using logarithms, the two parts of the equation have the following form:

$$\ln A_1(\omega) = \ln C_2 - \omega h_1; \quad C_2 = I.$$

That means, regarding large values of ω, the curve $\ln A_1(\omega)$ will be a straight line; its tangent of the angle of inclination is equal to the

the depth of the upper edge of the plate h. The cutting of this line with the x-axis is equal to C_2.

The equation, describing the phase, leads to the following expression:

$$\lim_{\omega \to 0} \phi(\omega) = \pi/2 - \Theta.$$

The angle of inclination of the plate follows by analyzing the phase spectrum. The depth of the lower edge results from the constant C_1, because all the other parameters are already known. Nevertheless, we can evaluate H and Θ without using the phase spectrum. The equation for $A_1(\omega)$ enables us to determine h and $I\rho$. Further, we get H, using the relation:

$$H = \frac{h}{\exp\left[\pi/I\rho\,(V_{min} + V_{max})\right]}$$

Now we calculate the angle of inclination of the plate as follows:

$$\Theta = \sin^{-1}\left[C_1 \frac{H - h}{C_2}\right].$$

Assumptions with regard to the inclination of the plate are possible, either on the basis of the measured anomaly or with the help of geological considerations. Then the following equation is valid:

$$H = h + C_1 \sin\Theta/I\rho.$$

The phase spectrum is a result of the destination of the origin of the coordinates. The spectrum is calculated, starting from the end of the profile. The phase spectrum is a straight line with regard to higher values of ω. The tangent of the angle of inclination defines the shifting of the origin of the coordinate system away from the centre of the profile. It is possible, too, to define the origin by analyzing the observed anomaly V (x), since the origin is situated not far from the abscissa of the minimum with a certain shift in the direction of the abscissa of the maximum.

The proposed algorithm was tested for the interpretation of SP anomalies by measuring the crossing of black schists with sulphurous mineralization in the government region of Pradesh in India. The discussed anomaly curve is represented by a profile with a lenght of 255 m; the distance between the single points is 1 m. Figure 15 shows the amplitude and the modified amplitude spectrum. The theoretical anomaly was calculated on the basis of the parameters, which were evaluated using the characteristics of the spectrum. The anomalies can be seen in Fig. 15. The calculated depth of the upper edge (15.9 m) agrees with the depth of 17 m, calculated by using the method of the analytical downward continuation of the field.

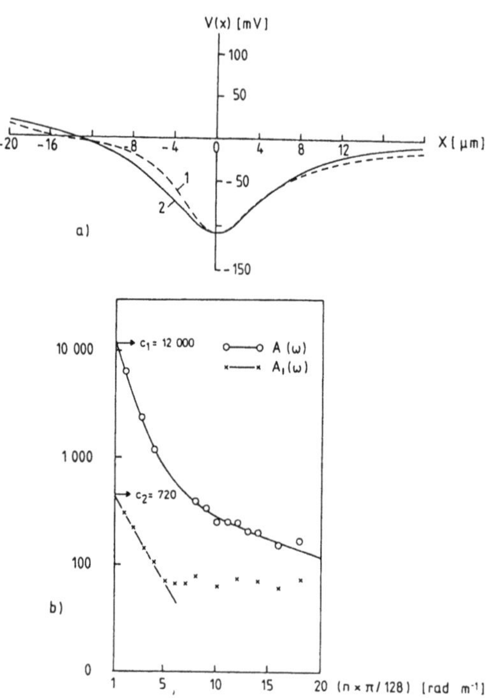

Fig. 15: SP anomaly, crossing a sulphurous body (a), spectra of the amplitudes. A (ω) and the modified amplitudes A_1 (ω) (b). 1 theoretical curve; 2 measured curve

Let us now consider the plate with an angle of inclination α and an unlimited striking in the direction of the y-axis. This plate passes through the xy-plane at the points (a cos α) resp. (-a cos α, -a sin α). Let us assume two linear poles, characterized by the current densities $\pm$I and situated in the vicinity of the upper and lower edge of the body. The potential, induced by two linear poles, may be calculated at each point of the space as follows:

$$V(x) = \frac{I\rho}{2\pi}\left[\ln\left\{(x-a\cos\alpha)^2 + (H-a\sin\alpha)^2\right\} - \ln\left\{(x+a\cos\alpha)^2+(H+a\sin\alpha)^2\right\}\right],$$

where ρ = specific resistivity of the surrounding medium; a = half of the width of the plate; H = depth to the middle point of the plate.

The Fourier transformation of the function V (x) may be written as follows:

$$V(\omega) = \int_{-\infty}^{\infty} V(x) \exp(-i\omega x)dx,$$

where ω = frequency.

The real part of R (ω) may be written as follows:

$$R(\omega) = \frac{2I\rho}{\omega} \cos(\omega a\cos\alpha) \sinh(\omega a\sin\alpha) \exp(-\dot{\omega} H);$$

If $\omega \to 0$, then:

$\lim_{\omega\to 0} R(\omega) = 2I\rho a\sin\alpha$ and for $\alpha \neq 0$, $R(\omega)$ becomes zero,

only if: $\omega_n a\cos\alpha = (2n-1)\pi/2$; $(n = 1, 2, 3, \ldots)$.

For $\omega = \omega_1$, we get $\omega_1 a\cos\alpha = \pi/2$ and $a\cos\alpha = \pi/2\omega_1$.
With regard to the limit value, it follows:

$$\alpha = tg^{-1} \frac{\lim_{\omega\to 0} R(\omega)}{I\rho} \omega_1/\pi;$$

$$a = \sqrt{\pi/4\omega^2 + \frac{(\lim_{\omega\to 0} R(\omega))^2}{4\pi^2\rho^2}}$$

The real part then becomes

$$\exp(-\omega H) = \frac{R(\omega)}{2I\rho \cos(\omega a\cos\alpha) \sinh(\omega a\sin\alpha)}.$$

Assuming R(ω) for the right side of the above equation and after derivation, we get:

$$-H = \frac{a \ln[R(\omega)]}{d\omega}.$$

10.2 Methods Based on the Calculation of Electrokinetic Effects

The above mentioned methods provide too little information about the nature of the source of SP anomalies. This is the main disadvantage. In this sense, electrokinetic considerations are more effective, because information on the physical processes without the earth can be obtained. Fitterman (1979) postulated that the development of SP anomalies in fracture zones is based on the electrokinetic effect. Mizutani (1976) and Fitterman (1979) reported that the electrokinetic phenomenon, induced by the electrical double layer on the boundary surface between the solid and the fluid phases as a consequence of water flow, is the main reason for SP anomalies, which are associated with earthquakes. Nourbehecht (1963) formulated the task concerning SP signals, which are produced by the electrokinetic mechanism in inhomogeneous media, and used the results for diagnosis of subsurface nuclear tests. Electrokinetic processes in an inhomogeneous medium must be regarded, if we intend to use the electrokinetic theory for interpretation of observed anomalies. Then, the field, induced by differences of pressure or temperature, is given; it is determined at the contacts between the spheres with different binding coefficients (for example with different binding coefficients of the filtration potential). Further, we assume that the primary source generates a constant field on the earth's surface. The distribution of temperature and pressure in the half-space is not described exactly, because it is not considered when calculating the pseudopotentials. As long as the anomaly is observed along the equipotential line of the pseudopotential, there are differences in the constant value only, with respect to these two anomalies. In these cases, we can calculate the filtration potential, the thermoelectrical effect and the effect of the electrochemical source. Fitterman (1983) developed an analytical solution pertaining to the potential on the surface, which is induced by a current parallel to a vertical plane, which is situated at any depth and separates spheres, characterized by different binding coefficients. This plane represents the fracture zone with respect to hydrothermal problems. It separates spheres, characterized by different thermoelectrical or electrokinetical binding coefficients and by the presence of the vertical component of the fluid flow.

For solving problems of the primary potentials, it is especially important to have the right conditions, which are valid for the boundary surface between the earth and air. The final result, i.e. the values of the electrical potentials, depends on these conditions. An optimal boundary condition will be a constant zero temperature with regard to problems dealing with heat transfer. On the surface, there will be the normal component of heat flow and if the rocks on the earth's surface do not have any zero-binding coefficients, flow sources will be induced inside the rocks. But the situation is quite different when any surplus of the pressure between groundwater level and surface becomes zero. Then the application of the non-zero boundary condition, concerning the pressure on the surface and problems of filtration potentials, leads to the fact that the normal pressure gradient does not become zero. This again causes a flow of the fluid through the boundary: earth surface-air.

For horizontal flowing of the fluid, the vertical pressure gradient must be zero. This is possible, assuming the air has an infinite low hydraulic permeability. The vertical zero gradient in the vicinity of the groundwater level may be approximated by a thin layer with a very low water permeability; this layer covers the water-saturated material in the vicinity of the groundwater level. Therefore, the flux, which must be modeled, becomes limited by water-impermeable layers, and thus variations in the groundwater level will not have any influence.

Figure 16 shows the voltage fields in the vertical plane x(o, h), induced by point sources of temperature and voltage in a homogeneous half-space with $C \neq 0$. With regard to problems in connection with pressure, the boundary condition means, that the temperature on the surface must also be zero. By comparing Fig. 16a and b, it can be seen, that merely a source of pressure on the surface generates an electrical anomaly. The surface current is parallel to the boundary earth-air. In this case, the sole induced electrical source becomes identical to the source of pressure at $\Delta P \neq 0$. With regard to heat problems, the induced electrical sources are situated inside the sources of temperature, if $\Delta T \neq 0$; of course they are also situated at that point on the surface, where the normal heat flow occurs. On the earth's surface, the electrical sources, induced on the boundary, compensate the influence of deeply situated sources. These two variants can be solved analytically. In these cases the total current becomes zero, because the electrical current levels the convection flow. Nourbehecht (1963) and Fitterman (1983) do not believe there is ananomaly on the surface, which is generated by a point source of pres-

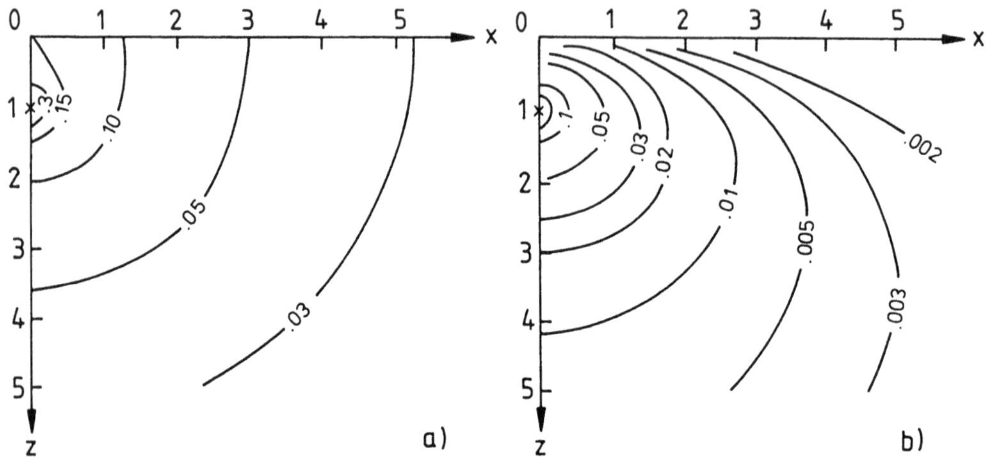

Fig. 16: a) Normalized voltage in the vertical plane (y=0) for a point pressure source in a homogeneous half-space. The boundary condition at the surface is 0-normal pressure gradient. The source of unit strength is at x=0, z=1 and distances are in units as a (Sill, 1983).
b) Normalized voltage in the vertical plane (y=0) for a point temperature source in a homogeneous half-space. The boundary condition at the surface is 0-temperature. The source of unit strength is at x=0, z=1 and distances are in units of a (Sill, 1983).

sure in a homogeneous half-space. The reason is the chosen boundary condition with respect to the zero pressure on the surface. In this case, the wish for a generalized potential, equal to the electrical potential, leads to non-acceptable boundary conditions. The more likely boundary conditions consist of a normal gradient equal to zero. So, the development of anomalies on the surface may occur, as already mentioned above. The models, used by Fitterman (1979, 1983, 1984) are based on boundary conditions, assuming that the primary potential on the earth's surface is zero. Thus, a vertical non-zero component of the heat or fluid flow is induced in the air. This may be valid for sources of temperature, but not for sources generated by fluid flow. The conception of Nourbehecht (1963), with regards to pseudopotentials and an equipotential field of the primary source on the surface, is valid for heat sources only, but does not help to solve problems connected with filtration potentials. Nourbehecht (1963) conceptualizes the generalized electrical potential as a linear combination of the usual electrical potential and other potentials (temperature, pressure, etc.). It is not favourable to measure the usual electrical potential, since SP anomalies are caused by binding potentials (primary potentials). The method for solving SP problems, proposed by Sill (1983), clearly separates the potential, and regards the primary potentials as the sources of the usual electrical potentials. This method is based on induced sources of flows, i.e. induced by the divergence of the convection streamings, generated by primary heat or fluid flows. Thus, the accent is transfered to the convection flows and its alternating effects on inhomogeneities, inducing the existence of natural potentials. This is another opinion with regards to the methods of Nourbehecht (1963) and Fitterman (1984), who have considered the potential of the primary flow and the binding coefficients between flow and voltage. So, the problem acquires a plausible physical content to explain the generation of SP effects.

Nourbehecht (1963) formulates a common solution of the problem, concerning streamings with alternating effects, as follows:

$$\Gamma_i = \sum_i L_{ij} X_j ,$$

where Γ_i = flows (charges, particles, heat); X_j = different forces (gradients of the electrical potential, pressure, temperature, etc.); L_{ij} = binding coefficients.

We now consider the streamings and potentials of the secondary electrical field, induced by primary heat and fluid flows. We may divide the relation of the primary flux into two equations, if there is only a minimal influence of the secondary electrical potentials upon the primary flow:

$$\Gamma_1 = -L_{11} \nabla\zeta , \quad j = \Gamma_2 = -L_{21} \nabla\zeta - \sigma\nabla\phi ;$$

where Γ_1 = primary flow; L_{11} = primary conductivity (permeability, heat conductivity, etc.); ζ = primary potential (pressure, temperature, etc.); j = total electrical current; L_{21} = the binding electrical conductivity; σ = electrical conductivity; ϕ = electrical potential.

The equation for the primary flow may be solved separately and may be used to solve the problem of the electrical current. The method of Nourbehecht (1963) solves the second equation with the help of the generalized (pseudo-) potential ψ, representing the sum of the electrical potential and the corresponding parts of the supposed potentials with regards to the primary sources:

$$\psi = \phi + L_{21} \zeta / \sigma$$

The method of Sill (1983) equalizes the summands of the equation and the convection flow, induced by the primary field (first summand), as well as that flow, which was influenced by the gradient of the electrical potential (second summand). To use the method of Sill (1983), we write the equation as follows:

$$j = j_K + j_l \quad \text{with } j_K = -L_{21} \nabla\zeta \quad \text{and} \quad j_l = -\sigma \nabla\phi .$$

If the external sources of the flow are absent and the conditions for a direct current $\partial\rho / \partial t = 0$ are valid, ∇j becomes zero and

$$\nabla j_l = -\nabla j_K = \nabla (L_{21} \nabla\zeta) = \nabla L_{21} \nabla\zeta + L_{21} \nabla^2 \zeta .$$

That means, sources of conducting currents are everywhere. Neither gradients of the cross-coupling coefficients (parallel to the primary flow; flow perpendicular to boundaries) nor external resp. induced sources of the primary flow are observed. The sources of conducting currents may be used to determine the resultant electrical potential. With regards to the summand $\nabla L_{21} \nabla\zeta$, it becomes evident, that gradients, oriented in the same direction, generate positive sources of the conducting currents; gradients in opposite directions, however, induce negative sources. We obtain two additional sources by substituting into the first of the two equations the expression $L_{21} \nabla^2 \zeta$. The first of the two sources $L_{21} \nabla \Gamma_1 / L_{11}^2$ does not become zero at the external sources of the primary field; the second one is equal to $L_{21} L_{11} \Gamma_1 / L^2{}_{11}$. The electrical source becomes negative for the case $L_{21} > 0$, if $\nabla \Gamma_1 > 0$ (source of the primary flow is positive). The second term will not be zero at boundaries, where the conductivity of the primary flow changes. These boundaries also localize the secondary or induced sources.

The sources of the generalized potentials are generated at boundaries, where the primary potentials are not zero and the coupling coefficients of the voltage change. The sources of the primary potentials in the convection current problem come from regions where there is a divergence of the convection current $L_{21}\zeta$. In this way the emphasis is transferred from the amplitude of the primary potential and the coupling coefficients of the voltage to gradients of the primary potential and the coupling coefficients of the current. There are some reasons, which demonstrate that the formulation of the problem would be better using convection currents.

The first is that the solution gives directly the real, unmeasurable electrical potential and not a combined potential as in the total potential approach. The second is that the source terms depend upon gradients of the primary potential, and these gradients are more directly connected to the physical generation of cross-coupling effects. With regards to flowing due to pressure, the velocity is given by the product of the permeability and the negative pressure gradient. The velocity of the pore fluid is related to a transfer of the surplus charge in the diffusion layer and this carrier flux denotes the convective current. Indeed, it is possible to formulate the problem of fluid flow (electrokinetic effects) in a new manner, using the terms of the velocity field, replacing $-\nabla\zeta$ by the vector of the velocity Γ_1 and the vector L_{21} by the cross-coupling coefficient of the velocity $L_{21}^{-1}L_{11}^{-1}$ (L_{11}^{1} = water permeability).

This replacement is justified, if the fluid flow is not related to the pressure decrease, but induced by heat convection. However, it is not justified to use the generalized potential, as in the case of more complicated problems of fluid flow. Then, we have to approximate the geometry of the flow, using appropriate boundary conditions. The method of Sill does not provide any advantages over the method of Nourbehecht, for solving single problems, using well-known analytical methods. Regarding more complicated problems the numerical solution developed by Sill (1983) offers, in comparison with the method of Nourbehecht (1963), no additional difficulties. For these problems we find a solution on the basis of the existing algorithm, with respect to the primary potential (fluid flow, heat, etc.). Using the solution for the primary potential and the model of the cross-coupling coefficients, the induced sources may be calculated. Afterwards the electrical potential, corresponding to the given electrical model, can be calculated.

Madden (1971) proposed an algorithm for calculating the direct current potential. It is appropriate for numerical modeling of the problem, using concevtion flows.

According to the algorithm of Madden, we give a brief review starting with the general potential equation $\Gamma = -L\nabla\zeta$ and $\nabla\Gamma = S$ (Γ = flow;

ζ = potential; S = source, L = electrical conductivity). If L does not depend on y (strike), we obtain for the Fourier transformation in the direction of y the following expressions:

$$-L(x\ ,z)\ \frac{\partial\zeta(x,\ \lambda,\ z)}{\partial\ x} = \Gamma_x(x,\ \lambda,\ z);$$

$$-L(x,\ z)\ \frac{\partial\zeta(x,\ \lambda,\ z)}{\partial\ z} = \Gamma_z(x,\ \lambda,\ z);$$

$$\frac{\partial\Gamma_x}{\partial x} + \frac{\partial\Gamma_x}{\Gamma z} + \lambda^2 L\zeta = S(x,\ \lambda\ ,\ z).$$

The system of these equations is approximated by a rectangular grid. Each point of the grid is described by following difference equation:

$$y_x(i,j-1)\ [\zeta(i,j-1)-\zeta(i,j)] + y_x(i,j)[\zeta(i,j+1)-\ (i,j)] + y_z(i-1),j)\cdot$$
$$\cdot[\zeta(i-1,j)-\zeta(i,j)] + y_z(i,j)\ [\zeta(i+1,j)-\zeta(i,j)] + y(i,j)\zeta(i,j) =$$
$$= S(i,j)\Delta x\ \Delta\ z.$$

$y_x = L\Delta\ z/\ \Delta\ x;\ y_z = L\Delta\ x/\ \Delta\ z;\ Y = \lambda^2\Delta x\ \Delta\ z;\ i = 1\ ...\ n;\quad j = 1\ ...m$

The system of equations consists of (n x m) grid points and is written in the form of a matrix as follows: $C\,\zeta = S$; C = matrix of the coefficients; ζ = vector of the potential at the grid points (n x m); S = vector of the grid point sources. The final solution regarding the space x, y, z results by inverse Fourier transformation of a suite of solution $\zeta\ (\lambda)$. The solution of the problem demands three models of the physical properties: the distribution of the resistivity in the primary flow L_{11}^{-1}, the cross-coupling coefficients of the voltage $C_{21} = \rho L_{21}$ and the specific electrical resistivity ρ . We can calculate the primary flow potentials ζ for the primary model with regard to each λ, if the distribution of the sources is known. After that, we calculate the electrical sources for each grid point with regards to the cross-coupling coefficients. Now we can evaluate easily the electrical potential, because we know the electrical sources and the distribution of the specific electrical resistivity. As a result of numerical transformations, we get the primary potentials ζ_n and the electrical potentials, both dimensionless:

Primary potential: $\zeta_n = \frac{L_{11}a}{I\ \zeta}$

Electrical potential: $V_n = \frac{\phi}{C_{21}}\ \frac{L_{11}a}{I\ \zeta} = \frac{\phi}{C_{21}}\ \frac{\zeta_n}{\zeta}$;

where a = factor of the scale in units of a length; $I\zeta$ = source of the primary flow (in units of $\Gamma_1 xS$). We obtain the real potentials ζ and ϕ by multiplication by the appropriate factor. We establish, as a result of numerical modeling with the help of the coefficients C_{21}, that the induced sources of the flow are reversed proportionally ($C_{21} = C_{21}/\rho$). Since the voltage is proportional to the current-resistivity product, the resultant model voltages depend only on resistivity ratios. That is, the same potentials will result for all models that differ only by a multiplication factor in all the model resistivities. All the models, given by Sill (1983), have dimensionless scales; the model parameters are given as specific resistivities and coupling coefficients of the voltage. Figure 17 shows the influence of the site of the pressure source, relative to the vertical boundary,

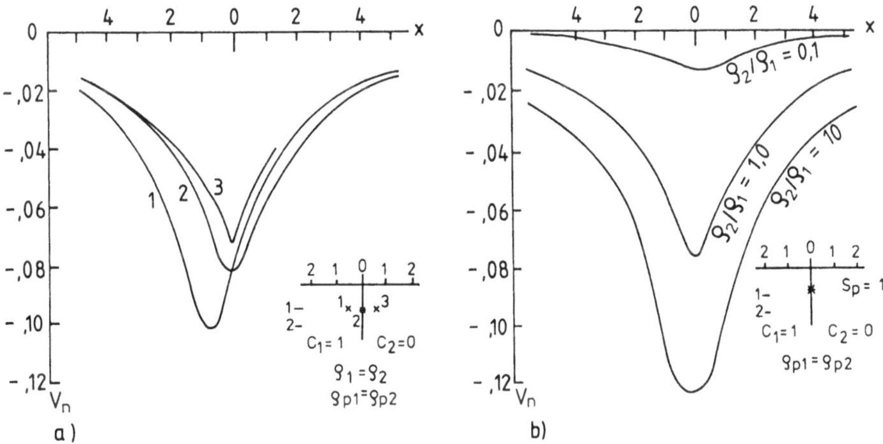

Fig. 17: a) Surface voltage (y=0, z=0) for a point pressure source and a vertical contact, showing the effects of variations of the source location with respect of the contact (Sill, 1983).
b) Surface voltage (y=0, z=0) for a point pressure source and a vertical contact, showing the effects of variations in the resistivity ratio across the contact (Sill, 1983).

which separates the media with different coupling parameters. Regarding curve 1 (source on the left side of the boundary), we note positive electrical sources, generated on the vertical boundary, wherever movement in the medium $C_1 = 1$ occurs. The sources diminish the amplitude of the potential in comparison with the homogeneous half-space (Fig. 16). In this case, the anomaly is symmetrical relative to the site of the pressure source. The anomaly becomes twice so low, as in the case of the homogeneous half-space, if the pressure source is directly situated at the boundary (curve 2), because there is a divergence of the flow from the right side of the boundary to the sphere with C = 0. Curve 3 (pressure source on the right side of the vertical boundary) describes a high anomaly, directly crossing the boundary; it must be reduced to a negative electrical source, induced at the vertical boundary, wherever a flow exists in the medium with $C_1 = 1$. Variations of the relation of the resistivities inside the two contacting media effect an increase or a decrease of the amplitude of the anomaly, induced by pressure, but do not change its shape (Fig. 17b). There are two possibilities for increasing the amplitude, assuming $\rho_2/\rho_1 = 10$. First, if $\rho_1 = 1$, $\rho_2 = 10$ and the current source, induced at the site of the pressure source, are the same as in the homogeneous medium: $\rho_2 = \rho_1 = 1$. The current flow from the right hand side ensues in the medium with a high resistivity and increasing amplitude. Secondly, if $\rho_1 = 0,1$ and $\rho_2 = 1,0$. The induced current source will be ten times more powerful in comparison with the homogeneous case ($\rho_1 = 1$), because $C = L/\rho$. The increase of the potential amplitude as well as the resistivity will be the same in this case. There is an analogous explanation, regarding the diminution of the amplitude, if $\rho_2/\rho_1 = 0,1$. The existence of overburden effects the asymmetrical dipolar shape of the anomaly, even if there is homogeneous resistivity (Fig. 18). The diminution of the positive maximum is effected by an induced source, at the horizontal contact being at a greater depth. Variations of the resistivity of the overburden effected changes in the shape of the amplitude (Fig. 18a). We can change the shape and the amplitude of the anomaly by alteration of the specific electrical resistivity, with respect to the contacting spheres, and by assuming a constant resistivity of the overburden. The anomaly will be monopolar and asymmetric, assuming the resistivity of the sphere with $C_1 = 1$ is lower than the resistivity of the sphere with $C_2 = 0$. A variation in the primary flow resistivities also influences the shape of the anomaly of heat conductivity. A decrease in heat conductivity is related to a decrease in the temperature and the gradient, regarding a homogeneous medium. This produces an increase of the quantity of the induced electrical sources, in the temperature point sources as well as in the right part of the denoted sediments. The resulting anomaly will be higher in comparison with the cáse of the homogeneous medium (Fig. 18a). Temperature and gradient will be lower than in a homogeneous medium, if one of the spheres is characterized by a higher electrical conductivity. The consequence would be a diminution of the quantity of the induced electrical sources and of the potential. Though the anomaly of adjacent spheres is unipolar, it becomes dipolar for sedimentary layers

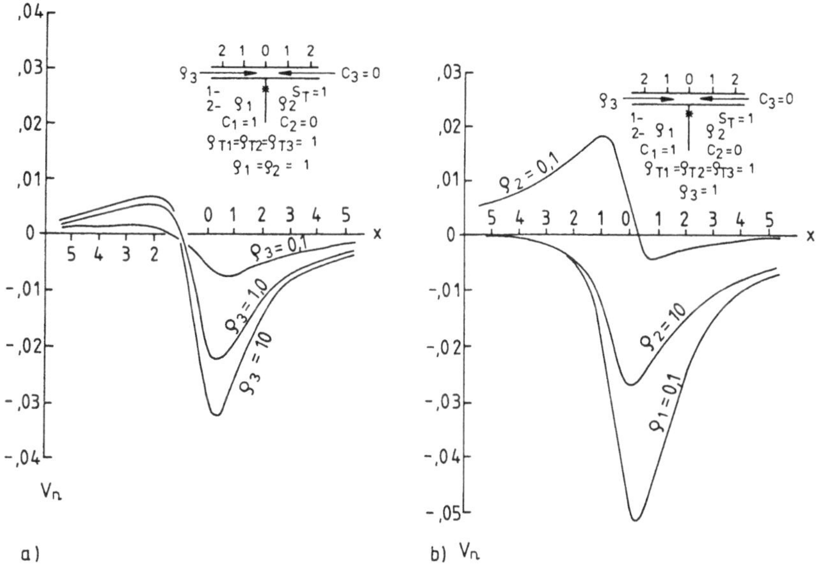

Fig. 18: a) Surface voltage (y=0, z=0) for a point temperature source and a vertical contact - overburden model, showing the effects of variations in overburden resistivity (Sill, 1983).
b) Surface voltage (y=0, z=0) for a point temperature source and a vertical contact - overburden model, showing the effects of variations in the quarter-space resistivities (Sill, 1983).

with better permeability, with regards to problems of a point source of pressure. The reason is avertical flow through the horizontal boundaries. Such a distribution of the fluxes corresponds to analogous distributions with regards to "heat problems" and to the coupled production of intensive electrical sources at the lower boundary of the overburden interface. Nevertheless, variations of the electrical resistivities of adjacent spheres change the dipolar shape of the anomalies. Let us now consider the alternating effect of the point source with two vertical boundaries, representing a dyke. The anomaly on the surface will be symmetrical, if the point source is situated at the centre of the vertical sheet (Fig. 19a; model 1). The reason is symmetry of the flows and the induced sources, with regards to the centre of the dykes. Referring to heat sources, the anomaly is positive, crossing the vertically oriented dyke, due to the positively induced sources on the boundary between the earth and the air; but there are negative "wings" at the sides, too. By shifting the temperature point source to the left side, the negative effect of the heat flow anomaly increases. The anomaly acquires a dipolar shape by gra-

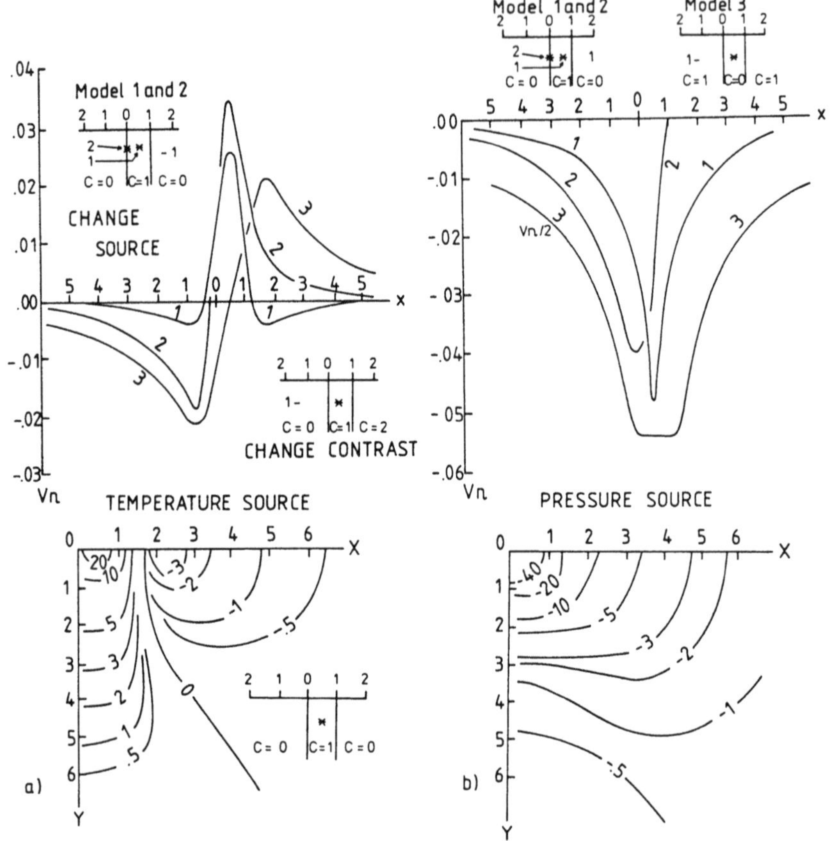

Fig. 19: a: TOP Surface voltage (y=0, z=0) for a point temperature source and dike, showing the effects of variations in the source locations and coupling parameters (Sill, 1983).
b) TOP Surface voltage (y=0, z=0) for a point pressure source and a dike, showing the effects of variations in the source locations and coupling parameters (Sill, 1983).
a) DOWN Contours of surface voltage (x, y-plane, z=0) for point sources and a dike. Point temperature source (model 1, TOP) (Sill, 1983).
b) DOWN Contours of surface voltage (x, y-plane, z=0) for point sources and a dike. Point pressure source (model 1, TOP).

dually in-creasing the parameter C (Fig. 19a; model 3). This conclusion corresponds completely with the conception of Fitterman (1984), that a dipolar anomaly is induced by a continuous increase of the thermoelectrical coefficients. We observe an anomaly with a reversed sign, changing the parameter C of the model 1, as follows: C = 0, inside the dyke; C = 1, outside the dyke. The sign of the anomaly does not change by shifting the pressure source, but shape and amplitude (Fig. 19b; model 3). Corresponding to the sheet flows, in this case, negatively induced sources are generated on the place $x = 0$, only. We get an asymmetric negative anomaly by shifting the pressure source to the left side of the dyke (Fig. 19b; model 2). The extended minimum on the left side ensures from the divergence of the source. By shifting the measuring points to the right edge of the dyke, the positively induced sources on the plane $x = 0$ level the influence of the negative electrical sources, situated at the point sources of the divergence of the flow. Finally, let us assume that the combination of the positive pressure source at the points $x = 0$ and the negative source at $x = 0$ produces a dipolar anomaly on both sides of the dyke. Commonly, each relative homogeneous fluid flow, crossing the vertical dyke, generates a dipolar anomaly.

On the maps of isolines, with respect to the potential distribution on the surface, which is influenced by a temperature source, the positive part of the dipolar anomaly is extended in the strike direction (y-axis). We observe a quick diminution of the unipolar anomaly in the strike direction, with regards to the pressure source (Fig. 19b). Perpendicular to this direction, the above mentioned diminution ensues more slowly. The extension of the positive heat anomaly in the strike direction can be explained easily from the standpoint of localization of the positively induced sources on the boundary between the earth's surface and air along the dyke. In the case of a pressure source, the anomaly decreases quickly in the strike direction, as a consequence of the disappearance of the negative point sources with the help of positive sources on two vertical boundaries. Kilty (1984) used the method for the interpretation of SP anomalies, crossing sulphurous orebodies, proposed by Nourbehecht (1963). He demonstrated that the mechanism of the SP must be observed as a result of the laws of thermodynamics, concerning the processes, which are in this case also characterized by non-balance. Regarding the phenomena between the orebody and the surroundings as a process of balance between chemical reactions, we should not expect any correlation between the composition of the orebody and the intensity of the measured SP anomaly. In practice, however, the amplitudes of the anomalies, crossing graphite, pyrrhotites and pyrites are higher than the anomalies, crossing galenites and other sulphides. Consequently, SP anomalies are not determined by an equalized potential on the surface of the orebody only, but by an additional potential, induced by streamings through the boundary between orebody and electrolyte. The assessment of this potential ensues analogous with the assessment of the magnetic scalar potential of a spontaneously magnetized body: the main difference is that the orientation of the SP field is normal to the surface of the orebody, under all circumstances. Sill (1983) refers also to the similarity between the SP-current density and the magnetic induction. The SP potential and the scalar potential satisfy the same differential equations by solving SP problems; therefore, programs of the magnetic modeling may be used for 2D- and 3D-problems and also for SP interpretation. It is only necessary to transform the corresponding boundary conditions and to calculate the single vector components of the magnetic field, because these components are used for the calculation of the scalar potential by integration.

11 Conclusions and Recommendations

One of the most important tasks at the present stage of the development of the society and the future, too, is related to an increase of the raw-material base of our country. The exhaustion of deposits, easily detectable, and the transition of prospecting for non-convectional types of deposits emphatically demand the development of methods and the modification of geophysical methods already known. Methods for investigating natural physical fields play an important role.

Natural galvanic cells, associated with the development of SP anomalies, may be induced by functional systems of the metasomatosis, including conditions for the development of different electrical processes. Electrochemical systems may be more or less ore-producing processes. The morphology of the electrical potential field, corresponding to a natural electrochemical cell (electrolyte), is established by the SP method. This enables us to develop more positive prognoses with regard to the establishment of aureoles of sulphurous, graphitic and radioactive metasomites.
The electrokinetic model, concerning the generation of SP anomalies, may be used for investigating the hydrodynamic mechanism of advancing ore-containing solutions and for detecting paleogeothermal gradients, associated with the generation of ore metasomatites. The movement of geothermal fluids, inside pores and joints, induces variations of the rocks, containing the fluids. Consequently, the thermoelectrical

coupling coefficients are changed. The importance of the coupling coefficients, concerning the zones of circulation of hydrothermal fluids, depends on the intensity of the metasomatic processes, on horizontal variations of temperature, on the pressure and on the mineralization of the pore fluids. Metasomatic processes are either an indicator or a reason, with regards to variations of the redox state, as diagnosed by the anomalies of the natural potential. Therefore, the SP method should be used for the direct or indirect establishment of ore structures, situated in zones of secondary hydrothermal changes of satisfied and magmatic complexes. The SP anomalies point out geochemical barriers of reduction, in this case, this means, sudden variations of the physicochemical conditions of the surroundings.

References

Apparo, A.; Rao G. (1983): The method of downward continuation in interpretation of induced polarisation data obtained with linear electrodes.- Geophys. Res. Bull., vol. 21, 3: 271-281.

Bacon, L.O.; Elliot, Ch.L. (1981): Redox chemical remanent magnetisation - a new dimension in exploration for sulphide deposits in volcanic covered areas.- Geophys., vol. 46, 8: 1169-1181.

Bhattacharya, B.B.; Biswas, D.; Kar, G.; Gosh, H. (1984): Geoelectric exploration for graphite in the Balangir district, Orissa, India.- Geoexplor., vol. 22: 129-143.

Bhattacharya, B.B.; Mahajan, D. (1984): Interpretation of mining geophysical data by downward continuation technique. Gerlands Beitr.- Geophys., 1: 12-22.

Bhattacharya, B.B.; Roy, N. (1981): A note on the use of a nomogram for self-potential anomalies.- Geophys. prosp., vol. 29: 102-107.

Blain, C.F.; Andrew, R.L. (1977): Sulphide weathering and the evaluation of gossans in mineral exploration.- Miner. soil. eng., vol. 3: 119-150.

Bogoslovsky, V.A.; Ogilvy, A.A. (1972): The study of streaming potentials on fissured media models.- Geophys. prosp., vol. 20: 109-117.

Bogoslovsky, V.A.; Ogilvy, A.A. (1973): Deformations of natural electric fields near drainage structures.- Geophys. prosp., vol. 21: 716-723.

Bolviken, B. (1981): The redox potential field of the Earth.- Origin and distribution of the elements, N 211: 649-664.

Corvin, R.F.; De Moulli, G.T.; Harding, R.S.; Morrison, H.F. (1981): Interpretation of self-potential survey results from the East Mesa geothermal field, California.- J. of geophys. res., vol. 86, B3: 1841-1848.

Corvin, R.F.; Hoover, D.B. (1979): The self-potential method in geothermal exploration.- Geophys., vol. 44: 226-245.

Cull, J.P. (1985): Self-potential and current channelling.- Geophys. prosp., vol. 33: 460-467.

Fitterman, D.V. (1979): Calculations of self-potential anomalies near vertical contacts.- Geophys., vol. 44, 2: 195-205.

Fitterman, D.V. (1983): Modeling of self-potential anomalies near vertical dikes.- Geophys., vol. 48: 171-180.

Fitterman, D.V. (1984): Thermoelectrical self-potential anomalies and their relationship to the solid angle subtended by the source region.- Geophys., vol. 49, 2: 165-170.

Fitterman, D.V.; Corvin, R.F. (1982): Inversion of self-potential data from Cerro Prieto geothermal field, Mexico.- Geophys., vol. 47, 6: 938-945.

Govett, G.J.S. (1976): Detection of deeply buried and blind sulphide deposits by measurement of H^+ and conductivity of closely spaced surface soil samples.- J. of geochem. explor., vol. 6: 359-382.

Govett, G.J.S.; Dunlop, A.C.; Atherden, P.R. (1984): Electrochemical techniques in deeply weathered terrain in Australia.- J. of geochem. explor., vol. 21: 311-331.

Ishido, T.; Mizutani, H. (1981): Experimental and theoretical basis of electrokinetic phenomena in rock-water systems and its applications to geophysics.- J. of geophys. res., vol. 86, B3: 1763-1775.

Ishido, T.; Mizutani, H.; Baba, K. (1983): Streaming potential observations using geothermal wells and in situ electrokinetic coupling coefficients under high temperature.- Tectonophys., vol. 91: 89-104.

Kilty, K.T. (1984): On the origin and interpretation of self-potential anomalies.- Geophys. prosp., 1984, vol. 32: 51-62.

Madden, T.R. (1971): The resolving power of geoelectric measurements for delineating resistivity zones within the crust.- In: Heacock, T.G. (Ed.): The structure and physical properties of the earth's crust.- AGU monograph, 14: p.95.

Mizutani, H.; Ishido, T.; Yokokura, T.; Ohnishi, S. (1976): Geophys. Res. Letters 3, 7, 365-368.

Murakami, H.; Mizutani, H.; Nabetani, S. (1984): Self-potential anomalies associated with an active fault.- J. Geomag. geoelectr., vol. 36: 351-376.

Murty, B.V.S.; Haricharan, P. (1985): Nomogram for the complete interpretation spontaneous potential profiles over sheet-like and cylindrical two-dimensional sources.- Geophys., vol. 50, 7:1127-1135.

Nayak, P.N. (1981): Electromechanical potential in surveys for sulphides.- Geoexplor., vol. 18: 311-320.

Nayak, P.N.; Saha, S.; Dutta, M.S.V.; Rao, R.; Sarker, N.C. (1983): Geoelectrical and geohydrological precursors of earthquakes in northeastern India.- Geoexplor., vol. 21: 137-157.

Nourbehecht, B. (1963: Irreversible thermodynamic effects in inhomogeneous media and their application in certain geoelectric problems.- Ph. D. thesis, M.I.T., Cambridge.

Olhoeft, C.R. (1981): Electrical properties of granite with implications for the lower crust.- J. of geophys. res., Vol. 86, B2: 931-936.

Petrovskij, A.D. (1928): The problem of a hidden polarized sphere.- Phil. Mag. and f. of Sci. 5, 27: 914-929 and 334-367, 7. Ser.

Rao, D.A.; Ram Babu, H.V.; Sivakumar Sinha, G.D.J. (1982): A Fourier method for the interpretation of self-potential anomalies due to two-dimensional inclined sheets of finite depth extent.- Pageoth., vol. 120: 365-374.

Rao, S.V.S.; Mohan, N.L. (1984): Spectral interpretation of self-potential anomaly due to an inclined sheet.- Current science, vol. 53, 9: 474-477.

Roy, A.; Chowdhury, D.K. (1959): Interpretation of self potential data for tabular bodies.- J. Sci. Eng. Res. 3, 1: 35-54.

Sato, M.; Mooney, H.M. (1960): The electrochemical mechanism of sulphide self potentials.- Geophysics, Tulsa 25, 1: 226-249.

Schiavone, D.; Quatro, R. (1984): Self-potential prospecting in the study of water movements.- Geoexplor., vol. 22: 47-58.

Sill, R. (1983): Self-potential modeling from primary flows.- Geophys., vol. 48: 76-86.

Sivenas, P.; Beales, F.W. (1982): Natural geobatteries associated with sulphide ore deposits. I. Theoretical studies.- J. of geochem. explor., vol. 17: 123-143.

Stern, W. (1945): Relation between spontaneous polarisation curves for depth, size and dip of ore bodies.- A.I.M.E. Geophysics: 189-196.

Streshnikov, G.B. & Kedrinsiy, I.A. (1965): Electrochemical solution of sulphide ores.- Int. Geol. Rev., 7(2):225-232.

Thornber, M.R. (1975a): Supergene alteration of sulphides, I. A chemical model based on massive nickel sulphide deposits at Kambalda. Western Australia.- Chem. Geol., 15: 1-14.

Thornber, M.R. (1975b): Supergene alteration of sulphides, II. A chemical study of the Kambalda nickel deposits.- Chem. Geol., 15: 117-144.

Wynn, J.C.; Sherwood, S.I. (1984): The self-potential (SP) method: an inexpensive reconnaissance and archaeological mapping tool.- J. of field archaeol., vol. II: 195-204.

Zablocki, C.I. (1976): Mapping thermal anomalies on an active volcano by the self-potential method, Kilauea, Havaii.- Proc. 2-nd UN symposium on the development and use of geothermal resources. San Francisco, vol. 2: 1299-1309.

AN ATTEMPT TO DETERMINE THE EARTH EMBANKMENT CONDITIONS BY RESISTIVITY INVESTIGATION

D. Arandjelovic[1]

1 Introduction

The attempt to use geophysical methods in investigating the conditions of the earth's embankments is based on reasonable assumption that the values and variations of individual geophysical parameters of the materials in the embankment are controlled by the variations and conditions of hydrogeological and geomechanical properties. In this context, the study of the resistivity is one of the most interesting of all physical characteristics of the embankment.

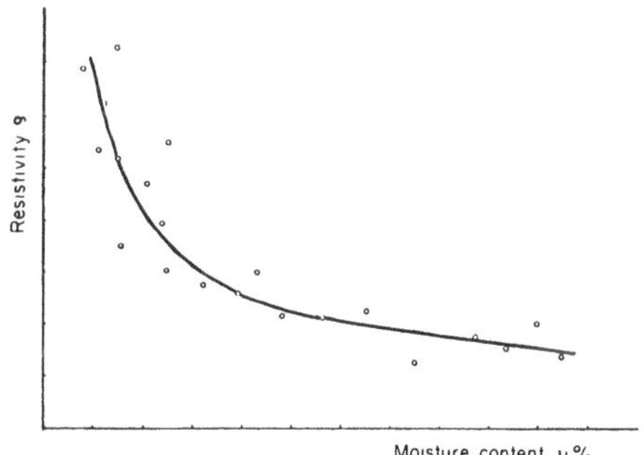

Fig. 1: Relationship between resistivity of the clay/sand embankment material and moisture content.

For the geophysical survey of an embankment, the geophysicist has the task of investigating a body which has a regular, constant and defining shape. Also, much is known about its lithology and hydrogeological situation. Such conditions are very favourable for the application of geophysical methods and correct interpretation of the data. Naturally, the embankment which is an artificial body, includes heterogeneous or unstable lithological and geomechanical zones of a larger or smaller extent and intensity. These zones may reasonably be expected to become the sources of geophysical anomalies, whose amplitudes will

[1] Geophysical Institute, Beograd, Karadjordjeva 48, Yugoslavia

Lecture Notes in Earth Sciences, Vol. 27
G.-P. Merkler et al. (Eds.)
Detection of Subsurface Flow Phenomena

be a function of the hydrologic regime both in the embankment and surrounding area.

Geophysical parameters of the embankment, particulary moisture content, are essentially controlled by the variation in the groundwater table, river water level and pluvial regime. The effect of moisture will be amplified in less compact, or geomechanically waeker zones. Compared to other geophysical parameters the resistivity is one of the most sensitive to the moisture content. Figure 1 shows the relationship between resistivity of the clayey-sandy embankment material to the moisture content obtained by laboratory measurements. Extremely high resistivity values are obtained for very dryed cores and, lowest resistivity values are gained for wet, or water-saturated cores.

It may be included from such analyses that systematic measurement and the study of resistivity (resistivity monitoring) throughout an embankment would have a high probability of locating zones with the highest moisture content. Naturally, these changes will be associated with changes in porosity or geomechanical properties (relative compactness) of the embankment.

2 Methodology of Geoelectrical Prospecting

The methodology of geoelectrical prospecting is quite specific. Bearing in mind that the investigating depth should be constant, and the lateral changes along the embankment are dominant, the method of electrical profiling was used, accepting the following methodological principles (Fig. 2):

1. Investigation should be detailed, at intervals of 5 m because lithological composition and geomechanical conditions may change over very short distances.

2. Investigation should be carried out at two depths of exploration, one up to 3 m and the other up to 5 m deep. Two depths of exploration are the minimum requirement for both qualitative and quantitative results of electrical profiling. The deeper investi-

gation should indicate changes under the foot of the embankment where the most intensive changes may be expected.

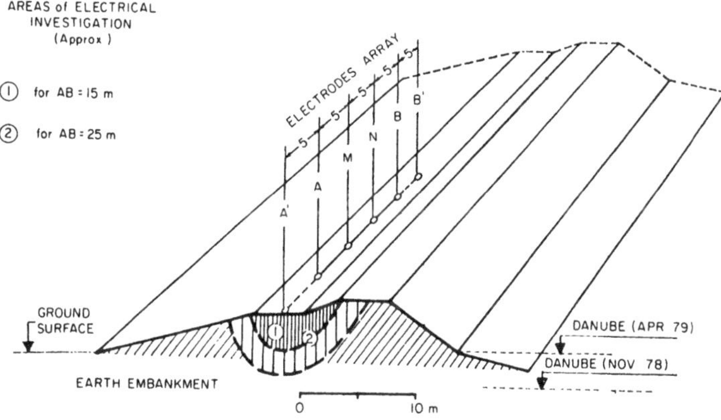

Fig. 2: Cross-section of an earth embankment and scheme of electrical exploration.

3. Investigation should be carried out during two different hydrological periods: a period of low and period of high water level, or low and high groundwater level. This is a most interesting and important methodological principle. The difference between extreme water levels in the explored area, or groundwater levels, may be several meters and may directly influence the hydrological state in the embankment. The state will also be partly a function of the lithology and geomechanical conditions. Systematic processing of the electrical profiling data while taking these factors into account may give very useful information.

4. For maximum comparability of data, the conditions of periodic electrical measurements should be as similar as possible. Measurements must be taken with similar accuracy using the same equipment and methodology.

3 Results of Electrical Investigation

Two typical stretches of embankment along the Danube ans Tisa rivers were investigated over a length of 5000 m. On the average the embankments were 5 m high and constructed a 100 years ago.

Surveying was carried out during two periods: in November 1978 when both river and groundwater level were low, and in April 1979 during the period of high river and groundwater levels. The differences in river and groundwater levels were about 4 m and 2 m, respectively. These levels oscilations were normally expected to be reflected in the values and variation of the resistivity of the materials in the embankment, and consequently to be a function of the lithological-geomechanical state of the embankment. A symmetrical arrangement of electrodes with double-depth exploration (A'B'= 25 m, AB = 15 m, MN = 5 m) was used in electrical profiling. Measurements were made at 5-m intervals.

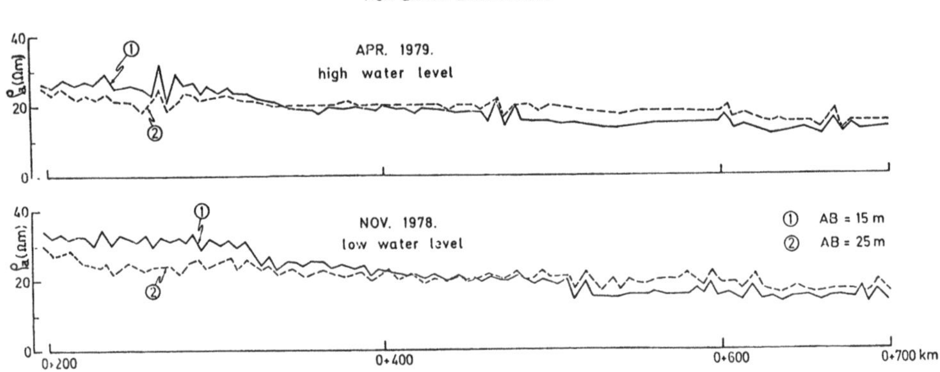

Fig. 3: Tisa embankment: resistivity profiles obtained in 1978 and 1979.

4 Embankment Along the Tisa River

Electrical profiling curves for both depths of investigations obtained in 1978 during low river water levels are similar and generally uniform (Fig. 3). Statistical analysis of the 442 apparent resistivity values obtained in 1978 for AB = 15 m gave values between 16 and 24 ohm-m in 80% of cases. For AB = 25 m, the values were between 16 and 28 ohm-m in 95% of cases. This extreme electrical homogeneity is taken to indicate lithological and geomechanical homogeneity. Survey data from 1979 (period of high water level) were similar. Somewhat lower apparent resistivity values for both depths of exploration were a normal consequence of an increased moisture content in the embankment.

Survey data from 1978 and 1979 are summarized in curves of apparent resistivity ratios with the corresponding histograms (Fig. 4). Analyses of the ratio of apparent resistivity is based on the supposition that if no change occurred in the physical properties of the embankment between periods of low and high water level, the electrical resistivity profiles would be virtually identical. But the results indicated certain anomalous changes which should be classified by significance for selective analysis.

Before separating and defining the anomalous phenomena, the normal background variation in resistivity ratio should be taken into consideration in addition to the measuring error. For AB = 15 m, in 74% of 439 cases analyzed the ρ_{78}/ρ_{79} ratio was between 1.00 and 1.25. The normal background variation and the measuring error may be considered within this range. Consequently, it is assumed that only apparent resistivity values outside this range should be considered anomalous. This amplitude of an anomaly is proportional to the deviation value from accepted range.

For AB = 25 m, in nearly 85% of 436 analyzed cases the ρ_{78}/ρ_{79} ratio was within range 1.00 to 1.25. The small number of anomalous values for AB = 25 m may be explained by the larger exploration depth.

The analyses of the data obtained suggests that, generally, the investigating stretch of embankment along the Tisa is in good condition. Slightly poorer conditions may be expected in some localities were significant deviations in the ρ_{78}/ρ_{79} ratio where registered for both depths of exploration. Some of these are marked by the letter B in the Fig. 4. However, the low anomalous phenomena suggest that they are of little importance.

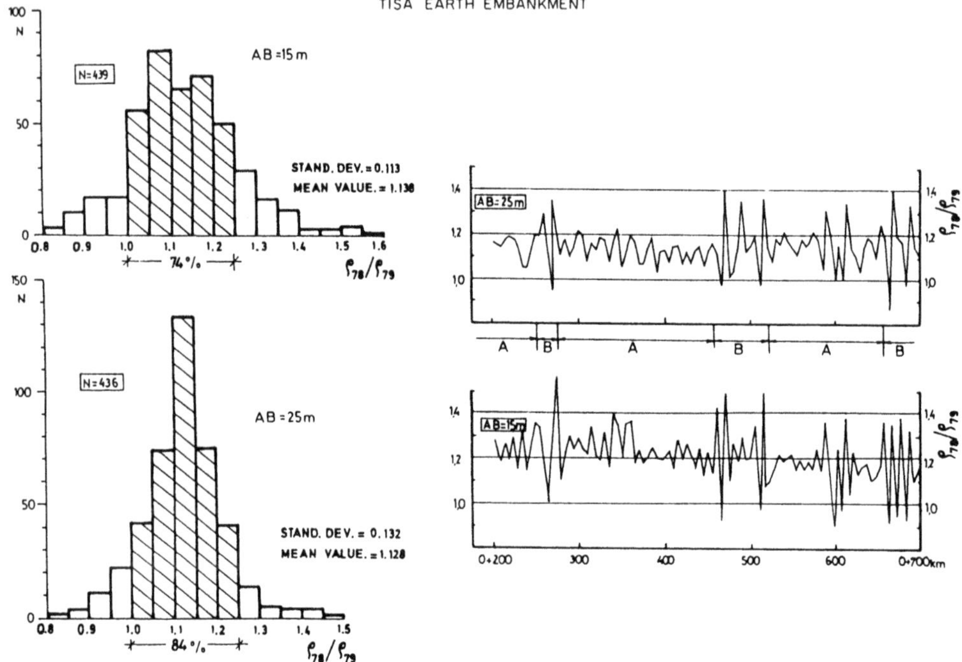

Fig. 4: Tisa earth embankment: histograms and profiles of resistivity ration ρ_{78}/ρ_{79} with prognosticated condition of the investigated earth embankment: A good condition; B relatively poor condition.

5 Embankment Along the Danube

Electrical resistivity profiles obtained in 1978 (period of low water level) are very pronounced both in the form and intensity of variation (Fig. 5). One can see that curves for AB =15 m and AB = 25 m are not very similar, indicating variable lithological and geomechanical conditions along the embankment.

Exploration results from 1979 (period of high water level) agree with the 1978 exploration data, but a much smaller range of apparent resistivity variations was obtained as a direct consequence of generally ioncreased moisture content in the embankment.

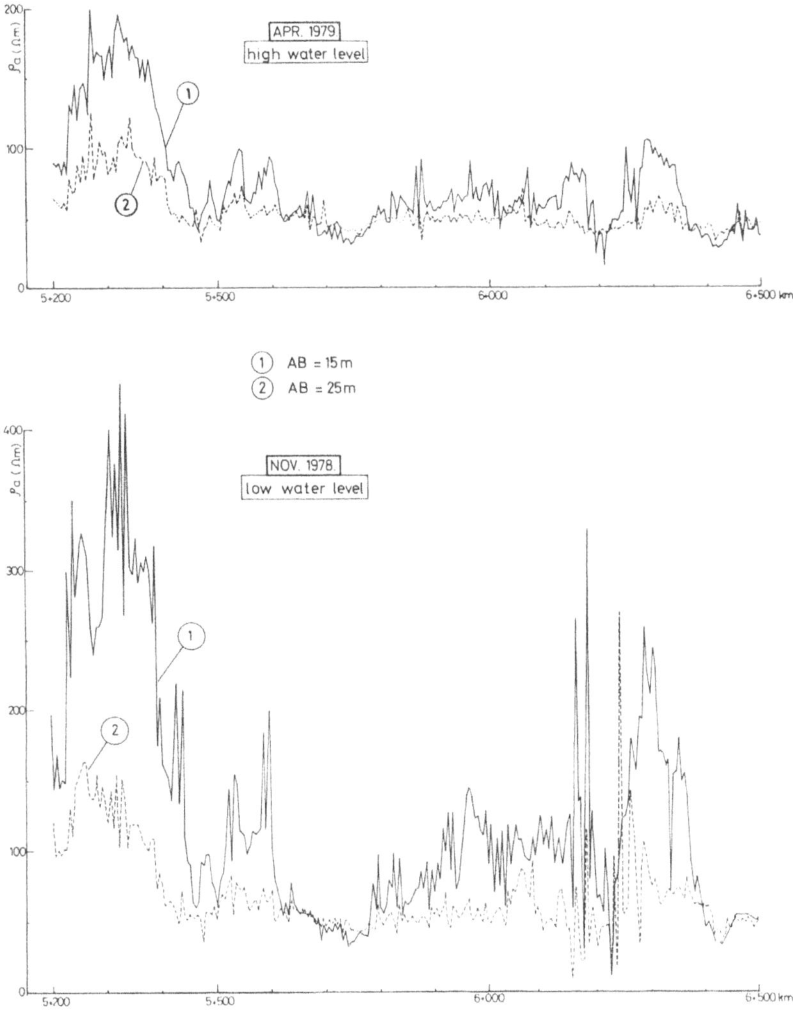

Fig. 5: Danube earth embankment: resistivity profiles obtained in 1978 and 1979.

The 1978 and 1979 exploration data are summarized in the ρ_{78}/ρ_{79} ratios and respective histograms (Fig. 6). Curves ρ_{78}/ρ_{79} for AB = 15 m and AB = 25 m are quite similar in shape. However, the curves for AB = 15 m compared to that for AB = 25 m indicates more and larger anomalies. This is a normal consequence of the exploration to a shallower depth which registers electrical properties in the embankment where changes in lithological-geomechanical properties will be greatest. The prediction of embankment conditions is marked by the letters A, B and C in Fig. 6.

The resistivity ratio values ρ_{78}/ρ_{79} are also plotted as a histogram. The wide and asymmetrical ranges of ratios indicate the considerable differences in physical properties of the embankment during the low and high water levels. The largest variations are likely to occur in localities where the lithological composition has changed or where geomechanical properties of the embankment have been weakened.

Consideration of ρ_{78}/ρ_{79} ratios is based on the same principles as accepted for the Tisa embankment. The anomaly background for the Danube embankment is greater, from 1.00 to 1.40. More than 60% of the analyzed values are within this range. The normal background variation and measuring error may be considered to fall within this range. Any ρ_{78}/ρ_{79} ratio value beyond this range may be considered anomalous. The greater the deviation, the weaker is the inferred geomechanical condition of the embankment.

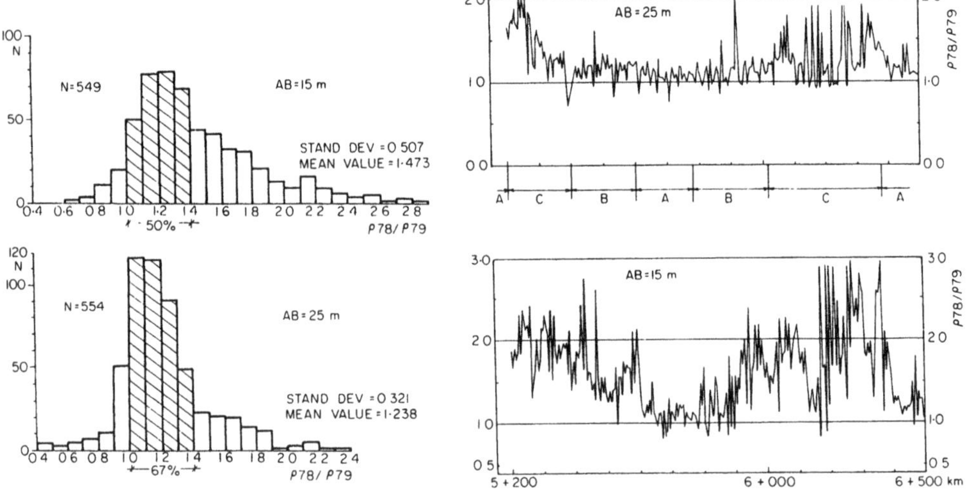

Fig. 6: Danube earth embankment: histograms and profiles of resistivity ratio ρ_{78}/ρ_{79} with prognosticated condition of the investigated earth embankment: A good condition; B relatively poor condition.

Comparison of the geoelectrical results obtained with the known geomechanical condition along the explored embankments shows us that:

1. Generally a satisfactory correlation between geoelectrical and geomechanical results are obtained.

2. Geoelectrical prediction of the embankment condition proved quite satisfactory in 60%, partly satisfactory in 15%, and unsatisfactory in only 25% of cases. In our opinion that prediction can be considered good.

6 Conclusion

The results of resistivity surveys give hope for a useful and efficient application of geoelectrical methods in study of embankment conditions. For this purpose, the methodology of geophysical examination, by introduction of new techniques, should be advanced. It is particulary important to try to establish as realiable a correlation as possible between geophysical parameters and corresponding geomechanical and hydrogeological properties of the embankment.

ASPECTS CONCERNING THE DETECTION BY GEOELECTRICAL METHODS OF SEEPAGE AT SOME ROMANIAN DAMS

T. Moldoveanu[1] and O. Suciu[1]

Abstract

This work presents a synthesis regarding the detection by geoelectric methods of natural potential and resistivity of the seepage that took place in some Romanian dams such as: Vida-Dobresti, Leşu, Racova, Mǎlaia and downstream of Rîul Mare.

1 Introduction

Among the geophysical methods that have found widespread applications in the field of hydraulic construction surveillance, the geoelectric investigation has an important role since its methods have been used to a larger extent (SP and resistivity methods).

Some of the aspects regarding the way the geoelectric investigation methods were used for the detection of seepage in some dams and dykes of Romania are presented later. The technical features of the dams investigated by geoelectric methods are presented in Table 1.

2 Results of Geophysical Measurements

2.1 Vida Dobreşti Dam

In 1969, when during the first impoundment, the storage capacity reached its maximum level, seepage was discovered on the left abutment

[1] Institute of Hydroelectrical Studies and Design, Bucharest, Romania

Lecture Notes in Earth Sciences, Vol. 27
G.-P. Merkler et al. (Eds.)
Detection of Subsurface Flow Phenomena

Table 1: Technical features of the investigated dams and dykes in Romania

Ref. No. Dam features	Name of dam					
	Vida-Dobresti Oradea	Leşu	Racova	Mălaia	Rîul Mare-aval Ostrovul	Păclişa
1. Year of completion	1969	1973	1965	1978	1987	1988
2. River	Vida	Iad	Bistriţa	Lotru	Riul Mare	Riul Mare
3. Dam Type[a]	E	R	G/E	G/E	G/E	G/E
4. Dam/dyke height (m)	15	61	20/15	30.50/ 17.50	30/20	30/20
5. Dam/dyke crest length (m)	70	180	80/7300	90.55/ 294	100/ 3500	100 2500
6. Dam/dyke volume ($10^3 m^3$)[b]	31	(B)560	(A)30/ (C)550	(A)45/ (C)196	(A)80/ (C)3000	(A)80 (C)800
7. Reservoir capacity ($10^3 m^3$)	400	28000	10000	3440	8000	8000
8. Purpose[c]	S	S, H, I	H	H	H, S, I	H

[a] Dam type: G = gravity; E = earth; G/E = concrete gravity and earth dykes; R = rockfill
[b] Dam/dyke volume: A = concrete; B = rockfill; C = earth.
[c] Purpose: I = irregation; H = hydroelectric; S = water supply.

of the dam, about 30 m downstream. The preliminary investigations led to the assumption that seepage mainly occurs at the contact between the impervious clay core and the grout cap on the left bank.

With respect to the detection of the actual seepage path, the following geoelectric methods were used:

1. The self potential method (SP);

2. The resistivity method using the procedures:
   a) electric profiling (with Wenner configuration);
   b) charged body method, known as "mise à la masse".

The geoelectric measurements (SP and electric profiling) were performed on a detailed network of profiles presented in Fig. 1.

The electric resistivity and SP measurements rendered evident a zone with minimal values situated close to the left dam shoulder. In order to solve the problem, the charged body method was used, drilling a hole near the mentioned zone. The resistivity measurements by this procedure were carried out in three stages: before pumping NaCl solution in the drillinghole (stage I), 2 h after pumping (stage II), and 5 h after the begin of the procedure (stage III). In all three stages, the measurements were carried out in the same points arranged on four concentric circles with radii of 1, 2, 5 and 7 m around the drillhole location. The measurements results are presented in Figs. 2, 3 and 4.

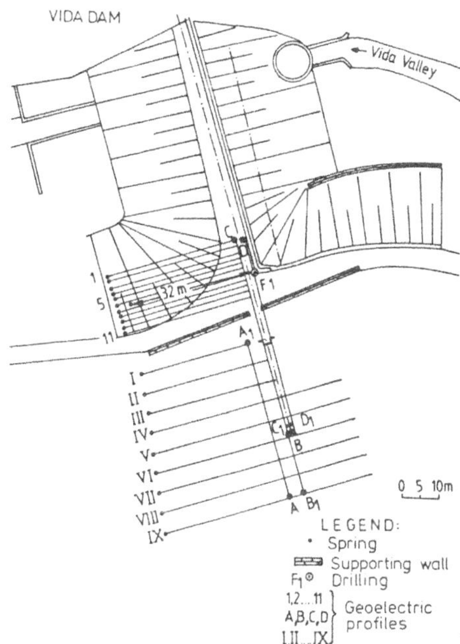

Fig. 1: Sketch with location of geoelectric profiles.

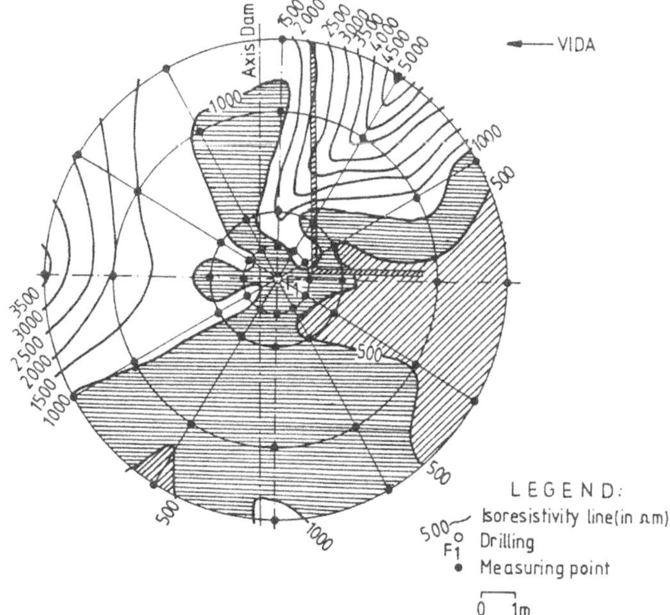

Fig. 2: Map with isoresistivities (charged body method)
Drilling F_1-time: 13,00; 03.04.1969

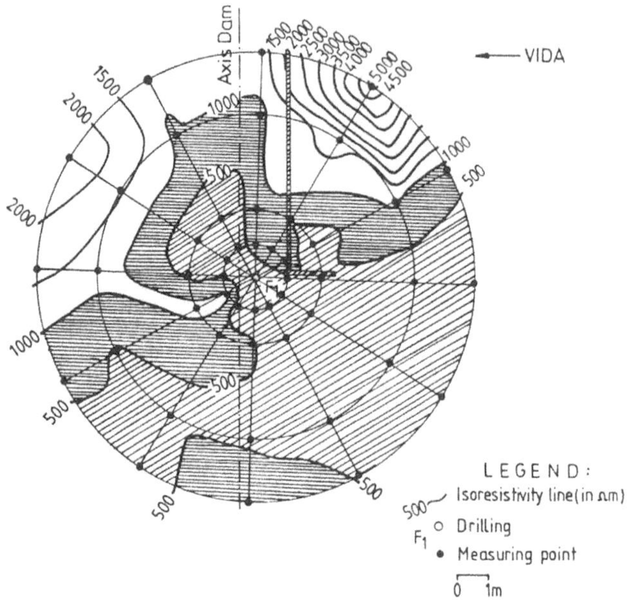

Fig. 3: Map with isoresistivities (charged body method)
Drilling F_1-time: 15,15; 03.04.1969

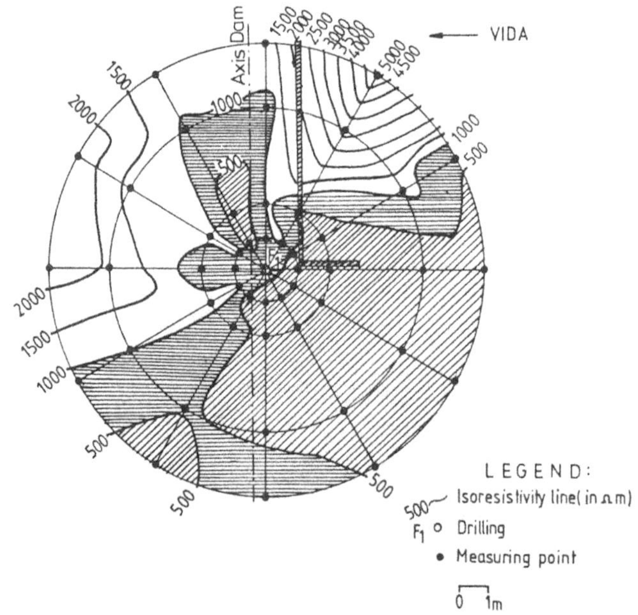

Fig. 4: Map with isoresistivities (charged body method)
Drilling F_1-time: 18,00; 03.04.1969

The comparative analysis of the obtained isoresistivity maps offer a clear image of water circulation through the dam body. The isoresistivity map drawn up in stage I presents the distribution of the conductivity zones (wetter) and the resistivity ones, before introducing salted water into the borehole. As can be noted, the conductivity zone conventionally delimited by the 100 and 500 Ωm isolines is located on the upstream shell and does not exceed the dam axis. In stage II the conductivity zone becomes wider and in stage III it exceeds the dam axis, extending downstreams to the left abutment. These results offer indications of the salted-water seepage through the clay core.

On profiles CC_1 and AA_1 located on the left bank, both resistivity measurements (by Wenner configurations; Fig. 5) and SP measurements (Fig. 6) were carried out twice, namely before introducing salted water into the borehole and after introduction.

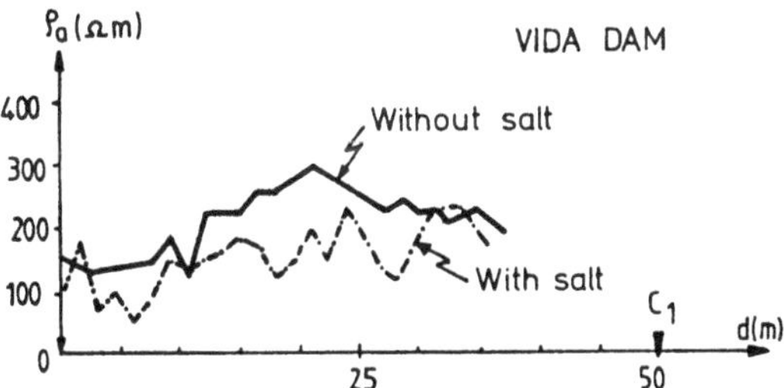

Fig. 5: ρ_a variation on the $C\text{-}C_1$ profile

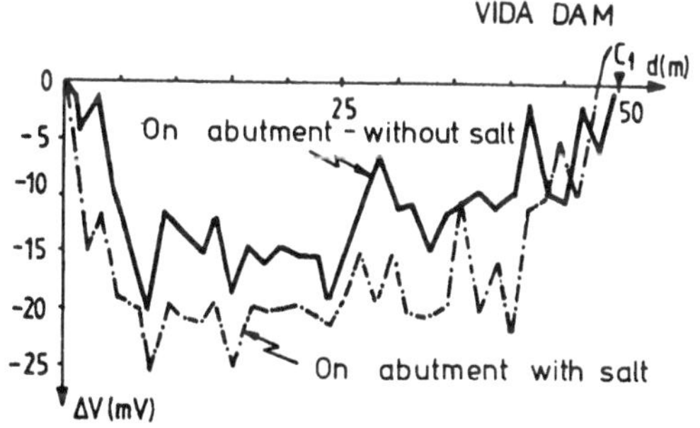

Fig. 6: SP variation on the $C\text{-}C_1$ profile

The SP values obtained on profile AA_1 after the introduction of salted water into the borehole, as compared to those obtained before introduction are much lower (on the average by 5 mV), at a distance of about 20 m starting from point A of the profile. Such a decrease of the SP is explained by the higher effect of potential determined by the salted water, indicating also a water circulation that bypasses the concrete embedding.

These results, obtained by means of several geoelectric methods, indicated clearly that water seepage in the reservoir occurs both through the clay core and through the left abutment, bypassing the concrete embedding. The water tightening works, carried out subsequently in the investigated zone by grout holes, reduced the seepage to a minimum value.

2.2 Leşu Dam

After starting the reservoir impoundment, in July 1973, seepage occured, which, as measured in two manholes at the downstream face toe, was kept relatively reduced (10-60 l/s), until the water level in the reservoir reached about 30 m below the crest level.

When exceeding this level by about 5 m, the seepage discharge increased rapidly, reaching 250 l/s in 2 weeks (7-20 November 1973).

The observations and measurements carried out regarding the evolution of levels in the downstream hydrogeological drillings and in the drain holes performed in the dam cutoff gallery led to the assumption that the seepage is produced through the right abutment.

To confirm this assumption, geoelectric measurements were performed by the resistivity method, using the gradient system (AB = fixed; MN = mobile). The position of the geoelectric profiles is presented in Fig. 7. The selection of the gradient system was imposed by the condi-

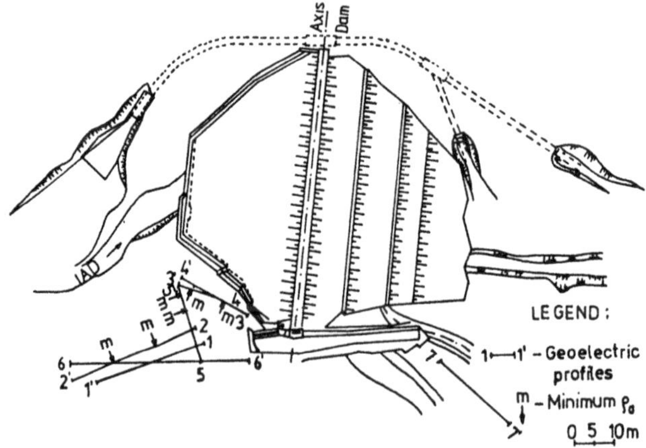

Fig. 7: Sketch with locations of geoelectric profiles in Leşu-Dam.

tions of a very rough relief in the right bank area.

It should be mentioned that the measurements were performed in two stages, namely at a low and high level of water in the reservoir.

The results of these measurements are presented in Fig. 8, which renders evident the zones of minimum apparent resistivities (ρ_a) that mark the seepage zones in the right abutment.

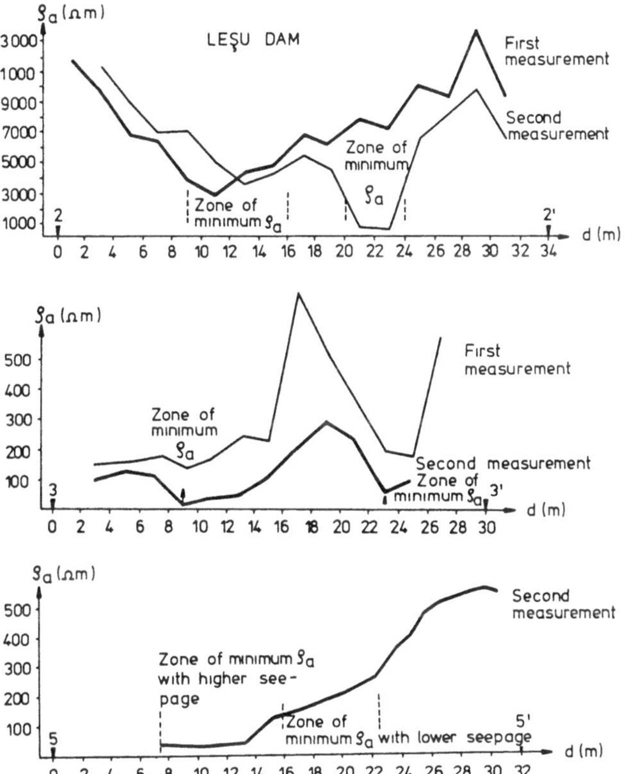

Fig. 8: Profiles with apparent resistivity variation obtained by gradient configuration

The reservoir drawdown, carried out in January - February 1974 confirmed the results of the geoelectric measurements. Thus, the zones of seepage penetration into the abutment were discovered. By means of the measurements mentioned above, the zones could be established where groutings were necessary to reduce and stabilize seepage.

2.3 Racova Dam

In 1974 the settlements were reactivated in the dam and in the dykes, and especially in the right bank dyke, in the zone of connection to the dam.

The occurence of seepage and the acceleration of settlements in the dyke required geoelectric measurements to localize the dangerous zones. The geoelectric measurements were performed on both dykes (right and left bank) by the resistivity method, and an electric profiling procedure using Wenner configurations of the following dimensions: a = 4, 8, 12 and 16 m.

The measurements were performed in two stages:

Stage I: December 1974;
Stage II: November 1975.

Figure 9 presents the results of the resistivity measurements per-

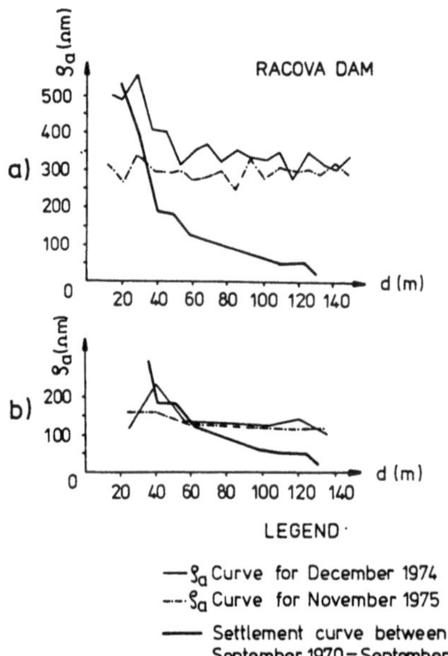

Fig. 9: Apparent resistivity variation on EP-3 profile, right bank dyke.
a) Wenner configuration AM=MN=NB=8 m;
b) Wenner configuration AM=MN=NB=16 m

formed with the Wenner configuration a = 8, 16 m on the right bank of the dyke (EP-3). On this profile (EP-3) a comparison of resistivities was achieved of stage II to stage I. It was found that in stage II the resistivity decreases relatively by 10 %, which reflects the increase of compaction and a slight increase of the moisture content. This interpretation has been correlated with the results of other categories of investigations. Thus, around meter 20 on this profile (Fig. 9), the permeability measured in the vicinity of the zone in the grout holes had higher values.

Also, a strong enough correlation was noted between the settlement curve, measured on bench marks replaced on the dyke, and the apparent resistivity curve corresponding to stage I (1974). Thus, the maximum values of settlements correspond to the maximum values of resistivities. The correlation is natural: the looser zones, easier to settle, generally present higher resistivity values.

The resistivity measurements carried out rendered evident the dyke sections that required periodic consolidation, water tightening and control works.

2.4 Mǎlaia Dam

In June 1982, two wet zones were noted on the downstream slope of the earthfill dam, situated 30 m from the first profile of piezometers and about 170 m in relation to the same profile towards the left abutment.

On detecting the mentioned zones, complex geoelectric measurements were performed in order to establish the seepage paths and to determine the drawdown curves in the seepage zones. The geoelectric measurements were performed by the following methods:

1. The self potential method (SP);

2. The resistivity method, using the procedures:
   a) vertical electric sounding (VES);
   b) electric profiling (by Wenner configurations).

The location of the geoelectric profiles is presented in Fig. 10.

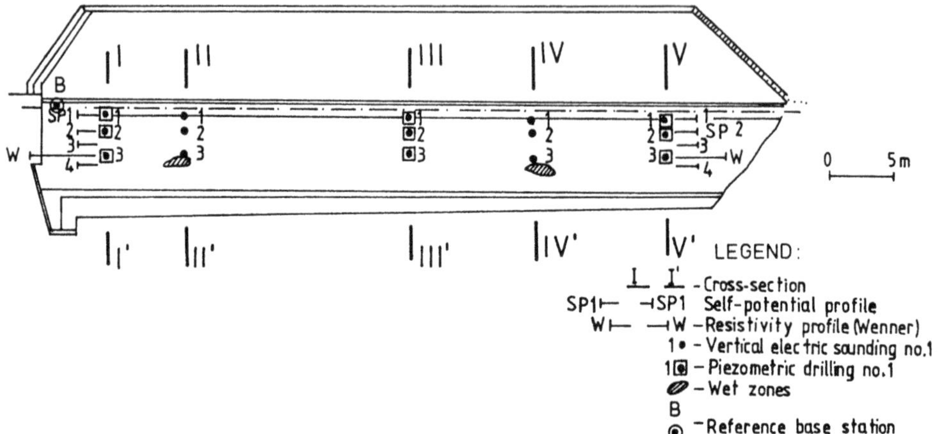

Fig. 10: Sketch with location of geoelectric profiles in Mălaia Dam

The SP measurements were carried out on the downstream face of the earth dam that was marked in a square network with the side of 10 m.

The results obtained are presented in the form of a map with lines with the same value of the SP (Fig. 11), and in the form of profiles of potential variation at crest elevation, at berm elevation and at two other elevations situated between the berm and the crest.

From the SP map two regions of electrofiltration potentials are evident: the upper part of the slope up to the crest, in which the isolines are almost parallel, and the lower part of the slope on a line in which closed contours of the isolines appear, delimiting the local seepage zones.

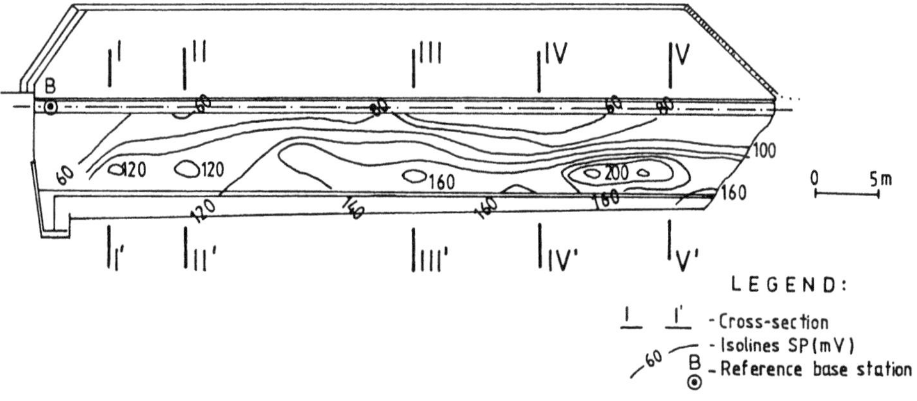

Fig. 11: Self potential map in Mălaia Dam

On the left abutment, the electrofiltration potentials present maximum values determined by the intense water circulation from the left abutment.

The drawndown curves were determined by VES in three cross-sections, from which sections I, III and V coincide with the piezometer profiles, and sections II and IV are between the piezometer profiles and pass through the detected wet zones.

On each cross profile at the dam axis, three VES's were placed on the downstream shell. The measurement results are represented in five vertical sections of apparent resistivities in directions transverse on the axis (Fig. 12).

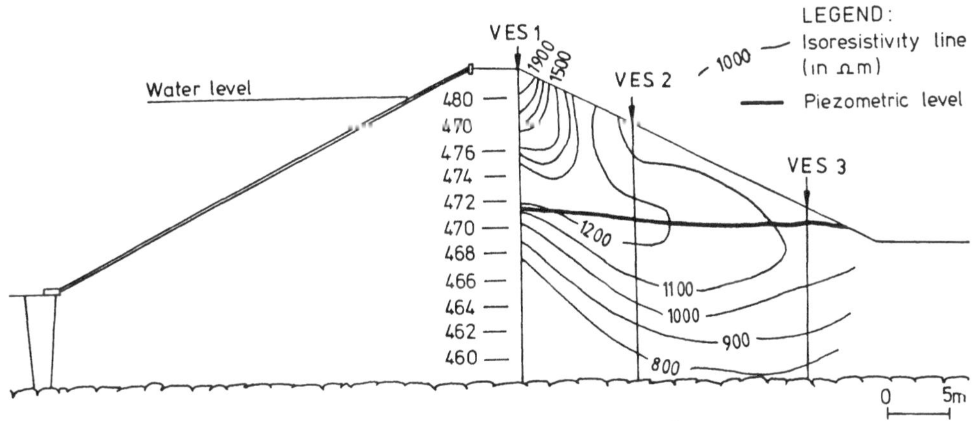

Fig. 12: Vertical resistivity cross-section in Mălaia Dam
VES = vertical electrical sounding

Also, an electric profiling by Wenner configuration (with AM = MN = NB = a = 6 m) was performed. The Wenner configuration was placed on the downstream shell in order to determine the increased moisture zones and the seepage zones that have not reached the slope surface.

Based on the vertical resistivity cross-sections, representing the variation of the resistivity with depth, the seepage lines on the five profiles were determined. These lines were drawn considering the limit between the maximum resistivity values corresponding to the aeration zone of the downstream shell and the minimum resistivity value corresponding to the saturation zone (Fig. 12).

From the analysis of the SP data and the apparent resistivity data (ρ_a), the following final conclusions were drawn:

1. The SP values range between 60 and 200 mV and present a maximum towards the left abutment, indicating intense water seepage through the abutment which supplies the springs downstream of the dam.

2. From the minimum resistivity values obtained by electric profiling using Wenner configuration zones with increased moisture content, between 20-30, 60-78 and 100-125 m were evident.

3. The springs that appeared on the downstream slope of the embankment dam, having discharges not exceeding 5 l/min, originate from the reservoir.

4. The springs that appeared downstream of the dam towards the left abutment, having discharges of about 100 l/min, originate partly from the precipitation water in the abutment and partly from the reservoir, bypassing the contact of the dam with the bank.

2.5 Downstream of Rîul Mare

Downstream of the Gura-Apelor dam, at present under contstruction, which will be the highest dam of Romania (h = 168 m), two other dams were designed and are now in the final construction stage: Ostrovul Mic and Pačlişa.

The ocurrence of massive seepage through the dykes of Ostrovul Mic reservoir led to the decision of emptying the reservoir for repairs. In order to detect the zones with cavities, slightly compacted ballast, as well as the seepage paths, geoelectric measurements were carried out in September 1987, when the reservoir was emptied. The geoelectric measurements investigated the ballast mass of the dyke, that is, the zone between the crest and the dyke base, with the aim of detecting the heterogeneities, the cavities, or the slightly compacted ballast but especially in order to achieve a reference stage for the surveillance of the variation in time of the compaction degree and the seepage rate.

Resistivity measurements by the electric profiling method were carried out along the right and left bank dykes of Ostrovul Mic, on about 800 m using Wenner configurations with four investigation depths: a = 8, 12, 16 and 20 m. The same method was used for the development investigations of Păclişa.

The resistivities values obtained are represented as apparent resistivity profiles and vertical resistivity cross-sections (Fig. 13).

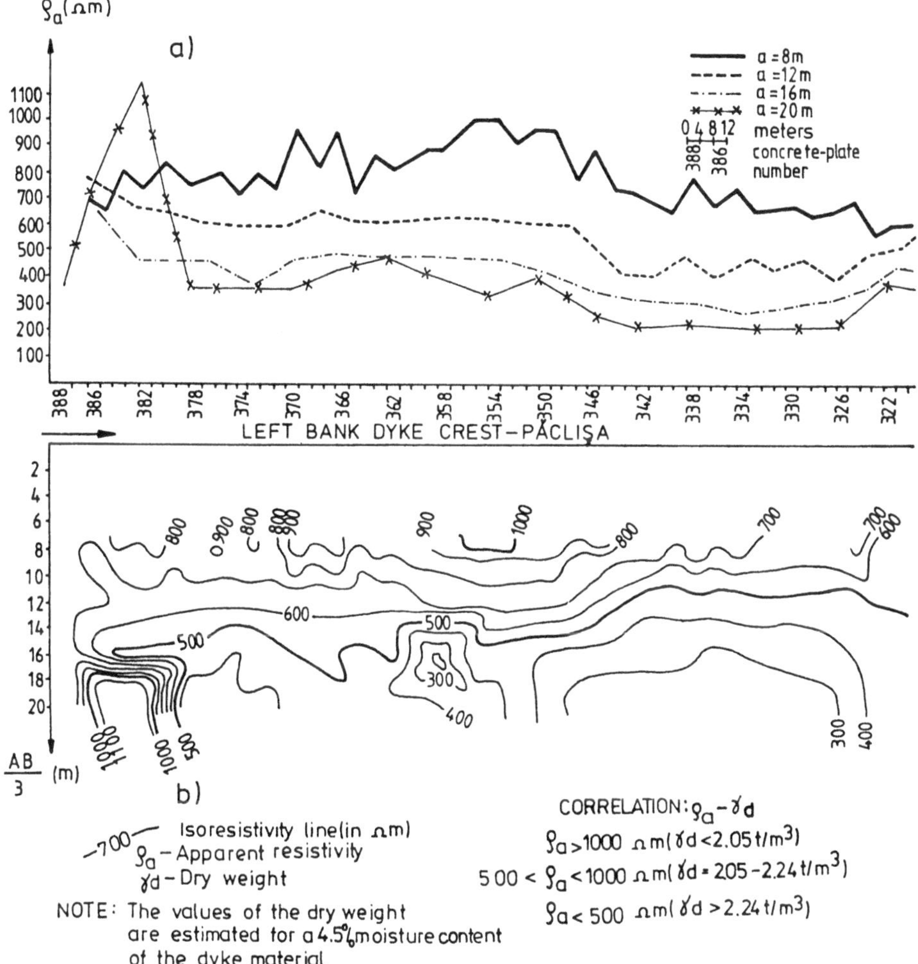

Fig. 13: Resistivity profiles and geoelectric section on left bank dyke.
a - electric profiling with Wenner-configuration
b - vertical resistivity cross-section

In the case of Ostrovul Mic, the measured resistivity values are influenced by natural moisture content of the dyke body material, the moisture content supplemented in the seepage zones, the initial compaction degree and the secondary differential settlements.

In the case of Păclişa, measurements were carried out before the first impoundment. Thus, the resistivity values are influenced only by the natural moisture content of the dyke body material and the initial compaction degree. For these dykes a correlation between resistivity and volumetric weight (Fig. 13) was expected to be found.

The measurements carried out on the dykes of the two projects in September 1987, with the reservoir empty, represents very important "zero" measurements and, together with the future measurements, will serve the control of the resistivity varition in time and for the detection of the zones of potential seepage.

The performance of measurements before reservoir impoundment can also help in the detection of the zones with an inadequate compaction degree and where remedial measures are necessary before starting the first impoundment.

3 Conclusions

The SP and resistivity methods are two basic systems of geoelectric investigations, capable of revealing the existence of a seepage flow. They have been confirmed by convincing results obtained by their application in the cases of the mentioned dams and dykes. Thus, the seepage zones were rapidly and efficiently detected and subsequently subjected to treatment.

The fact that the seepage was reduced to minimum values as a consequence of such remedy applications proved that the detected seepage zones were real ones.

The results obtained by the investigations performed in the modelling tanks in the laboratory and on site show clearly that the existing connection between the underground flow and the distribution of the electric field can be measured at soil surface.

From the experience grained through problems successfully solved up to the present, it can be estimated that seepage detection in dams and dykes was made with a higher efficiency and precision when geoelectric measurements were carried out by both self potential and resistivity methods.

References

Constantinescu, P.; Moldoveanu, T.; Stefǎnescu, D.; Vîjdea, V.; Visarion, V. (1979): Eng. Geophys, Ed. Tech., Bucharest.

Merkler, G.P.; Blinde, A.; Armbruster, H.; Döscher, H.D. (1985): Field investigations for the assessment of permeability and identification of leakages in dams and dam foundations.- 15th Congr. Large dams, Lausanne.

Merkler, G.P.; Moldoveanu, T. (1970): Some examples of the applicability of geophysical measurements in solving certain engineering geological problems.- 16 Proc., 2nd Int. Congr. Rock mechanics, Belgrade.

Moldoveanu, T. (1969): Results of geoelectric measurements performed in Vida-Dobresti Dam.- ISPH Arch., Bucharest.

Suciu, O. (1982): Electrometric measurements for seepage surveillance in Mǎlaia Dam.- ISPH Arch., Bucharest.

Suciu, O. (1988): Downstream Riul Mare development, Clopotiva-Hateg Section. Geophysical measurements on Ostrovul Mic storage, 1st stage, empty reservoir.- ISPH Arch., Bucharest.

MAIN TOPICS OF THE ROUND TABLE DISCUSSION

H. Militzer[1]

Regarding to the main subject of the symposium, the round table discussion dealt with the following:

1. Some general aspects of the teamwork between geophysicists and construction engineers;

2. Some perspective trends in developing single methods as well as complex cooperation.

The common motive of the discussion was the conviction that the importance and application of geophysical surveys are increasing rapidly. This is also valid for the investigation of subsurface structures, for the determination of petrophysical parameters of the soils and rocks in situ as well as for longterm observation of the system construction site/building.

Important aims of the investigations include the determination of the depth and morphology of the considered construction site, as well as faults, cavities and other inhomogeneities. Thus on the one hand, geophysical investigations quickly and economically support the choice between several alternative sites and, on the other hand, they procure precise information required for a fixed site.

In determining the physical properties of rocks in-situ, some of these parameters may be derived directly from geophysical measurements, e.g. bulk density or water content. For engineering the most important geophysical parameters are electrical resistivity, seismic wave velocity and thermal conductivity. Of course some of the derived properties must be modified, usually according to semiempirical constitutive relationship, for example the elastic moduli or the jointing of rocks. Some parameters, for example magnetic susceptibility, enable a quanti-

[1] Bertolt-Brecht-Str. 9, GDR-9200 Freiberg

Lecture Notes in Earth Sciences, Vol. 27
G.-P. Merkler et al. (Eds.)
Detection of Subsurface Flow Phenomena

tative interpretation of magnetic anomalies obtained from traversing dolerite dikes.

Investigations concerning the observation of the system construction site/building, amongst others, comprehend control measurements for determing the state of the existing buildings, e.g. the water permeability of dams and dikes as a function of time or as influenced by shocks and vibrations with regard to their bearing capacity.

However, it must be emphasized that geophysical investigations neither replace direct methods such as drillings, trenching, etc. nor laboratory measurements for determining geotechnical or hydrological relevant parameters. Furthermore, direct engineering-technical measurements, for example of settlements, should be generally continued in conjunction with geophysical investigations. Geophysical methods may be considered as a means of interpolation between, and extrapolation from, borehole data. By careful planning, the number of borings required for an adequate definition of building site conditions can be remarkably reduced and the locations for boreholes exactly defined; therefrom optimal knowledge with respect to the subsurface structures and parameters can be obtained. At the same time by geophysical mapping we avoid misleading interpretations of geological situations, for instance in faulted ground or in search of cavities, buried channels, etc. are often evident from borehole data alone.

But a single geophysical method may not provide enough information on subsurface conditions to be used alone. Each method typically responds to the different physical characteristics of earth materials, and only the correlation of data from different methods may provide the most meaningful results. Certain combinations of methods may be useful under one set of subsurface conditions, but offer little or no information in others. The above mentioed facts presume a good technical understanding between construction engineers, engineering geologists and geophysicists. They should be aware of the mutual possibilities and limits of efficiency of the other disciplines and should have the derive for cooperation.

One of the participants of the meeting expressed his opinion concerning these topics as follows: "As far as I can see, the clients who come to see you are interested in results and these results are prepared by the geophysicists and by means of their methods applied to geomechanical or geotechnical problems. The question is, what can I do with it? In that case I feel that the necessity of interdisciplinary cooperation between engineers, geologists, mineralogists and others is urgently needed (Wilt)".

Furthermore, one of the main problems pertains to the transfer of the parameters determined in the laboratory to real ground conditions. Despite much remarkable progress over the last years this problem also needs enhanced cooperation. Someone explained: "We have problems with discrepancies in the data between laboratory measurements and field data. Occasionally we even perhaps don't know how to define hydologic and geoelectric properties, if we go from the laboratory to nature. Therefore, we also have discrepancies between the results of model calculations and the results of field measurements. That is also the reason for modeling only very simple structures (Brauns)".

Undoubtedly the symposium induced a strong impulse concerning the demand for cooperation and international exchange of experience.

One of the participants of the meeting expressed his opinion as follows: "Formerly we have had a lot of groups working in isolation around the world, for example on the problems of electrodes and we have had people in the USA as well as in the European countries - especially in the USSR - many exotic investigators ... but few of their results have been published. This symposium, however, has been extremely valuable in letting all these different groups know what the others are doing. The same thing occurs with field procedures. There is no fixed established field procedure for SP measurements as for the other geoelectric methods. People in different countries use entirely different techniques and do not believe particulary in the other people's work. A great contribution of this symposium will be, to bring all this together, putting it into one volume, where we can examine each other's techniques, look for the strengths and weaknesses, and possibly combine the best of all these methods to produce an internationally acceptable technique for SP application (Corwin)".

Concerning methodological orientations there were very interesting and important papers dealing with questions of devices, measuring techniques and future interpretation. The main topics included:

1. Self-potential measurements;

2. Geothermical measurements;

3. Complex geophysical measurements.

With regards to self-potential measurements it was proved that long-term observations of flow phenomena demand an extremely high precision with respect to the stability of the elctrodes. Only thereby is it possible to register independently real streaming potentials and time-variable variations of the mineralization of subsurface water as well as the porosity as a consequence of variations in stratification resp. the degree of the dissolved mineral components. Some comments emphasized that this question too is more a question of exchange of experience than of new developments. In discussing the problems of interpretation a main topic included questions of analogous and numerical models.

In connection with the research project a big laboratory channel was developed (6 x 2 x 1.5 m) as well as a large model dam (H = 3.5 m; L = 20 m; V = 600 m^3) in a huge open pit in the area of BAW. These models are the result of cooperation between the University of Karlsruhe and the Institute of Federal Waterway Engineering, Karlsruhe. Thus it was possible to model fluid streamings and to quantify the relationships between hydraulic, electrical and thermic fields on the basis of numerous objective measurements.

Some of the participants underlined the international happening of the above mentioned dam model and the big experimenatal channel, they had been constructed by the University of Karlsruhe and the BAW. In comparison, there are only a few practically useful numerical models until now. These models are mainly 2D-models and are used for projection as well as for interpretation of field measurements. With regards to this situation it was logical that some remarks emphasized the demand for more attention to 3D-modeling for better approximation of real events and petrophysical situations, including the relationships between flow phenomena and physical fields. "But modeling needs good information about the primary flow and the geoelectrical properties (Sill)".

Demanding more and better data simultaneously means to consider the relationship between SP-effects and temperature, and in particular, pressure. "A lot of experiments we talked about here are very dependent on temperature and if we model the interesting situations over the summer and winter we will understand this better (Morgan)".

Additionally, laboratory investigations were suggested up to temperatures greater than 100 C and high pressures. Accordingly new and interesting results are expected, which may explain streaming effects and SP-fields under thermic conditions better than at the present time.

The discussion concerning questions of geothermics was mainly related to problems of imaging infrared techniques. Instructive examples were presented concerning the effective application of infrared scanning as well as infrared imaging using cameras. These cameras are positioned at fixed points in the field and are useful for detecting leakages in dams and dikes. The infrared radiation measuring technique is not only a very economic tool, but is also appreciated by clients, since real pictures of the surface temperature of the objects can be produced. But this fact can easily become dangerous, when the clients do not know and understand the limits of the method, thus believing that it is good for everything. "We have almost perfect techniques to measure the barrage image of the surface in different wavelength ranges and I think that this is a very economic method, which in such a short time produces hundreds of kilometers of information about the dams along existing waterways. On the other hand, the interpretation of these images is still a problem. It is difficult to pinpoint the reason for a sudden anomaly (Kappelmeyer)". With regards to this point of view questions of overlapping of surface temperatures, microclimate, sun radiation and the influence of streaming groundwater were discussed. It was thus proven that the infrared method is of little use as a single method, but can be used alone when the radiation anomaly can be clearly interpreted. The necessity of combining this technique with surface methods became evident by the remark "...what you need are many studies ... or you have to know a lot of things about the dam, about the underground, about the history, the hydrology, the climate ... in some cases the studies are not enough. After accumulating all these studies you have to discuss what you are going to do with the surface methods ... you have to think about what you want to know ... you have to decide at what time the measurements are to be made. But in any case you need more methods (Armbruster)".

"Another possibility to accelerate the learning process with regards to the application of geophysical tools in engineering and especially in detecting leakages in dams and dikes as well as solving environmental problems is to publish more case histories. Case histories are useful tools for explaining the complex application of geophysical methods, especially under geotechnical aspects and other points of view, including the complex interpretation (Corwin)".

In summary all the participants of the round table discussion underlined the high scientific level and the practical relevance of the symposium concerning questions of a stable function of waterway buildings and constructions for environmental protection. The exchange of experience and results was excellent for further investigations. This was also valid with respect to solving basic problems, the theory of the methods, for developing devices, for data processing and interpretation as well as for diminishing the influence of noise effects.

For the future the participants agreed to the following proposals for the continuation of the investigations:

1. In principle, such symposia should be repeated at intervals of approximately two years. Thus, the international exchange of results should be organized in such a manner that progress can be accelerated and the expenses can be minimized for the participating countries, for example by the common use of expensive equipment.

2. With respect to the problems of technical devices at this time, efforts should be concentrated on developing electrodes with good, long-term stability for measuring the temperature as well as the apparent specific electrical resistivity directly in the vicinity of the electrodes.

3. Methodological terms of references should be developed with the aim of minimizing the the noise level of geoelectrical measurements, of recognizing the geological and petrophysical situation, of observing telluric and straying currents as a basis of arrays in the future for the areas in question.

4. The devices of arrays should be standardized with regards to their main parameters.

5. The numerical modeling becomes more and more important for the interpretation of field measurements as well as for projection and effective practical investigations. Therefore, the development of models with regards to the real petrophysical situation, becomes an important task.

6. The modern infrared technique has opened new possibilities to detect and analyze near-surface flow phenomena. To improve the accuracy of interpretation it is increasingly necessary to advance the complex cooperation with other geological, geophysical, geochemical, petrophysical and further earth bound disciplines (Hötzl, Merkler, Militzer).

(These conclusions are the main part of the magnetic tape recording registered during the round table discussion).

Excursion

SOME DETECTION METHODS USED FOR THE STORAGE RESERVOIR AT IFFEZHEIM, FRG

H. Armbruster, Bundesanstalt für Wasserbau
Kußmaulstr. 17, D-7500 Karlsruhe

1 Reasons for Building (History)

There are many reasons for building. Let us begin with the history. For 150 years the Rhine crossed the Niederterasse (alluvial plane) with several arms, with a maximum depth of 30 m and a maximum width of about 6 km. During the low water period there were numerous islands (about 2000 between Basle and Mannheim), and during the high water period the whole area was under water, two or three times a year. There have always been two particular occasions when there is high water up to the present day: The Advents-Hochwasser (the flood of advent) and the flood of Pfingsten (Whitsun). This means: the first snow in the Alps melts during the first warm period before Christmas time, the melting water passes through the Lake of Constance (Bodensee) and fills, together with the rainfall in the Black Forest, the river Rhine. After springtime most of the snow melts in Switzerland, Austria and France . The snow of the German part of the Alps melts during the first warm period. We await the next flood at the end of springtime, approximately during the middle of May.

Now back to the history: during these periods many tragic events have happened (the flood has destroyed the houses and the agricultural efforts of a small population). Colonel Tulla (later Professor of the Technical High School) planned and constructed a new bed for the River Rhine (Fröhlich 1975), much shorter than the old one (about 23 %, the length is shorted from 350 km to 270 km from Basle to Mannheim). The work (the new deep bed and the dikes along the river in the hinterland against the periodic floods) was completed between 1817 and 1868. This was the first period of the regulation of the river Rhine.

Lecture Notes in Earth Sciences, Vol. 27
G.-P. Merkler et al. (Eds.)
Detection of Subsurface Flow Phenomena

The second period (from about 1870 until 1914), named after the engineer HONSELL, was necessary because of the shipping traffic. The traffic needs a particular water depth even during the low water period, therefore, the 250-m bed of the Rhine was compressed with the aid of sills, diversions, and stone packages. The third period of regulation started about 1900, and was initiated from the idea of obtaining energy from the water. There were many Russian/German plans, but they were destroyed during the First World War. At the end of this war, according to the Versailles Treaty, France was given the right to use the water power of the river. The government of France decided to use the power by transporting the whole discharge of water to the French side with a special channel (canal d'Alsace) parallel to the river. Only the remaining discharge of about 15 m^3/s was normally on the old axis and also the high-water occured twice a year. In this way, three storage reservoirs were built with all installations for power and shipyards on the French side (in the new channel).

Because there was usually only a small discharge in the river, the groundwater table fell in the range of some meters (maximum 7 m). Therefore we speak of a desert in the Rhine valley, due to the change of vegetation on the German side of the Rhine. After the Second World War, when Germany and France became first political and later on personal friends, both governments began to consider a common solution for the groundwater/water problem. They decided in 1956 to choose another form of constructing storages, the so-called Schlingenlösung (sling resolution) and also decided to support the groundwater while constructing fixed sills (Feste Schwellen) in the old river. With this solution a new, special channel was not planned, but a new route selection for about half the length of the river. The power plant station is always on the French side in the shorter, new channel. For the German side the contract included the construction of some sills for raising the groundwater table. Between 1956 to 1970 four of these stations were built (Graewe, 1975).

But soon, also during construction of the buildings, the main problem of the river and the erosion of the bed was recognized and in 1969 the two governments decided to construct a new system, called "barrages in the river" (Flußlösung). Two barrages were constructed, a third one near Karlsruhe was in discussion. If it were possible to avoid the erosion of the bed, then it would not have been necessary to build either the third or other barrages. The problem of avoiding erosion was discussed over a relatively long time: At the end of every storage reservoir the river takes the bed load from the nonimpounded river and starts the next erosion step so that the problem, starting with Tulla's regulation and following with the construction of power plants, is only to be solved by constructing new barrages. So hydraulic scientists, especially those from the Federal Waterway Engineering and Research Institute finally determined that the erosion can be stopped:

1. Either by constructing barrages through to Mainz, where the velocity of the river is too small and the bed is too hard (natural barrier of rock) for erosion;
2. Or by giving back the load artificially to the bed of the river (Geschiebezugabe, Felkel et al. 1977).

This last Method is still under development but for about 10 years the erosion has been stopped by giving the bed load back to the river, which is taken off from big pits along the river but not from the river itself.

In conclusion, the barrage reservoirs in the river Rhine, the last of which was called "Staustufe Iffezheim" (Martens et al., 1975), are necessary for the following reasons:

1. Flood: regulation in the 19th century;
2. Shipyard: regulation until the beginning of the 20th century;
3. Groundwater supporting: constructions of sills or analogous constructions

2 Constructions

For every damming of a navigable river the following constructions are necessary:

1. Locks with one or two lock chambers for shipyards;
2. Weirs for the regulation of the water table, especially during the high water period;
3. Power stations for energy;
4. Dams along the river;

5. Constructions for the regulation of former crossings like streets and former mouths of small rivers;
6. Dams across the valley, only needed at special constructions.

The Staustufe Iffezheim (built in 1974 ÷ 1977) belongs to that type of dam, which needs all the preceding constructions (Fig. 1). The main dimensions are

- two locks with 270 m length, 24 m width;
- one weir with six fields of 20 m, that means a width of 120 m for a discharge of about 7000 m^3/s;
- one power station with four turbines for about 1100 m^3/s, average difference of water about 11 m, power output of about 100 MW;
- dams along the river of about 2 X 24 = 48 km with lateral trenches;
- dam across the valley, length about 250 m, height about 12 m.

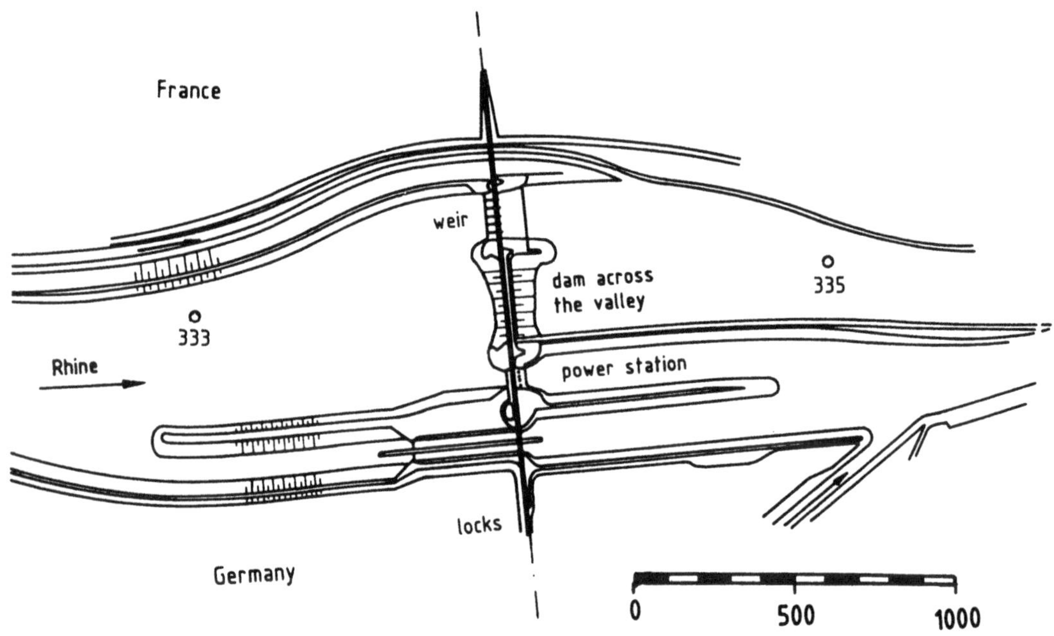

Fig. 1: Staustufe Iffezheim (barrage of Iffezheim)

3 The Constructions and Their Seepage Problems

3.1 The Main Buildings

3.1.1 The Complex Locks/Power Station

This complex, on the right bank of the river, was built in a common excavation pit (about 300 x 600 m) (Fig. 2) combined with low dams against the high water of the running river.

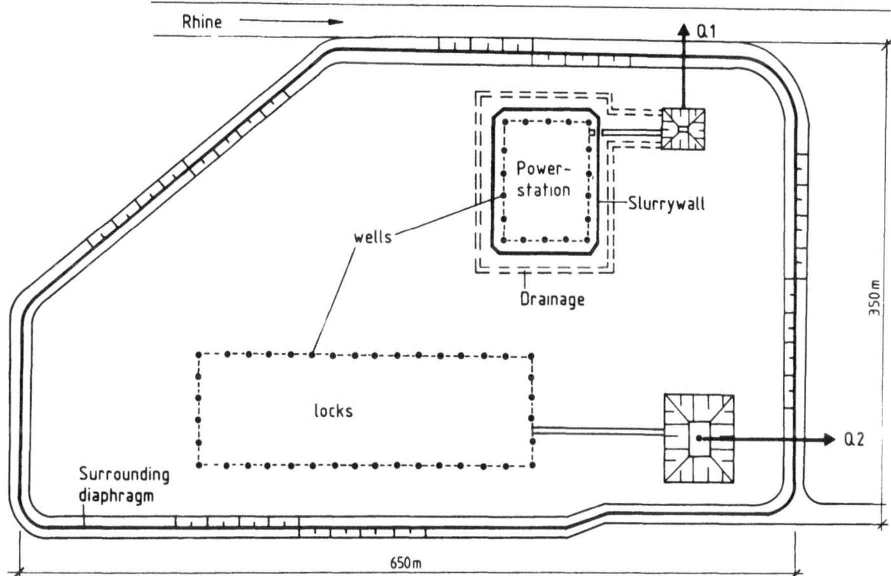

Fig. 2: Sealings and groundwater lowering of the locks and the power station

The transmissivity of the soil was very high, the lowering of the water by about 10 m on average, and by 15 m at the deepest points, was only reached with the help of a diaphragm along the big pit and another deeper diaphragm around the deepest part of the power station. These diaphragms have two functions:

1. The surrounding diaphragm is only necessary during building to reduce the discharge of seepage. If there are leakages, the problem is normally not a security problem, only a problem of energy costs.

2. The diaphragm around the deeper parts of the power station and the part of the diaphragm along the river (later near the main dam)

also have the function to reduce the later uplift. Therefore, leakages can cause serious problems later.

Both problems occured and had to be solved. The first problem was very important, because there were two different owners of the constructions (locks belong to the Ministry of Traffic, the power station to the electricity foundation RKJ). The combination of sealing and lowering at different heights or partly lowering and different times to start pumping cause many serious problems in finding a solution if the sealings are defect or the soil highly permeable.

The second problem initiated numerous hydraulic measurements and also a test with coloured tracers. In the end, after finishing work, the diaphragm was partly excavated and partly replaced by a new one. But until today some problems have remained. Apart from the tracer test, no other measurements were done, apart from the hydraulic ones with many observation wells. The pressure under a fine sand layer caused many problems for the pumps, the drainage system as well as for the stability of the slopes. In the deeper part of the pit the whole system of water lowering had to be changed during work. The discharge of pumping was about 5 m^3/s for more than 1 year.

3.1.2 The Complex Weir

This complex belonged to the ministry and therefore the legal problems were less serious. Fortunately, also, the constructional problems were less serious because the surrounding diaphragm (ca. 200 x 350 m; Fig. 3) in the flood protection dam was shorter and the sealing element in the inner part of the excavation pit was constructed in a two-phase procedure of slurry walls, which means: a concrete wall. Also, the foundations were less deep and therefore the deep, fine sand was less dangerous.

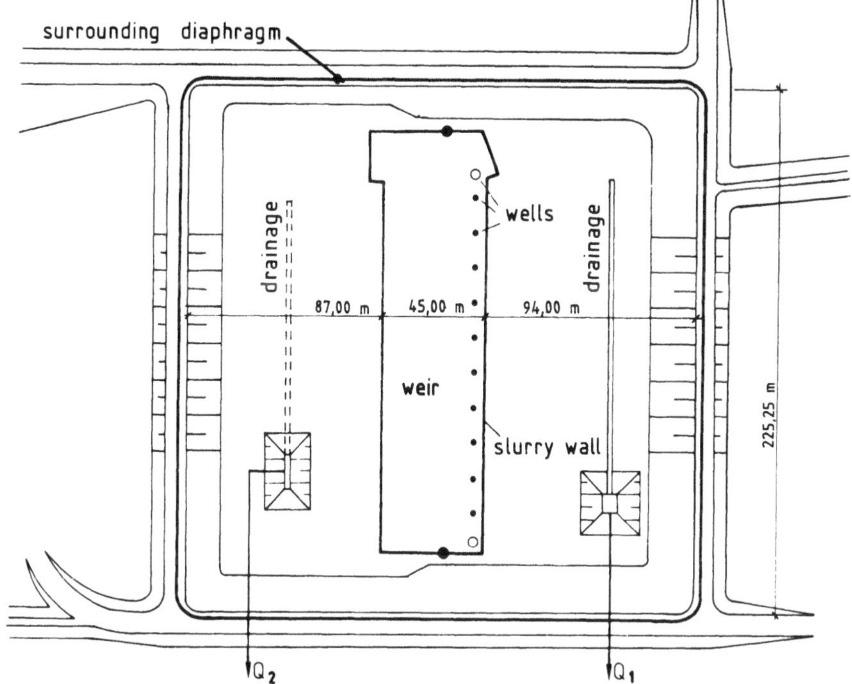

Fig. 3: Sealings and groundwater lowering of the weir

Beyond this, the problems were well-known from the power-plant station and tension in the sand layer could be reduced by the wells. The hydraulic measurements were taken regularly (every day, sometimes twice a day); no other measurements have been taken.

3.1.3 The Dams Along the River

These dams (Fig. 4) have a diaphragm sealing system called "Schmalwand".

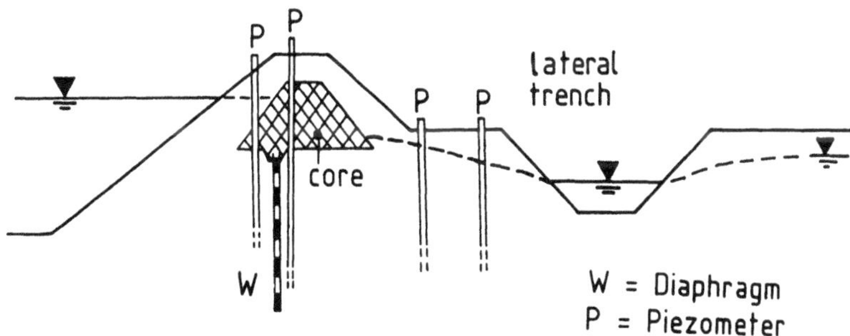

Fig. 4: Sealing system of the lateral dams

This method was at this time a new mode of action and the machines were only able to function down to a depth of about 14 m. Besides, the ground (gravel) was very hard to generate. So the head of the diaphragm was on the original surface of the ground, the new dams were constructed later on. The connections between the diaphragm and the later core of the dam were carried out by excavating the head of the diaphragm, cleaning it and enveloping it with the core material. The core was often an economy core (Sparkern), that means a thin core, depending on the quantity of available material. Along the dam a lateral trench was made for two reasons:

1. First, some small rivers had to be diverted, so that they were not impounded. So the trench (ditch) can also be considered a small river (maximum of about 30 m^3/s on the German side).

2. The trench is a drainage system for the toe of the dam, i.e. a control system, where measurements of discharge can be executed.

During work and also after completion, no real control could be done. Only at high water after construction, but before raising the water table, some leakage effects were seen. Therefore, a few control sections were installed.

3.1.4 The Dam Across the Valley

This dam (Fig. 5) was the last construction, because it was filled directly into the running river. On one night the rock-filled, downstream dam locked the river and some auxiliary dams had to be removed, then the water streamed through the weir and the open locks.

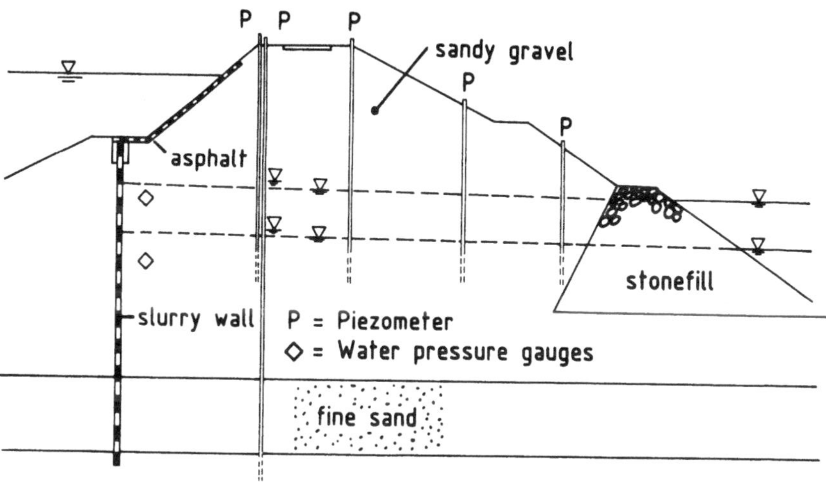

Fig. 5: Sealing system of the dam across the valley

The rocks are filled with sand and gravel material, which penetrate with the stream in the void space of the fill. The vertical sealing of the dam, a one-phase diaphragm, is constructed in the upstream berm at a depth of about 25 m. The material of the dam is an inhomogeneous mixture of sandy gravel and gravel, which was fitted from the bottom of the Rhine and taken from pits in the surroundings. The final sealing of the upstream slope of the dam consists of asphalt concrete. The sealing could be partly controlled with some hydraulic gauges and a drainage system behind the head of the diaphragm, which has three controlling outlets. Observation was possible from the beginning, because there was always a difference between the tailwater and headwater level.

4 The Seepage Problems of the Staustufe Today

4.1 The Complex Locks Power Stations

This complex is calculated with a precalculated water pressure under and behind parts of the construction. These pressures depend on the sealing systems and the effect of their drainage systems. So the pressure must be controlled by water pressure cells and observation wells. In addition, the temperature of the soil is measured to determine the dependence between pressure and temperature. Until now there have been no problems. Also, the settlements and the movement of some fixed points of the construction have been measured to prove the stability of the percolated soil materials.

4.2 The weir

The weir also is calculated with precalculated pressure sizes under the piers of the weir construction. This pressure has been measured with observation tubes extending up inside the piers since the beginning. Movement measurements are made and also temperature measurements in the tubes for controlling the seepage potentials.

4.3 The Dams Along the River

These dams (48 km) are hardly under control. One side is controlled by the German Ministry of Traffic, the other by the French Ministry (therefore, the frequency of collecting control data is not the same). On the right (German) side of the embankments, which is only regarded here, many anomalies are detected, especially by observation (once a week). Every anomaly is considered and observed with hydraulic and temperature measurements and often equipment (observation wells) is installed. Most of the anomalous areas (wet areas or stronger outlet to the lateral trench) are at the crossings of the former bed of the Rhine (before Tulla) and the present axis. Here, a gravel- filled, old valley first brings the higher transmissivities of the deeper soil under the end of the diaphragm and secondly the very inhomogeneous underground causes many failures in the vertical sealings. These two conditions often created too many possible reasons for the increased discharge of the seepage, so that the constructor of the sealing element could only seldom be persuaded that he was at fault. Many hydraulic measurements in the trench and in the piezometers are carried out. The thermal measurements in observation wells are available for a long period (measured in the water-filled tubes), the surface temperatures have been measured twice (1981/1987), also some geoelectrical and self-potential measurements have been carried out. The result of the observations and measurements forced the owner in some cases to improve the sealing construction (new diaphragm, for example, or a new filter construction at the slope of the lateral trench). Then normally many measurements could be taken before and after the repairs.

4.4 The Dam Across the Valley

This dam was controlled with the drainage system and few observation wells. Four years after the dam was put into operation, the asphalt sealing broke (in summertime) over the end of the drainage system. A great amount of water (about 150 l/s) flowed through the pipe of the controlling drainage system at a high velocity, making it very difficult to find the leak. Finally, it was found with the aid of a diver and numerous shiploads (gravel) were dumped over the hole. After the repair (injection of the leakage and closing of the whole drainage system) a large number of observation wells (more than 40) were installed. Since then the hydraulic and thermal fields have been measured weekly for years, including two flights with the multispectral scanner (MSS). With these measurements one can see the distribution of the hydraulic and thermal fields and the way they change over the years.

References

Fröhlich H. (1975)
Die Geschichte des Oberrheinausbaus. Wasserwirtschaft 65, H.9, FRG

Graewe H. (1975)
Die Notwendigkeit einer Rheinstaustufe bei Neuburgweier.
Wasserwirtschaft 65, H.9, FRG

Felkel K., Kuhl D., Steitz K. (1977)
Naturversuche mit künstlicher Geschiebezugabe zwecks Verhütung der Sohleneintiefung des Oberrheins, Wasserwirtschaft 67, H.5, FRG

Martens W., Klose H. (1975)
Die Staustufe Iffezheim. Wasserwirtschaft 65, H.9, FRG

THE BLACK FOREST OBSERVATORY SCHILTACH

H. Mälzer[1]

The observatory was established near Schiltach/Black Forest from 1970 to 1972 and belongs to the Universities of Karlsruhe and Stuttgart. It is located in an abandoned silver and cobalt mine founded in 1770. The observatory is operated by the Geodetic and Geophysical Institutes of Karlsruhe University and the Institute of Geophysics of Stuttgart (Emter et al., 1975). The buildings and the construction have been financed by the Volkswagen Foundation (Stiftung Volkswagenwerk) and the instrumentation has been provided by the universities, the participating institutes and the German Research Society (Deutsche Forschungsgemeinschaft.

The site of the observatory is about 5 km north of the Kinzig valley (Fig. 1) to minimize man-made noise from traffic, industry and rivers.

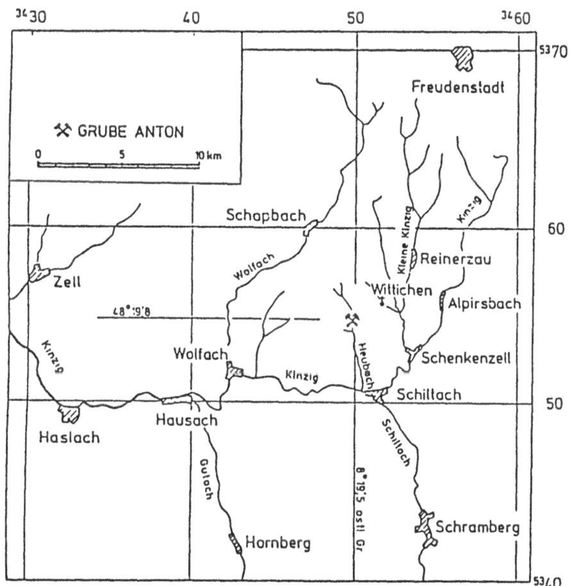

Fig. 1: Site of the observatory

[1] Geowissenschaftliches Gemeinschaftsobservatorium, Universitäten Karlsruhe/Stuttgart; Heubach 206, 7620 Wolfach, FRG.

Lecture Notes in Earth Sciences, Vol 27
G.-P. Merkler et al (Eds.)
Detection of Subsurface Flow Phenomena

The surrounding rocks are granites. The interior part of the tunnel system containing several instrument vaults is blocked off by an air lock, which avoids fast exchange of air and fast pressure fluctuations near the instruments (Fig. 2). This part of the mine is from 400 to 700 m horizontally away from the entrance and from 140 to 170 m below the surface of the earth. This provides very good temperature stability, which is mandatory for long-term geophysical measurements. The analog signals are brought to an electronic vault in front of the air lock, where some of the data are digitized by a microprocessor. All data (analog and digital) are then brought out to the lab house in front of the mine by cables where they are recorded on chart recorders or on magnetic tape. The lab house contains the offices of the personnel (two scientists, two engineers), the recorders, work shop, the computer (RAYTHEON 704-minicomputer) and other facilities. A garage houses a power back-up system which is able to provide power to instruments, recorders and computer in case of powerful failures. This system consists of an inverter, which runs continuously and which receives its power input from the mains, a set of batteries or a diesel generator. About 200 m higher on the slope a small amagnetic cabin houses instruments to measure the variations of the earth's magnetic field.

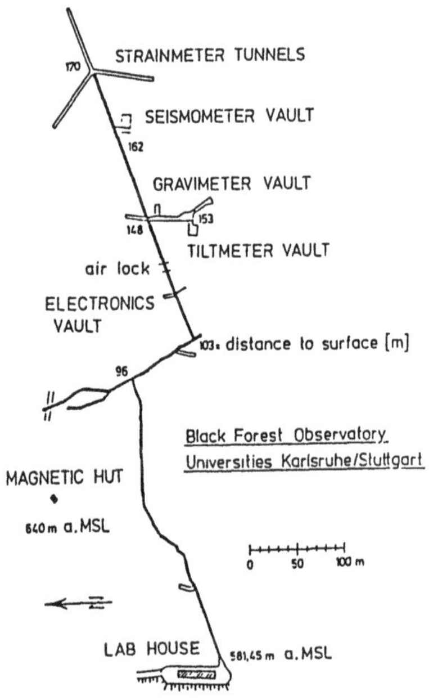

Fig. 2: Interior part of tunnel system.

The sites of the instruments have been precisely surveyed by the geodesists to obtain coordinates and azimuths. The coordinates of the tiltmeter vault are the follows:

The purpose of the observatory is to measure and monitor motions and deformations of the earth in as wide a frequency band as possible. The spectrum of the geodynamic signals ranges from the short-period body wavea (40 Hz) or near earthquakes over long-period body and surface waves from teleseism and the free oscillations of the earth excited by giant quakes to the earth tide deformations caused by the gravitational attraction of the moon and sun having periods of 12 h and longer. The observatory must be considered as an experimental station and research facility, where new instruments and techniques can be compared to established ones. Noise studies help to improve these techniques, therefore, metereological variables are monitored continuously and their effect on instruments or the earth's crust are a matter or intensive research.

The observatory is integrated into the seismic networks around the Upper Rhinegraben and Swabian Jura seismic regions for the study of local earthquakes. It will be integrated into the German Regional network of broadband seismic stations and into the worldwide GEOSCOPE network and very long-period digital seismic stations in cooperation with the Institute of Geophysics at the ETH Zürich, Switzerland. Strong cooperation in very long-period research exists with the working group "Geodesy/Geophysics" in Germany, the Institutes of Geophysics and Planetary Physics of the University of California at Los Angeles an San Diego. The observatory also contributes to Antarctic research by cooperating with the above and the German Alfred Wegener Institute for Polar Research in Bremerhaven.

At present the observatory is equipped with the following geophysical sensors: Short-period seismographs (three components, Geotech S-13) broadband seismographs (three components, Wielandt, with very long-perios channel), one LaCoste-Romberg gravimeter ET 19 with electrostatic feedback, one Askania borehole pendulum (two components), two Hughes level-bubble tiltmeter (two components each), one long baseline fluid-tube tiltmeter (differential pressure type), several 10-m Invar wire strainmeter, one Askania magnetic variograph, a set of three component variometers, strainless steel and $Pb-PbCl_2$ self-potential lines (60-m horizontal and 40-m vertical), and several metereological sensors. Two Askania gravimeters and one Askania borehole pendulum participate in field experiments in Denmark, Antarctica (Georg-von-Neumayer station) and California (Pinon Flat-Crustal Deformation Observatory).

Major research interests at present are the study of local elastic effects on earth tide tilts and strains, the analysis of dispersed globe-cylindrical seismic surface waves, the measurement of frequencies and attenuation of free oscillations of the earth, the effects of bariometric pressure variations on the data, the possibility of tidal influences on the statistics of tectonic and volcanic earthquakes and most recently the resonance due to the earth's core in the diurnal tidal band.

Reference

Emter, D., Kiesel, H., Mälzer, H., Schlemmer, H., Zürn, W. (1975): Wissenschaftliche Arbeiten am Geowissenschaftlichen Gemeinschaftsobservatorium Schiltach.- Dtsch. Geod. Komm., Reihe B, Heft 211; München.

www.ingramcontent.com/pod-product-compliance
Ingram Content Group UK Ltd.
Pitfield, Milton Keynes, MK11 3LW, UK
UKHW061655190726
13853UKWH00008B/2224